ICRRM 2019 – System Reliability, Quality Control, Safety, Maintenance and Management

D1827175

Vinit Kumar Gunjan · Sri Niwas Singh ·
Tran Duc-Tan · Gloria Jeanette Rincon Aponte ·
Amit Kumar
Editors

ICRRM 2019 – System Reliability, Quality Control, Safety, Maintenance and Management

Applications to Civil, Mechanical and Chemical Engineering

 Springer

Editors
Vinit Kumar Gunjan
Department of Computer Science
and Engineering
CMR Institute of Technology
Hyderabad, Telangana, India

Tran Duc-Tan
Electronics and Telecommunication Faculty
University of Engineering and Technology
(UET), Vietnam National University
Hanoi, Vietnam

Amit Kumar
BioAxis DNA Research Centre Pvt Ltd
Hyderabad, Telangana, India

Sri Niwas Singh
MMM University of Technology
Gorakhpur, Uttar Pradesh, India

Gloria Jeanette Rincon Aponte
Department of Computing
University Cooperativa de Colombia
Neiva, Colombia

ISBN 978-981-13-8506-3   ISBN 978-981-13-8507-0   (eBook)
https://doi.org/10.1007/978-981-13-8507-0

This Springer imprint is published by the registered company Springer Nature Singapore Pte Ltd.
The registered company address is: 152 Beach Road, #21-01/04 Gateway East, Singapore 189721, Singapore

ICRRM 2019

Proceedings of International Conference on Reliability, Risk Maintenance and Engineering Management (ICRRM) 2019

About Book

This proceeding includes selected articles from the International Conference on Reliability, Risk Maintenance and Engineering Management (ICRRM) 2019, held in Pune, India in 2019.

It offers in-depth literature and cutting-edge research papers on the latest developments in reliability in civil and chemical engineering, accelerated life testing, big data and IoT applications in R&M, business process improvement, design optimization using R&M techniques, discrete event modeling and simulation, risk analysis and management, diagnostics and prognostics, FMEA, R&M applications in supportability, R&M applications in service, International Conference on Reliability, Risk Maintenance and Engineering Management, and related areas.

ICRRM is one of the most prestigious conferences conceptualized in the fields of system reliability and management. The event witnessed industry professionals, academicians, governmental agencies, and universities to cover a broad range of perspectives, practices, and technical expertise during the conference. Over 300 delegates attended this conference.

Organizing Committee

Honorary Chairs

Sri Niwas Singh (Vice-chancellor)	MMM University of Technology, Gorakhpur, India
Vinit Kumar Gunjan	CMR Institute of Technology, Hyderabad, India
Tran Duc	University of Engineering and Technology, Vietnam
Gloria Jeanette Rincon Aponte	University Cooperativa de Colombia, Neiva, Colombia

General Chair

Rachayya Arakerimath	G.H. Raisoni College of Engg. & Management, Pune, India

Co-chair

Ajay Dahake	G.H. Raisoni College of Engg. & Management, Pune, India

Publication Chair

Rupali Patil	G.H. Raisoni College of Engg. & Management, Pune, India

Publicity and Design

Adik Yadao
Sandeep Patil

Registrations and Finance

Sanjay Mitkari   G.H. Raisoni College of Engg. & Management,
                                Pune, India
Sneha Kuralkar   G.H. Raisoni College of Engg. & Management,
                                Pune, India

Hospitality

Saurabh. Gupta   G.H. Raisoni College of Engg. & Management,
                                Pune, India
Girish Joshi   G.H. Raisoni College of Engg. & Management,
                                Pune, India

Website

Abdul Pathan   G.H. Raisoni College of Engg. & Management,
                                Pune, India
Dnyneshwar Jade   G.H. Raisoni College of Engg. & Management,
                                Pune, India

Easy Chair Management

Pramod Kathamore   G.H. Raisoni College of Engg. & Management,
                                Pune, India
Shaikh Serif   G.H. Raisoni College of Engg. & Management,
                                Pune, India

Conference Proceedings Coordinator

Santosh Hiremath   G.H. Raisoni College of Engg. & Management,
                                Pune, India
Prasad Dhumal   G.H. Raisoni College of Engg. & Management,
                                Pune, India

Workshop/Seminar Coordinator

Dipak Patil   G.H. Raisoni College of Engg. & Management,
                                Pune, India
V. Thorat   G.H. Raisoni College of Engg. & Management,
                                Pune, India

Technical Program Committee

Poonam Hiwal	University of Cambridge, Cambridge, UK
Sruti Das Choudhury	University of Nebraska-Lincoln, Lincoln, NE, USA
Sonia Sanchez-Cuadrado	Complutense University, Madrid, Spain
Amit Kumar Pandey	Aldebaran Robotics, France
Joy Deep Mitra	Michigan State University, USA
Krishna Raj	HBTU Kanpur, India
Pradeep Kumar	University of KwaZulu-Natal, Durban, SA
Nitin Chanderwal	University of Cincinnati, USA
Anabel Fraga	University of Carlos III, Madrid, Spain
R. Gowri	Petroleum University, Dehradun, India
T. R. Lenka	NIT Silchar, Assam, India
Pankaj Pal	NIT Uttarakhand, India
Govind Gupta	IIIT Raipur, India
Balwinder Raj	NIT Jalandhar, Punjab, India
Naushad Alam	AMU, Aligarh, India
Manoj Kumar Majumder	IIIT Raipur, India

Invited Speakers

Prof. Dr. Sc. Dina Simunic—University of Zagreb, Croatia
Dina Šimunić was born in Zagreb in 1963. She received B.Sc. and M.Sc. degrees in electrical engineering from the University of Zagreb, Faculty of Electrical Engineering and Computing (FER), Zagreb, Croatia, in 1985 and 1992, respectively. She received Dr. Sc. at Graz University of Technology, Austria, in 1995. Since 1991, she is with the Department for Wireless Communications at University of Zagreb, Faculty of Electrical Engineering. She was Guest Professor in research laboratories "Wandel & Goltermann" in Germany and "Motorola Inc.," USA, in 1996. In 2007, she was promoted to full professor. In Croatia, she served as Vice President of Telecommunications Council. In European Commission, she acted as Vice Chair of COST for precompetitive research in information and communication technologies. She is President of Technical Committee for telecommunications at Croatian Standards Institute. She is Leader of the group "Strategy toward the development of materials for education about standardization, as well as identification of common requirements" in ITU. She participated in five and led three scientific projects of Croatian Ministry of Science, Education and Sports. She was Leader of the WP6 of EU FP7 Project: eWALL. She was a collaborator of the research project, financed through structural funds of EU, Ministry of Science, Education and Sports: ICTGEN (Information and Communication Technology for Generic and Energy-efficient Solutions with Application in e-/mHealth). She

published more than 100 papers in journals and conference proceedings in the fields of secure routing protocols against denial-of-service attack in mobile ad hoc networks, cognitive radio systems, radiofrequency simulations, and biomedical effects of electromagnetic fields She is Senior Member of IEEE. She participated in 10 international program committees of scientific conferences, and she serves as Member of the editorial board of International Scientific Journal JOSE and Reviewer of Wireless Personal Communications and many other international scientific journals. She served as Editor in Chief of the Journal of Green Engineering. She is Leader of scientific Green Engineering Laboratory, GEL.

Topic of Talk: Smart City Communications
Abstract. The world is developing rapidly, with the fast-growing population. The current prediction is that by 2050 two-thirds of the population will live in cities. This fact puts a lot of pressure on life in the city, especially related to infrastructure. Therefore, the idea of smart city and its innovative infrastructure is becoming a reality. The communication layer is a very central part of the smart city concept, and it is located between service layers and physical layers. The choice of a communication network depends very much on the role, planned applications in the given scenario, as well as on their security, interoperability, lifetime, reliability, quality of service, consumed power, etc. Different technologies will be discussed, including traditional cellular networks (GSM/GPRS/UMTS/LTE), Bluetooth/BLE, RFID, NFC, 802.11, WBAN 802.15.6, NB-IoT, LPWAN, eMTC, EC-GSM-IoT, and 5G IoT.

Tadahiko Murata—Kansai University, Suita, Japan
Murata received the Ph.D. from Osaka Prefecture University in 1997. His research area includes soft computing, multi-objective optimization, and social simulation. He joined a number of international conferences as Program Committee Member and experienced Program Chairs in IEEE-sponsored conferences. He founded Technical Committee on Awareness Computing, IEEE SMCS, and chairs the TC. Since 2005, he directed Policy Grid Computing Laboratory to enhance research area among computational intelligence and computer science, economics, political science, sociology, and psychology. He joined Research Institute for Socionet work Strategies at Kansai University in 2008. He was invited by Computational Institute of University of Chicago in 2010 and 2011. He published more than 50 papers in journals and conference proceedings in the fields of Neuro, Fuzzy, Caossystem, genetic algorithm, social system engineering, combinatorial optimization problems, computer science, and optimal control of economic systems.

Title of Talk: Towards the Realization of Real-Scale Social Simulations
Abstract. In this talk, we show challenges in developing or realizing real-scale social simulations (RSSSs) and show several attempts to cope with them. RSSSs are simulations for the real world with communities and environments. There are four challenges in realizing RSSSs: (1) Modeling, (2) Attributes, (3) Computation, and (4) Analysis. In (1) Modeling, we should model decision-making processes for

entities in RSSSs. Here, "entities" include residents and their organizations. We should determine the scale of entity in the RSSS. We should also include interaction processes between entities and environments. Here, "environments" include physical and legal conditions existing outside entities. As for (2) Attributes, precise or specific information of entities and environments is required for developing RSSSs. While we can develop some decision-making process for an entity, its decision depends on its age, income, educational background, or some other attributes. We need specific values for those pieces of information. Even if personal information is registered or sensed in cyberphysical world, those private data should be protected appropriately. A technique of synthesizing populations is one of the solutions. Synthetic population is generated based on the statistics of the real world. Using the technique, the population that has the same statistical characteristics with the real world is available for RSSSs. Using such populations, we can see what events can happen in a world with the same statistical characteristics. When we simulate RSSSs with computers, the number of humans in a simulation becomes up to 7.5 billion humans on the planet with their organizations. In order to simulate such a huge people and organizations within a practical time, we should employ high-performance computers. Challenges in (3) come from parallelization, distribution, and reproducibility in the computation of RSSS.

Dr. Ali Razban—Purdue University Indianapolis, Indiana
He received M.S.E. degree in EECS Department at the University of Michigan, Ann Arbor, MI, in 1991. His areas of specialization are controls and signal processing. He received doctorate degree from the University of London, London, UK, in 1994. His areas of specialization are control, robotics, and automation. Since 1989 to 1994, he worked as Research Associate, Department of Mechanical Engineering, Imperial College, London, UK. In August 1999, he joined as Adjunct Professor, Department of Mechanical Engineering, Binghamton University, Binghamton, NY. From 2010 to 2016, he joined as Senior Lecturer, Department of Mechanical Engineering, Indiana University Purdue University (IUPUI). He was promoted as Assistant Director in 2011 and Director of the Bachelor of Science in Energy Engineering (BSEEN) Program in 2016 Industrial Assessment Center, Indiana University Purdue University (IUPUI).

Topic of the Talk: The Current and Future State of World Energy and Energy Management
Abstract. The presentation will explore topics related to the current and future world energy consumption. Important topics that will be discussed relate to shifts in energy sources, the role of energy efficiency, and renewable energy trends in the energy market. Furthermore, the effects of how energy efficiency has played a role in reducing energy consumption in developed countries will be covered. The implementation of energy-efficient measures could offset increases in energy demand in the future. The issue of how energy-efficient measures can impact future global energy demands will be addressed. The presentation will also explore the current state of renewable energy along with trends and projections for the direction

of renewable resources in the near future. Finally, the presentation will address financial implications of the power market in its current state and address the economic factors associated with expected shifts in the global energy profile in the future.

Dr. Gajjela Satyanarayana Reddy—National University of Singapore

He received master's degree in physics from the School of Physics, University of Hyderabad, in 2005. His area of specialization is nanomaterials science and engineering. He received doctorate degree from University of Singapore, Singapore, in 2013. His area of specialization is development of low-cost photovoltaic devices for solar energy harvesting. Since 2013 to till now, he worked as Postdoctoral Research Fellow in the Department of Mechanical Engineering at NUS Singapore. For the past several years, he has been involved in the development of safe, reliable, and environmentally friendly lithium-ion and sodium-ion battery technologies. His research work mainly focuses on developing promising components of the batteries which enables improved performance and a better environmental profile and offers greater risk mitigation. He was a key member of the teams which invented a nonflammable high energy density sodium-ion battery and low-cost high-power Li-ion battery. The former technology promotes the safety of batteries in commercial applications, while the later technology part of the solution to lowering batteries' environmental impacts.

His expertise is in novel materials development, energy efficiency improvement, and product and process development, new technology evaluation, industrial experience in manufacturing batteries for power electronics, automotive industries, and grid storage. He published more than 20 papers in international journals and conference proceedings in the fields of energy storage and renewable energy conversion. He was Organizer of Young Investigators Forum, CMCEE2018 Singapore. He is Active Member of MRS Singapore and American Ceramic Society, USA.

Mr. Kapil (Krishna) Baidya—AGM, Tata Motors, Pune, India

Mr. Kapil (Krishna) Baidya received M.E. degree in mechanical engineering from National University of Singapore in 2000 and B. Tech degree in mechanical engineering from IIT Madras in 1993. Since 2009, he is with Tata Motors Limited as Assistant General Manager (Dev)—Advanced Engineering. From 2008 to 2009, he worked as Joint General Manager in Essar Engineering Ltd. He also worked as Senior Program Manager in Flextronics, Chennai, from 2007 to 2008. KLA-Tencor Principal Mechanical Design Engineer during December 2006–July 2007 (8 months) Chennai Area, India. In VES Singapore, he worked as Senior Mechanical R&D Engineer from 2002 to 2006. In 2002, he joined as Senior System Design Engineer in Advanced System Automation, Singapore, from 2000 to 2002. He also worked as Graduate Engineer Trainee to Senior Engineer in Essar Steel Ltd, Surat, from 1993 to 1998. He worked in various fields like electric and hybrid vehicles, fuel cell bus covering the design of fuel cell power, as well as hydrogen storage and delivery system under the grants from DSIR, Design-Driven Cost Reduction (DDCR) Project for the existing midrange HP printers, etc.

Topic of the Talk: Industrial Automation for Safety, Risk Management and Reliability

Abstract. With the Industrial Revolution in Europe and America during the seventeenth and eighteenth centuries, the manufacturing was moved from home to factories. The demand for goods increased rapidly, so as the size of the factories without considering the safety and the working environment; so the accidents in the factories increased at the pace of industrialization. It is the industrial automation with the use of control systems, computers, or robots, and information technologies for handling different processes and machineries in the industry mitigated the need of volume production and the risk on human lives. In true sense, industrial automation is the second step beyond mechanization in the scope of industrialization. The advent of industrial automation increased the quality and flexibility of manufacturing process, thereby reducing the production cost and consistent high-quality products with minimal human intervention. With industrial automation, the factories have been able to work round the clock throughout the year without any break. Industrial automation made the production line safe for the employees by deploying robots to handle hazardous conditions. The main drawback to implement industrial automation is its high initial investment; also, the cost of switching from a human production line to an automatic production line is very high. It also involves in training employees to handle this new sophisticated equipment and that makes it further expensive. Industrial automation has recently found more and more acceptance from various industries because of its huge benefits, such as increased productivity, quality, and safety at low costs.

Dr. Harold D'Costa—Intelligent Quotient Security System, India
Dr. Harold D'Costa is President of Cyber Security Corporation and CEO of Intelligent Quotient Security System. IQSS is featured in top 10 cybersecurity companies in the country in matters related to cybersecurity, cyberforensics, and digital evidence. He is also working as Advisor in Law Enforcement Agencies. He worked as Senior Trainer (judicial officers—cyberlaw, digital evidence, and cybercrime). He worked on more than 4,000 cases of cybercrime from 7 countries and trained more than 35,000 cops and 2,000 judicial officers in cyberforensics, cyberlaw, digital evidence, and cybercrime investigation. He was a resource person in 15 international conferences and was also part of global cybersecurity summit in India. He is a regular writer in leading newspaper and an expert in cybersecurity in various TV channels. He had the distinction to train Bangladesh Judges on Cyber Forensics organized by National Judicial Academy. He also worked on cybercases related to Bollywood, MNC, banks, telecom, education, media, gas subsidies, etc. He was instrumental in developing and implementing the first e-delivery of question papers for Mumbai University. The system was widely appreciated and now is de facto in many universities in different states of the country. He was on the panel

of Maharashtra Police Academy to develop curriculum and courseware on cyber-crime and digital forensics for corporates. He is consistently getting a feedback score of 100% in training officials from Mantralaya and large government institutes. He developed courseware and curriculum for YCMOU on Diploma—Cyber security. He is associated with National Judicial Academy, NIA, Maharashtra Police Academy, Goa Police, CID, Central for Police Research, Detective Training School, Maharashtra Intelligence Academy, Judicial Officers Training Institute, YASHADA, in training senior officials and providing consistent support in cybersecurity defense. He has co-authored 14 books in cybersecurity space.

.

Contents

Mathematical Study for Determining the Time of Evaporation of Water Desalination System

A. B. Auti[1(✉)], J. Ratna Raja Kumar[1(✉)], Nisha Abhijeet Auti[2(✉)], and Usha Mandadapu[1(✉)]

[1] Genba Sopanrao Moze College of Engineering, Pune 411045, Maharashtra, India
autiabhijeet24578@gmail.com
[2] BSCOER, Pune, India

Abstract. In this work, a mathematical model was developed to predict the amount of wáter to be evaporated depending up on the intensity of solar radiations, size and material of the concentrator and absorber. A parabolic concentrator Is used as a heating source. Solar energy is focused by the concentrator on an absorber which contains salt water. Water gets heated and converted to steam. The steam is then condensed by the condenser to get purified water. These results were utilised to develop a mathematical model which predicts the time of evaporation of the impure water. The model was validated with experimental results obtained for different months. The correlation coefficient was calculated and found to be 0.93.

Keywords: Dimensional analysis · Regression analysis ·
Parabolic concentrator · Solar · Buckingham's theorem

1 Introduction

The design, optimization and fabrication of solar desalination system have been presented. Mathematical model is developed using two main tools, dimensional analysis and regression analysis. Dimensional analysis provides a mathematical tool for analysing and making model laws [1]. The process of modeling carried out in this work is divided into two main stages – dimensional analysis and regression analysis [2].

2 Selection of Variables

Rate of evaporation, (m) of the given quantity of water in specified time (t) is the primary unknown quantity. So it is chosen as the first variable [3]. Amount of solar radiation (I) and the quantity of water (Q) are the important physical quantities which are crucial in deciding the evaporation rate [4]. They are the variables chosen next. Other factors which affect the process are Emissivity (ε) of the absorber material, Reflectivity (ρ) of the concentrator, Specific heat capacity (C_P), Density (ρ_1), Absorber area (A_{abs}) [5].

© Springer Nature Singapore Pte Ltd. 2020
V. K. Gunjan et al. (Eds.): *ICRRM 2019 – System Reliability, Quality Control, Safety, Maintenance and Management*, pp. 1–5, 2020.
https://doi.org/10.1007/978-981-13-8507-0_1

2.1   Representation in Terms of Fundamental Physical Quantities with the Units

It is necessary in dimensionless analysis to represent dimensions of all the factors in terms of fundamental quantities, M, L, T and θ_1 in a table known as dimensional matrix (Table 1).

Table 1. Physical quantities in terms of fundamental dimensions

	I	m	Q	θ1	ε	Aabs	ρl	d	CP
M	1	1	1	0	0	0	1	0	0
L	2	0	0	0	0	2	−3	1	2
T	−3	−1	0	0	0	0	0	0	−2
θ	0	0	0	1	0	0	0	0	−1

2.2   Solution of Equations in Terms of Independent Variables and Final Development of Dimensionless Products

Now, homogeneous linear algebraic Eqs. 1 to 4 are written in terms of variables a to g whose coefficients are the numbers in the cells of the dimensional matrix corresponding to those variables.

$$a + b + d + f = 0 \tag{1}$$

$$2c + e - f = 0 \tag{2}$$

$$-a - 3b - 2f = 0 \tag{3}$$

$$-f + g = 0 \tag{4}$$

Equations are solved in terms of independent variables and the solution is written in the form of solution matrix. Values of the dependent variables d, e, f and g are expressed in terms of 3 independent variables a, b and c using Eqs. 1 to 4 as shown below.

$$d = -a/2 + b/2 \tag{5}$$

$$e = -2c + a/2 + 3b/2 \tag{6}$$

$$f = -a/2 - 3b/2 \tag{7}$$

$$g = f \tag{8}$$

By considering the determinant at the left hand side of the matrix of solutions, it is perceived that the rank of the matrix is invariably equal to the number of rows. Matrix of solution as given has a rank 3. So the rows are linearly independent

$$\Pi 1 = k\,(\Pi 2)b1 \times (\Pi 3)c1 \tag{9}$$

3 Forming and Solving Regression Equations

The dimensionless variables constructed by a dimensional analysis are now used as the variables of the regression to form a correct dimensional relationship, one that is homogeneous. Any coefficient in a fitted model is also dimensionless and will not change if the units of measurement are changed.

$$\log \Pi_1 = \log k_2 + a_1 \log(\Pi_2) + b_1 \log(\Pi_3) \tag{10}$$

$$\text{Let } Y = a + bX_1 + cX_2 \tag{11}$$

Three equations for three can be generated as under,

$$Y = N(a) + b \sum X_1 + c \sum X_2 \tag{12}$$

$$\sum X_1 Y = \sum X_1(a) + b \sum X_1^2 + c \sum X_1 X_2 \tag{13}$$

$$\sum X_2 Y = \sum X_2(a) + b \sum X_1 X_2 + c \sum X_2^2 \tag{14}$$

3.1 Experimental Data

Experiments were conducted in different months and with different quantities of water in absorber. The first test was conducted with 2 kg mass of water using the concentrator and cookers of diameters 0.3 m, 0.25 m and 0.27 m made of stainless steel and aluminium respectively. The observations are given in Table 2.

Table 2. Readings for 2 kg

Month	I (W/m^2)	m (kg/sec)	A$_{abs}$ (m^2)	ρ_1. Cp (J/m^3 K)	θ (^{0}C)	Observed time (sec)
Dec	580	0.00027	0.070734	4000000	28	7407
Dec	560	0.00026	0.070734	4000000	27	7700
Dec	660	0.00032	0.049112	2430000	28	6250
Dec	654	0.0003	0.057293	2430000	28	6671
Dec	660	0.00033	0.070734	4000000	29	6061
Mar	690	0.00031	0.070686	2430000	30	6461
Mar	800	0.00037	0.057293	2430000	30	5405
Mar	794	0.000364	0.049112	2430000	32	5494
Mar	820	0.00037	0.057293	2430000	30	5405
Mar	825	0.00038	0.049112	2430000	32	5263

Similar tests were conducted with 5 kg mass of water using the concentrator and cookers of diameters 0.3 m, 0.25 m and 0.27 m made of stainless steel and aluminium respectively. The constants c and k are determined using the set of observations such that overall error in t is minimized and −1 and 1/1236.3 are found to be suitable.

$$t = \frac{123.6X \in A_{abs}(Q^{0.395} - 1)}{\rho(I^{0.21}d^{0.22})(y\theta)^{0.185}} \, hours \tag{15}$$

All the required parameters can be calculated from the above equation

4   Validation

Time of evaporation was estimated for input data such as product of ρ_1 and c_p equal to 4000000 J/m^{3}K, dish diameter 2.3 m and absorber area 0.07 m^2. Solution obtained is written below. The observed time for the above analysis was 6061 s. The accuracy of the generated equation is found to be 93%. Correlation coefficient was determined for experimental and calculated values using Karl Pearson's formula (Mujumdar, 2006) (Fig. 1).

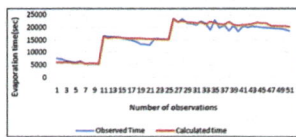

Fig. 1.   Experimental and calculated time of evaporation

The developed model shows that time of evaporation decreases with increase in the size of concentrator, intensity of solar radiations, surrounding temperature, emissivity of the absorber and reflectivity of the material. The time of evaporation decreases with decrease in the size of the absorber and the quantity of water to be evaporated.

5   Conclusions

In this study a mathematical model was developed to find quantity of water that was to be evaporated and time required for evaporation. The model was developed on the basis of almost 50 experiments conducted by varying different parameters. Equation involves parameters such as, size and material of the absorber, size of the concentrator, reflectivity of the concentrator and atmospheric conditions. The model was validated with experimental results value with 93% accuracy. The correlation coefficient was calculated and found to be 0.93.

References

1. Albahloul, K.E., Zgalei, A.S, Mahgjup, O.M.: The Feasibility of the Compound Parabolic Concentrator for Solar Cooling in Libya, Thermal Conversion of Solar Energy ICRE8-5, 9–20 (2010)
2. Beltran, R., Velazquez, N., Alma, C.E., Daniel, S., Guillermo, P.: Mathematical model for the study and design of a solar dish collector with cavity receiver for its application in stirling engines. J. Mech. Sci. Technol. **26**(10), 311–332 (2012)
3. Grigonienė, J., Mindaugas, K.: Mathematical modelling of optimal tilt angles of solar collector and sunray reflector. Energetic **55**(1), 41–46 (2009)
4. Patil, R.J., Awari, G.K., Singh, M.P.: Performance analysis of Scheffler reflector and formulation of mathematical model. VSRD-TNTJ **2**(8), 390–400 (2011)
5. Boric, M., Velimir, S.: Design of a stationary a symmetric solar concentrator for heat and electricity production. In: Proceedings of the Fourth IASTED International Conference, 2–4 April 2008. Power and Energy Systems, Langkawi (2008). ISBN CD: 978-0-88986-732-1

Estimation of Aggregate Characteristics Using Digital Image Processing

Madhuri N. Mangulkar[1(✉)], Suddhasheel Ghosh[2],
and Sanjay S. Jamkar[3]

[1] Marathwada Institute of Technology, Aurangabad 431001, Maharashtra, India
mangulkarm@yahoo.com
[2] Jawaharlal Nehru Engineering College,
Aurangabad 431001, Maharashtra, India
[3] Government College of Engineering, Aurangabad 431001, Maharashtra, India

Abstract. Keeping in mind indirect and direct methods used for the quantification of aggregates, Digital Image Processing (DIP) based system for the measurement of volume, equivalent volume diameter, sphericity, roundness index of coarse aggregate particles is developed. The system is calibrated using standard objects such as marbles, coins and then used for the measurement of coarse aggregate particles having varied characteristics. The system is applied to four types of aggregates. Analysis of the various shape characteristics shows that sphericity is the most effective measure. The developed system is successfully implemented to study aggregate characteristics.

Keywords: Aggregate · Shape · Roundness index · Particle packing · Digital image processing (DIP)

1 Introduction

Aggregates occupy bulk of the volume of concrete. A proper packing of aggregates with binding material ensures reduction in voids and thereby better performance of concrete. Therefore, precise evaluation of characteristics of ingredients of concrete is highly essential. The effect of aggregate characteristics such as size, shape and grading is more pronounced in the case of high strength (HSC) and high performance concrete (HPC) due to very low values of water to cementitious material ratio. Aggregate characteristics have a significant impact on the cost effectiveness of concrete [1, 2]. Attempts are continuously being made for the objective quantification of characteristics of aggregates. The methods used for this purpose can be broadly classified as Indirect and Direct methods.

The manual test methods for characterization of aggregates are tedious, labour intensive, time consuming and prone to human errors. Digital-image processing is a powerful computer-based method for studying and has been important tools in many diverse fields. In the earlier digital image processing based methods, the aggregate shape is characterised through images of three orthogonal faces and other opposite faces are assumed to be identical. Actually the aggregate particle shape is highly irregular. It is essential to have DIP system which overcomes the effect of shadows in

© Springer Nature Singapore Pte Ltd. 2020
V. K. Gunjan et al. (Eds.): *ICRRM 2019 – System Reliability, Quality Control,*
Safety, Maintenance and Management, pp. 6–12, 2020.
https://doi.org/10.1007/978-981-13-8507-0_2

order to increase accuracy in the measurements. In the present study a digital image processing based system for the quantification of morphological properties of coarse aggregate particles is proposed.

2 Literature Review

IS 383 [3] has classified the shapes of the aggregates based on visual inspection in to rounded, irregular or partly rounded, angular and flaky. ASTM D3398 [4], BS 812 [5], IS 2386 Part 1 [6], ASTM D 1252 [7], ASHTO [8], ASTM D 4791 [9], ASTM D 5821 [10] are the manual methods for measuring shape characteristics are provided by These methods are laborious, time consuming and approximate. Therefore the researchers started exploring the use of new technologies to accurately and rapidly measure aggregate shape characteristics. These methods can be mainly categorised as Tomographic, Laser scanning and Digital image processing (DIP) based techniques. Garboczi [11], Kim et al. [12] and Illerstorm [13] have studied laser scanning based techniques to capture the entire shape of the aggregate. Kwan et al. [14], adopted DIP to analyse the size and shape in terms of sphericity, shape factor and convexity ratio of coarse aggregate using DIP. The digital image processing based methods available in literature consider maximum three faces of the aggregate for analysis, assuming remaining three opposite faces to be similar. Actually the aggregate shape is highly irregular. Based on the literature review, it is decided to carry a systematic study for most effective shape parameter for distinguishing various shapes of aggregates and to system be designed, developed and implemented which can capture the dimensions of an aggregate particle and characterise the sample with precision.

3 Shape Characteristics of an Aggregate

The shape characteristics of an aggregate have been traditionally expressed in terms of sphericity, elongation, flatness and shape factor. The various formulae for determining these characteristics for a given triplet (dl, di, ds) are given in Table 1.

Table 1. Shape characteristics of an aggregate with triplet (dl, di, ds)

Factor	Formula	Reference
Shape factor	$\frac{dl \times ds}{di^2}$	Barksdale et al. [18]
Flatness	$\frac{di}{ds}$	Barksdale et al. [18]
Elongation	$\frac{dl}{di}$	Kuo et al. [19]
Sphericity	$\sqrt[3]{\frac{di \times ds}{dl^2}}$	Kuo et al. [19]

4 DIPAM: An Imaging System for Aggregates

For imaging the aggregates from various angles, we design the DIPAM (Digital Image Processing based Aggregate Measurement) system. The proposed system consists of (a) Conveyor, (b) Turntable, (c) Illumination arrangement, (d) Image acquisition and (e) Control system integrated with each other. Conveyor arrangement facilitates movement of aggregate particle from a sample to the required position where image is acquired and back to the original position after image acquisition is over. Image acquisition system consisted of three Logitech Pro 9000 web cameras having 8 Megapixel resolution placed in an orthogonal manner (bottom, top and front). In order to acquire images of more faces of the aggregate, a turntable is added to the system. It is mounted on the shaft of the stepper motor which rotates the aggregate moved on a horizontal semi-transparent milky sheet by a predefined angle and enables the acquisition of more than three images of the aggregate. Illumination arrangement consisted of LED module which takes care of required illumination. In order to eliminate shadow effect the system is designed in such a way that camera and LED light both face each other, and camera exposure is set to maximum with minimum gain. A semi-transparent milky acrylic sheet is mounted in between the camera and the focus light. The sheet gets illuminated uniformly providing white background for acquiring the image of target aggregate. The control system controls the activities of (a) Stepper motor for Turntable, (b) DC motor for linear movement of conveyor along with Aggregate Tray, (c) Optical sensor of Encoder Disk, (d) Home position limit switch. The entire system is interfaced with the computer through MATLAB (Mathworks Inc.) software. The integrated Digital Image Processing based Aggregate Measurement (DIPAM) System is shown in Fig. 1.

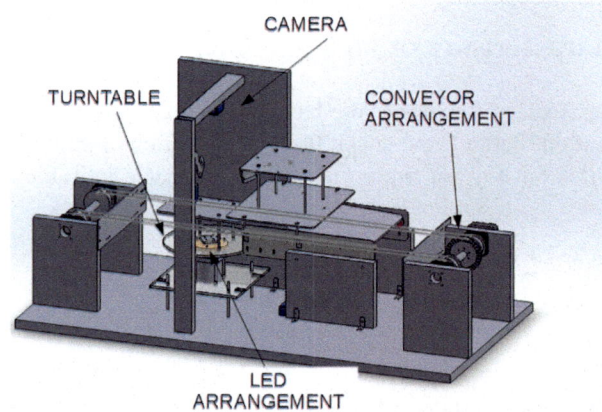

Fig. 1. Digital image processing based aggregate measurement system

In order to measure the various characteristics of an aggregate, it is placed on the aggregate tray mounted on a conveyor. The conveyor arrangement moves the tray linearly to the first position (in front of bottom camera) where the first LED lamp illuminates the aggregate. The bottom camera captures the first image. The aggregate tray then moved to a second position on the turn table where a second image is captured by the top camera. The third camera at the front also captures an image simultaneously. It also captures images by turning the tray through 90, 180 and 270°. The turntable rotates the tray by 270° in opposite direction and conveyor brings it to the original position. The steps are repeated till the image acquisition of all the aggregate particles in the sample is completed.

The system is calibrated using standard objects of known dimensions such as coins, rectangular prisms and marbles. The dimensions of the objects are measured manually using standard vernier calliper. Digital images of these objects are acquired by placing the objects one at a time, at a distance of 10 cm from each camera. Six images are taken at an interval of 5 s. Each camera in the system provided an image of dimensions 1600 pixels x 1200 pixels. The images are processed using MATLAB through the following steps: (a) conversion of RGB images to gray scale, (b) noise filtering, (c) detection of boundary, and (d) computation of longest (dl), intermediate (di), and shortest (ds) dimensions of the object. Based on the results, it is observed that 1600 pixels correspond to 62.4 mm i.e. 0.039 mm/pixel.

4.1 Analysis of Shape Characteristics

In this study, four different categories of aggregates are tested, viz. elongated, rounded, angular and cylindrical. 30 aggregate pieces of each of the types are passed through the imaging arrangement (DIPAM), and the measurements of their longest, intermediate and shortest dimensions were obtained.

4.2 Results of Coarse Aggregate Characteristics Using DIPAM

Ten samples of coarse aggregate particles of Type-A, B, C and D from each size fractions are sampled through a standard process. DIPAM system described in the previous chapter is used for measuring longest, intermediate and shortest dimensions of the particles. The results of the same are presented in columns 3, 4, and 5 of Tables 2, 3, 4 and 5 for each type of coarse aggregates. Each result presented in the table is the average of measurements of each particle three times using DIPAM. These dimensions of some randomly selected particles from the samples of various types of aggregates are also measured using Verniers caliper. Shape measures of aggregates namely, sphericity, shape factor, elongation ratio flatness ratio and equivalent diameter of the sphere having same volume as that of the aggregate particle are determined based on the measured values of longest, shortest and intermediate dimensions.

4.3 Computation of the Shape Measures of Coarse Aggregate Particles

The shape measures such as sphericity, shape factor, elongation ratio and flatness ratio are determined as per the relations presented in Table 1. The average results of these shape measures are given in columns 12 to 15 of Tables 2, 3, 4 and 5.

The shape factor is calculated using formulae presented in 3rd row of Table 2. The weighted average shape factor presented in Tables 2, 3, 4 and 5 for Type-A aggregate lies between 0.39 to 0.59. For Type-B, C and D aggregate it lies between 0.69 to

0.75,0.72 to 0.79 and 0.62 to 0.72. From the results it is observed that shape factor cannot clearly distinguish all the four types of shapes.

The elongation ratio is calculated using formulae presented in 4th row of Table 1. From the results presented in Tables 2, 3, 4 and 5, weighted average elongation ratio for Type-A aggregate lies between 1.29 to 2.98. For Type-B aggregate it lies between 1.59 to 2.00. For Type-C aggregate it lies between 1.57 to 1.82 while for Type-D aggregate it lies between 0.9 to 1.02. From these results it is evident that even elongation ratio also cannot clearly distinguish all the four types of shapes.

The Flatness ratio presented in 4th row of Table 2 indicate that for Type-A aggregate it lies between 2.15 to 3.74, for Type-B between 1.66 to 2.06, for Type-C between 1.59 to 1.87 and for Type-D between 1.36 to 1.58. These results also represent that flatness ratio cannot clearly distinguish all the four types of shapes.

It can be seen from Tables 2, 3, 4 and 5 that weighted average sphericity for elongated aggregates ranges from 0.4 to 0.62. The range for angular, cubical and rounded aggregates is observed to be approximately between 0.62 to 0.71, 0.66 to 0.78 and 0.87 to 0.93 respectively. This clearly indicates that sphericity is able to fairly distinguish between the four classes of aggregates. Therefore sphericity is used in the proposed particle packing theory to account for particle shape and size.

Table 2. A: Shape characteristics of aggregates

Aggregate type	Sieve size fraction in mm	Average dimensions in mm as per DIPAM			Equivalent diameter of sphere (Dp) (mm)	Shape measures by DIPAM			
		Longest (dl)	Inter- mediate (di)	Shortest (ds)		Spher- icity (ψ)	Shape factor (SF)	Elon- gation ratio (ER)	Flatness ratio (FR)
1	2	3	4	5	11	12	13	14	15
Type-A- Elongated	4.75– 12.5	16.10	5.40	4.30	8.25	0.45	0.46	2.98	3.74
	12.5–16	27.40	12.75	10.88	17.86	0.57	0.58	2.15	2.52
	16–20	43.70	23.00	12.50	26.61	0.53	0.39	1.90	3.50

Table 3. B: Shape characteristics of aggregates

Aggregate type	Sieve size fraction in mm	Average dimensions in mm as per DIP			Equivalent diameter of sphere (mm)	Shape measures by DIP			
		Longest (dl)	Inter- mediate (di)	Shortest (ds)		Spher- icity (ψ)	Shape factor (SF)	Elon- gation ratio (ER)	Flatness ratio (FR)
1	2	3	4	5	11	12	13	14	15
Type-B Angular	4.75– 12.5	9.20	5.43	5.28	6.31	0.70	0.75	1.70	1.74
	12.5–16	20.40	11.83	11.50	13.84	0.69	0.74	1.73	1.77
	16–20	27.80	16.30	15.70	18.94	0.69	0.74	1.71	1.77

Table 4. C: Evaluation of shape characteristics of aggregates

Aggregate type	Sieve size fraction in mm	Average dimensions in mm as per DIP			Equivalent diameter of sphere (mm)	Shape measures by DIP			
		Longest (dl)	Inter-mediate (di)	Shortest (ds)		Spher-icity (ψ)	Shape factor (SF)	Elon-gation ratio (ER)	Flatness ratio (FR)
1	2	3	4	5	11	12	13	14	15
Type-C Cubical	4.75–12.5	9.40	5.25	5.20	7.88	0.68	0.74	1.79	1.81
	12.5–16	22.90	14.48	14.10	20.74	0.73	0.77	1.58	1.62
	16–20	26.30	14.48	14.10	21.72	0.67	0.72	1.82	1.87

Table 5. D: Evaluation of shape characteristics of aggregates

Aggregate type	Sieve size fraction in mm	Average dimensions in mm as per DIP			Equivalent diameter of sphere (mm)	Shape measures by DIP			
		Longest (dl)	Inter-mediate (di)	Shortest (ds)		Spher-icity (ψ)	Shape factor (SF)	Elon-gation Ratio (ER)	Flatness ratio (FR)
1	2	3	4	5	11	12	13	14	15
Type-D Rounded	4.75–12.5	7.82	5.83	4.75	8.98	0.84	0.70	0.90	1.36
	12.5–16	15.83	15.75	10.83	19.95	0.83	0.62	0.95	1.58
	16–20	20.95	19.90	12.65	22.26	0.85	0.68	0.95	1.43

5 Conclusions

From the results of the experimental work discussed in the previous section we may safely draw following conclusions,

1. Proposed Digital Image Processing based Aggregate Measurement system (DIPAM) is capable of measuring longest, intermediate and shortest dimensions of each coarse aggregate particle in a sample with precision.
2. The results of the weighted average values of shape factor for Type-A, B, C, and D coarse aggregates are observed to be between 0.39 to 0.59, 0.69 to 0.75, 0.72 to 0.79 and 0.62 to 0.72 indicating that shape factor cannot clearly distinguish between coarse aggregates of varied shape characteristics.
3. The results of weighted average values of elongation ratio and flatness ratio also depict insensitivity of these parameters towards coarse aggregate shape.
4. The results of weighted average sphericity for Type-A, B, C, and D coarse aggregates ranges from 0.4 to 0.62, 0.62 to 0.71, 0.66 to 0.78 and 0.87 to 0.93 respectively representing higher sensitivity towards shape.

References

1. Jamkar, S.S., Rao, C.B.K.: Index of aggregate particle shape and texture of coarse aggregate as a parameter for concrete mix proportioning. Cem. Concr. Res. **34**, 2021–2027 (2004)
2. Lee, J.R.J., Smith, M.L., Smith, L.N.: A new approach to the three-dimensional quantification of angularity using image analysis of the size and form of coarse aggregates. Eng. Geol. **91**, 254–264 (2007)
3. IS 383:1970: Specification for Coarse and Fine Aggregates from Natural Sources for Concrete. Bureau of Indian Standards (1971)
4. ASTM D 3398-97, Standard Test for Index of Aggregate Particle Shape and Texture, West Conshohocken, USA (1997)
5. BS 812 Part-1, Methods for determination of particle size and shape, British Standards, UK (1975)
6. IS: 2386 Part-I, Methods of test for aggregates for concrete (Particle size and shape), Bureau of Indian Standards, New Delhi, India (1963)
7. ASTM D 1252-03: Standard Test Methods for Uncompacted Void Content of Fine Aggregate (as Influenced by Particle Shape, Surface Texture, and Grading). Annual Book of ASTM Standards. American Society for Testing and Materials, West Conshohocken, 04 Feb 2003
8. AASHTO PROVISIONAL STANDARDS, TP 56-03. Uncompacted Void Content of Coarse Aggregate (As Influenced by Particle Shape, Surface Texture, and Grading). American Association of State Highway and Transportation Officials, Washington, D.C. (2003)
9. ASTM D4791, Standard test method for flat and elongated particles in coarse aggregate, Annual Book of ASTM Standards, West Conshohocken, Pa, 04 March
10. ASTM D 5821-95. Standard Test Method for Determining the Percentage of Fractured Particles in Coarse Aggregate. Annual Book of ASTM Standards, American Society for Testing and Materials, Philadelphia, PA, 04 March 1999
11. Garboczi, E.: Three-dimensional mathematical analysis of particle shape using X-ray tomography and spherical harmonics: application to aggregates used in concrete. Cem. Concr. Res. **32**, 1621–1638 (2002)
12. Kim, H.K., Haas, C.T., Rauch, A.F., Browne, C.: Dimensional ratios for stone aggregates from three-dimensional laser scans. J. Comput. Civ. Eng. **16**, 175–183 (2002)
13. Illersetrom, A.: A 3-D laser technique for size, shape, and texture analysis of ballast. Master's thesis, Royal Institute of Technology, Stockholm, Sweden
14. Mora, C.F., Kwan, A.K.H.: Sphericity, shape factor, and convexity measurement of coarse aggregate for concrete using digital image processing. Cem. Concr. Res. **30**, 351–358 (2000)

Failure and Vibration Analysis of Blower Assembly Used for Ventilation System for Life Enhancement

Mudassar A. S. Shaikh$^{(\boxtimes)}$ and R. R. Arakerimath

Department of Mechanical Engineering,
G. H. Raisoni College of Engineering and Management,
Wagholi, Pune 412207, India
mudassarshaikh58@gmail.com

Abstract. This paper presents failure and vibration analysis of blower used for ventilation system based on application of computer aided engineering (CAE). The objective of research is to enhance the life of blower assembly which is used to perform ventilation. In this paper we are going to perform the structural and modal analysis of blower assembly in Ansys workbench 2019 and compare the results with IBC 2015 code to check whether the deformation of structure is within the limit or not. The various mode shapes are studied from the CAE results and correspondingly we avoided the resonance on to the structure which is mainly due to the excitation frequency. In this project we are going to analyse the blower assembly under mechanical conditions. The objective of our project is to reduce the weight of blower assembly; here we are going to analyse the mechanical blower and provide the solution/recommendation accordingly based on FEA results.

Keywords: Turbo blower · CAE · Structural analysis · Dynamic analysis · Vibration

1 Introduction

Most gathering plants use fans and blowers for warmth ventilation and cooling, require a breeze stream for pharmaceutical and air process industry. Fan systems is vital to keep creating shapes working, and involve a fan, an electric motor, a drive structure, courses, piping, stream control contraptions, and circulating air through and cooling equipment (channels, pipe, cooling circles, condenser, evaporator etc.).

Ventilation frameworks generally introduced by focal frameworks incorporate supply and fumes fans; serve for ventilation of settlement and other than convenience regions with air with synchronous ventilation of capacity batteries and for air cooling and sanitization from hurtful and smelling contaminations.

Cooling frameworks are exhibited by nearby, compartment gathering and single conduit frameworks. These frameworks are utilized to give agreeable conditions as far as air temperature and moistness for the team in settlement zones and other convenience regions, air filtration in galleys, arrangement rooms, and sterile zones and furthermore for air blending in compartments.

© Springer Nature Singapore Pte Ltd. 2020
V. K. Gunjan et al. (Eds.): *ICRRM 2019 – System Reliability, Quality Control, Safety, Maintenance and Management*, pp. 13–20, 2020.
https://doi.org/10.1007/978-981-13-8507-0_3

Fan and blower decision depend on upon the volume stream rate, weight, kind of material dealt with, space imprisonments, and capability. Fan efficiencies of blower are not equivalent to blueprint to design moreover by sorts. Fans fall into two general characterizations: outspread stream and center stream. In diffusive stream, wind current adjusts course twice - when entering and second when leaving (forward twisted, backward twisted, inclined, extended). In center stream, air enters and leaves the fan with no adjust its course (propeller, tube significant, vane center). The genuine sorts of outward fan are extended forward twisted and in turn around twisted. Winding fans are present day workhorses in light of their high static weights and ability to deal with seriously sullied air (Fig. 1).

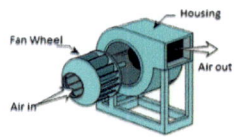

Fig. 1. Blower housing

2   Objective

- To study operations of blower under mechanical conditions.
- To reduce the weight of the blower assembly.
- To develop solid model of impeller with CATIA V5.
- Static analysis of blower assembly using ANSYS WORKBENCH 19.
- Modal analysis of blower assembly using ANSYS WORBENCH 19.
- To validate the simulated results to avoid failure.

3   Modeling of Blower

A simulation model for static and dynamic analysis of the blower is required to predict the stresses, deformation and mode shapes of blower assembly. The 3D model and finite element models of the bower assembly are made by using CATIA and ANSYS WORKBENCH 2019.

3.1   Geometric Modeling

The blower is composed of many components, as shown in Fig. 2. The blower structure, electric motor, lifting lugs, rotating shaft and the impeller. CATIA geometry files of the blower components are converted to STEP format and these format files are then imported into ANSYS WORKBENCH 2019. Each converted component is modeled as a rigid body that has mass, the center of mass and the moment of inertia. Figure 2 shows the important components and complete geometry of the blower assembly.

With the end goal to do a limited component investigation, the model we are utilizing must be partitioned into various little pieces known as limited components. Since the model is partitioned into various discrete parts, FEA can be depicted as a discretization method. In basic terms, a scientific net or "work" is required to complete a limited component investigation. In the event that the framework under scrutiny is 1D in nature, we may utilize line components to speak to our geometry and to do our investigation. In the event that the issue can be depicted in two measurements, a 2D work is required. Correspondingly, if the issue is intricate and a 3D portrayal of the continuum is required, at that point we utilize a 3D work. Region components can be triangular or quadrilateral fit as a fiddle. The determination of the component shape and request depends on contemplations identifying with the multifaceted nature of the geometry and the idea of the issue being displayed.

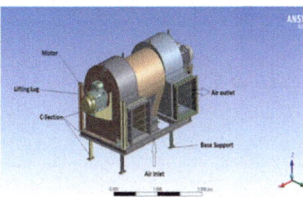

Fig. 2. 3D model of blower

3.2   Finite Element Model of the Blower

The meshing is performed for finite element analysis by using the geometry model of the blower with ANSYS WORKBENCH 2019. Through finite element analysis using ANSYS WORKBENCH, the modal parameters such as stresses, deformation, mode shape and natural frequency are calculated (Fig. 3).

Fig. 3. Mesh model

Element Quality
Skewness = 0.9998, Jacobian Ratio = 29.24.

4 Material Properties and Input Data

4.1 Material Properties

See Table 1.

Table 1. Material properties

Sr. no.	Material	Young's modulus Mpa	Poisons ratio	Density Kg/m^3
1.	Structural steel	2E + 05	0.3	7850
2.	Aluminium	71000	0.33	2770
3.	Nitrile rubber	5.5	0.499	1000

4.2 Input Data

Motor Rating: 37 Kw
Motor RPM: 985 rpm
Fan air flow: 12–13.2 m^3/sec

5 Analysis

5.1 Structural Analysis

Boundary Condition
The Blower model is provided with fixed support at four bottom base plate and standard earth gravity. The motor mass and impeller mass is provided as point mass applied to the bolting position (Fig. 4).
Moment = 358.68 Nm, Internal Pressure = 1063 Pa.

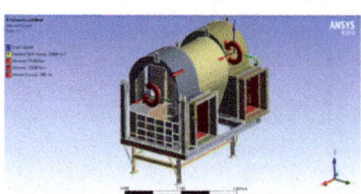

Fig. 4. Boundary conditions

Result
Total Deformation and Equivalent Stress (Figs. 5 and 6).

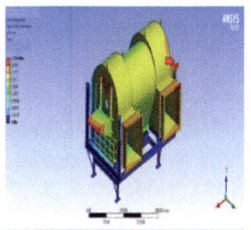 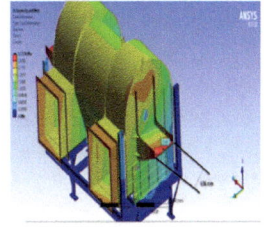 

Fig. 5. Total deformation   **Fig. 6.** Equivalent stress

5.2   Modal Analysis Mesh

See Fig. 7.

Fig. 7. Mesh model for modal analysis

Boundary Condition
The Blower model is provided with rubber stiffness at four bottom location and standard earth gravity.

Stiffness in X direction = 3625.178 N/mm
Stiffness in Y direction = 3658.2 N/mm
Stiffness in Z direction = 2771.9 N/mm

The motor mass and impeller mass is provided as point mass and applied to motor bolting location. As vibration isolator are used here hence vibration does not get transferred to the base structure, therefore we are just considering the assembly above the vibration isolator (Fig. 8).

Fig. 8. Boundary condition

Result
Mode Shapes (Figs. 9, 10, 11, 12, 13, 14, 15, 16*).*

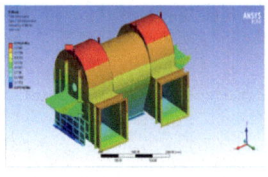

Fig. 9. Mode shape 1

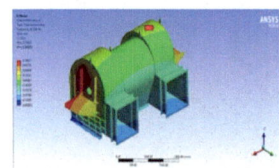

Fig. 10. Mode shape 2

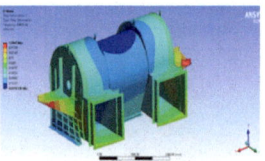

Fig. 11. Mode shape 3

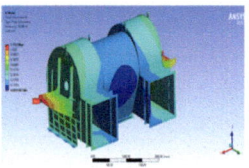

Fig. 12. Mode shape 4

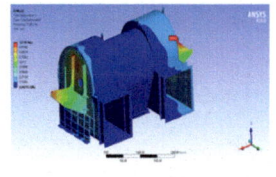

Fig. 13. Mode shape 5

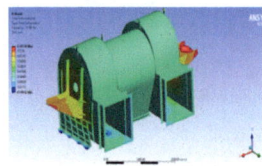

Fig. 14. Mode shape 6

Fig. 15. Mode shape 7

Fig. 16. Mode shape 8

Mode shape graph and table
The frequencies coming on the blower assembly and its values are given below (Figs. 17 and 18).

Fig. 17. Frequency vs. mode shape graph

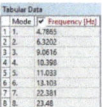

Fig. 18. Frequency on each mode shape

6 Frequency Calculation/Resonant Frequency Range Calculation

The RPM of motor as per given input is 985 rpm

The formula to find frequency is to divide the RPM by 60,

Frequency = 985/60 = 16.416 Hz

The resonant frequency range which with assembly should not operate is given by, 20% of excitation frequency

Therefore, we need to consider 20% in both plus and minus direction to find the resonant frequency range which is given as,

16.416 * 20% = 3.2833, The 20% of 16.416 is 3.2833

Hence, 16.416-3.2833 = 13.1327, 16.416 + 3.2833 = 19.6993

Hence the range of resonant frequency is **13.1327 to 19.6993 Hz**

7 Conclusion

1. From the structural analysis

(A) To pass the structure from stress point of view, the maximum allowable stress into the structure needs to be lower than the value $(0.6S_{yt} = 0.6 * 210 = 150$ Mpa). As we can see the stresses generated in the model is **52.985 Mpa** which is less than **150 Mpa**.

(B) To pass the structure from deformation point of view the maximum allowable deformation into the structure needs to be lower than the value L/300 as per IBC2015.

Where L = Length of the member on which maximum deformation is coming.

Maximum allowable deformation = L/300 = 636/300 = **2.12 mm.**

The Maximum deformation generated in the structure = 2.7176 − 2.5249 = **0.193 mm.**

As we can see the maximum deformation in the model is **0.193 mm** which is less than the allowable value **2.12 mm**.

2. From Modal Analysis

The maximum and minimum frequency coming from analysis is 4.7865 Hz and 23.48 Hz and no frequency is coming between the ranges of resonance i.e. 13.1327 to 19.6993 Hz.

Therefore from the above results of structural analysis and modal analysis we can assure that the structure is safe from rigidity point of view.

As the frequencies are not coming in the range of resonance we can conclude that at the time of working the structure will produce less vibration and noise.

So from all these points we can conclude that the life of blower assembly increases as the vibrations, deformation, stresses generated in the blower assembly are well within the limits.

References

1. Matta, K., Srividya, K., Prakash, I.: Static and dynamic response of an impeller at varying effects. Int. J. Appl. Sci. Eng. Manag.
2. Nizami, M.J., Sunman, R., Reddy, M.G.B.: Evaluation of static and dynamic analysis of a centrifugal blower using FEA. Int. J. Adv. Trends Comput. Sci. Eng. **2**(7), 316–321 (2013)
3. Dadhich, M., Jain, S.K., Sharma, V., Sharma, S.K., Agarwal, D.: Fatigue (Fea) and Modal Analysis of a Centrifugal Fan. Grob Design Pvt. Ltd., Jaipur, Rajasthan, India Apex Institute of Engineering and Technology, Jaipur, Rajasthan, India FET Agra College, Agra, Uttar Pradesh, India
4. Itha, V., Rao, T.B.S.: Static and Dynamic Analysis of A Centrifugal Blower Using Fea. Mechanical Engineering Department, Nimra Institute of Science and Technology, Ibrahim-patnam, Vijayawada, Andhra Pradesh, India
5. Iratkar, G.G.: Structural analysis, material optimization using FEA and experimentation of centrifugal pump impeller. Int. J. Adv. Res. Ideas Innovations Technol.
6. Mane, P.R., Firake, P.L., Firake, V.L.: Finite element analysis of M.S. impeller of centrifugal pump. Int. J. Innovations Eng. Sci. **2**(9) (2017)
7. Tare, D., Bhagat, V., Talikoti, B.: Static and dynamic analysis of impeller of centrifugal blower. Int. J. Innovative Res. Sci. Eng. Technol.

Quality Improvement Through Soil Stratum in Non-mechanized Treatment System for Wastewater

Sagar Mukundrao Gawande[✉] and Dilip D. Sarode

Department of General Engineering, Institute of Chemical Technology,
Mumbai 400019, India
gawande.sagar@gmail.com, dd.sarode@ictmumbai.edu.in

Abstract. The improvement in wastewater quality cited at prime importance because the quantities of the water available in surface and underground sources are depleted at very fast rate from last several years. This depletion is going on due to rise in water demand for development of industrial zones and fulfillment of growing population. The direct disposal of untreated wastewater from domestic and industrial activities contaminates the available quantity of surface and subsurface water beyond its acceptable limits. The consumption contaminated water not only leads to various water borne diseases but also pose difficulties in industrial processes. The soil has sufficient voids and permeability which can be used as medium to trap the available solid contaminates from the wastewater. The soil quality and characteristics will helpful to develop the quality improvement system to trap out the solids from wastewater. The void ratio and concentration of solids are important parameters to be considered in development of experimental stratum so that it will avoid the contamination of water bodies and improve the quality of wastewater. The motivational objective of this study is to develop the economical and efficient natural quality improvement system for wastewater treatment by locally available resources especially in rural areas where the improper sanitation practices are in existence. Authors are involved in water quality improvement for Ausa Lake by DWAT (Decentralize Wastewater Treatment) system followed by Reverse Osmosis (RO) unit under DST (Department of Science & Technology, Govt. of India) project at Ausa town of Marathwada region of Maharashtra.

Keywords: Quality improvement · Non-mechanized · Natural system · Safeguarding · Local resources

1 Introduction

In recent years the study area already faced the water stressed conditions and currently the availability of fresh water is reducing at alarming rate. An attempt is being made to develop a model by constructing a wastewater quality improvement system for the study area which will avoid the contamination of the lake. The study includes installation of water treatment unit so as to supply water to the town and surroundings after treated lake water passed through the Reverse Osmosis unit, so that it will achieve

© Springer Nature Singapore Pte Ltd. 2020
V. K. Gunjan et al. (Eds.): *ICRRM 2019 – System Reliability, Quality Control,*
Safety, Maintenance and Management, pp. 21–27, 2020.
https://doi.org/10.1007/978-981-13-8507-0_4

considerable savings in budget of local governing bodies and government authorities for supplying water by tankers under stressful condition. The paper will describe the total water availability in India to address the upcoming water scarcity and soil types and their pH with grain sizes to understand the interaction between voids in soil and solids in wastewater.

1.1   Water Availability

The Indian Territory every year receives rainfall of an average of 4,000 BCM (Billion Cubic Meters) out of which 48% ends in Indian River system from which only 18% can be utilized due to shortfall of storage basins and infrastructure facilities. Average annual rainfall in India is 1170 mm, in desert areas it is 100 mm and in Cherrapunjee 10000 mm. About 1047 BCM water is evaporated and runoff reducing the available water to 1953 BCM whereas the usable water is only 1123 BCM [1].

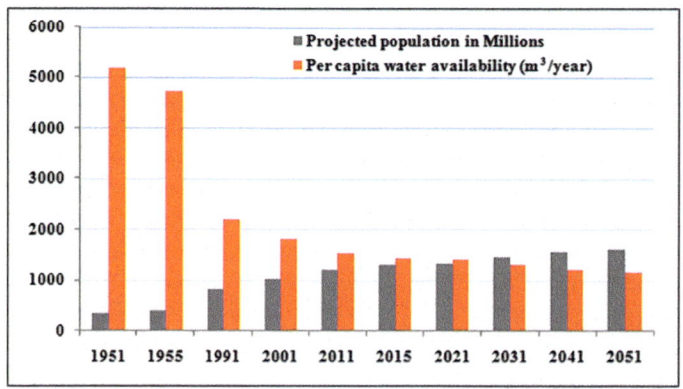

Fig. 1. Per capita water availability in India

The water availability in India is going to be a serious challenge, it is expected the population will be increase to 1.66 billion in the year 2050 and will face water stressed conditions. The Water Stress Indicator (WSI) is defined as per capita availability of water if it is less than 1700 cubic meters (m^3) similarly, while it is falls below 1000 m^3 capita^{-1} is expressed as a Water Scarcity Condition (WSC) [2] from Fig. 1 it is observed that in the year 1951 the per capita availability was 5178 m^3 year^{-1} and in year 2021, 2031, 2041 and 2051 the water availability will be 1421, 1306, 1225 and 1174 respectively which decreases at the rate of 72.55%, 74.77%, 76.34% and 77.32% indicating water stressed conditions.

1.2   Generation and Reuse

Table 1 describes the wastewater generation in India since year 1988 to year 2012, in the year 1990 the quantity of municipal wastewater generated was 4.9×10^9 m^3/year whereas in year it was 15.459×10^9 m^3/year which means around 31.71% increase in last 20 years. The higher discharge of untreated wastewater in year 2011 from primary treatment was 11.03×10^9 m³/year which is 36.11% more compared with year 1990. After secondary treatment about 62.52% wastewater discharge in year 2008 which is lesser than 81.28% in the year 1990 due to treatment through the available treatment facilities. It is estimated that 22900 Million Liters per Day (MLD) of domestic wastewater is generated from urban centers and 13500 MLD of industrial wastewater [3].

Table 1. Wastewater generated and treated in India during year 1988–2012 [5]

Description of particulars and quantity of municipal wastewater in $(10^9$ m³/year)	Year of assessment				
	1988–1992	1993–1997	1998–2002	2003–2007	2008–2012
Produced municipal wastewater	4.9 (1990)	6.684 (1995)	10 (2001)	9.583 (2004)	15.45 (2011)
Treated municipal wastewater	0.917 (1990)	—	1.631 (2001)	—	4.416 (2011)
No. of municipal wastewater treatment facilities	—	—	—	498 (2003)	498 (2009)
Not treated municipal wastewater	3.983 (1990)	5.188 (1995)	8.369 (2001)	7.012 (2004)	11.03 (2011)
Not treated municipal wastewater discharged (secondary water)	3.983 (1990)	5.188 (1995)	8.372 (2001)		9.66 (2008)

The reuse of wastewater may be evaluated through the evaluation of wastewater reuse potentials with total use of available fresh water. The wastewater reuse is small compared with total fresh water, but it is expected to increase significantly (Table 1) in coming years which can be use in agricultural, industrial processes, firefighting, aquaculture, domestic use, wetland creation and aquifer recharge, [4] fire frightening, flushing of toilet, cleaning of public toilets and sewers, and other municipal uses.

2 Types of Stratum

Soil is a blend of living as well as non-living matters which develop on the earth's surface. The climate, time, and biodiversity including the human activities are major factors which affects the formation of soil whereas the nature of soil in a place is basically influenced by such factors like climate, natural vegetation and rocks [6]. In India historically the soil was classified as Urvara (fertile) and Usara (sterile) [7]. The major classification of Indian soils tabulated as in Table 2.

Table 2. Range of distribution and types of soils in India with their pH [6]

Sr. no.	Type of soil	pH range	Range of distribution
1	Alluvial soil	6.5–8.4	River valley of Ganga and Brahmaputra, plains of Uttar Pradesh, Uttaranchal, Punjab, Haryana, West Bengal, Bihar
2	Desert soil	7.6–8.4	North Gujarat, South Punjab, and Rajasthan
3	Black soil	6.5–8.4	North Karnataka, Telengana, Rayalsema, Maharashtra, Kathiawar
4	Red soil	5.5–7.5	South of Kerala, Tamilnadu, Karnataka, Chota Nagpur
5	Grey and brown	7.6–8.5	Rajasthan & Gujarat
6	Laterite soil	Less than 5.5	Hills of Assam, Summits of Karnataka-Kerala, Eastern Orissa
7	Mountain soil	5.0–6.5	Sikkim, Uttaranchal, Himachal Pradesh, Jammu & Kashmir

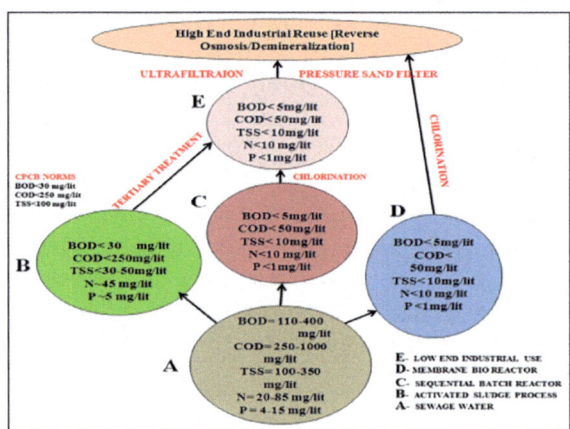

Fig. 2 Showing reuse of raw sewage and its characteristics after various treatment [8, 9]

3 Soil and Wastewater Composition

The organic matter of the soil includes both living organisms and dead organisms. A soil comprises mainly four components as Mineral matter, Organic matter, Air, and water. In an average soil sample is 45% minerals, 25% water, 25% air, and 5% Organic matter. The amount of air and water filling the pore space in the soil varies depending on the amount of precipitation, location, and water holding capability of the soil shown in Fig. 3.

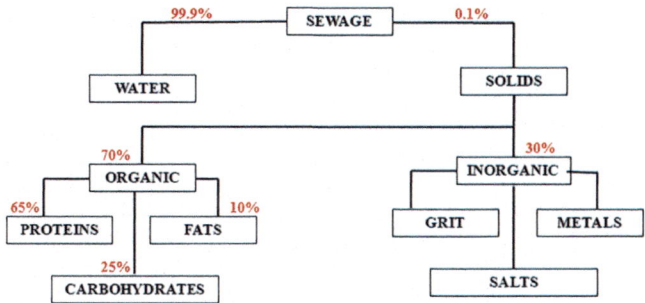

Fig. 3. Showing composition of wastewater (Sewage) [10]

The typical sewage or wastewater has 99.9% of water and only 0.1% of solids out of which 30% are inorganic matter and 70% organic matter which has 65% proteins with fatty matter 10%, the carbohydrates are available in wastewater around 25% as shown in Fig. 3.

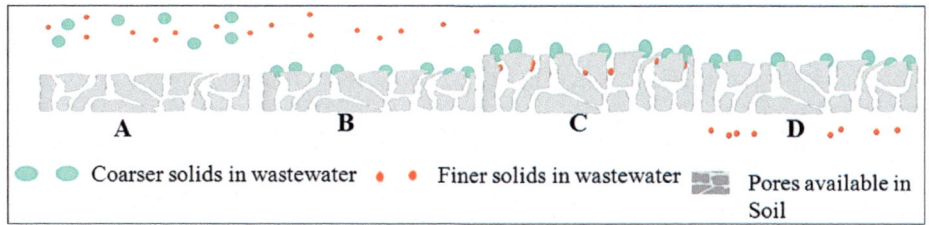

Fig. 4. Quality improvement of wastewater and voids in soil

4 Solids of Wastewater

In India the suspended solids standard is ≤ 100 mg/l. The classification of solids based on percentage of constituents considering Total Solids 100% in which about 30% are non filterable i.e. Suspended solids whereas remaining 70% are filterable [11]. Industrial and domestic activity decides the characteristics of wastewater in a town. Figure 2 giving the details of wastewater characteristic before and after treatment of raw wastewater using different treatments. In Indian conditions the temperature of wastewater varies from 15 °C to 35 °C and pH of raw sewage is in the ranges from 5.5 to 8.0. The size of the particles in the wastewater range from settleable (e.g. >100 µm), supra colloidal (e.g. 1–100 µm), colloidal (e.g. 0.08–1 µm), and soluble (e.g. <0.08 µm) [12]. Contaminates of interest in wastewater generally varies in the size range from less than 0.001 to 100 µm (1 µm = 0.001 mm). The material that comprises the Biochemical Oxygen Demand (BOD) and Suspended Solids (SS) are smaller than 50 µm [13].

Table 3. Indian standard soil classification system and their corresponding permeability

Type of soil	Class size	Class	Grain-size range	Permeability k in (cm/sec)
Very coarse soils	Boulder size	–	>300 mm	–
	Cobble size	–	80–300 mm	–
Coarse soils	Gravel size (G)	Coarse	20–80 mm	–
		Fine	4.75–20 mm	100
	Sand size (S)	Coarse	2–4.75 mm	10^0 to 10^{-1}
		Medium	0.425–2 mm	10^{-1} to 10^{-2}
		Fine	0.075–0.425 mm	10^{-2} to 10^{-3}
Fine soils	Silt size (M)	–	0.002–0.075 mm	1×10^{-5}
	Clay size (C)	–	<0.002 mm	10^{-7} to 10^{-9}

5 Pores and Solids Relation

Water that travels into and through soil is cleaned by physical, chemical, and biological processes. When pollutants carried by water get trapped in the small pores of the soil, they get physically cleaned. Nearly all soils have slight chemical charges which attract and capture chemicals with the opposite charge for instance; soils (especially clayey soils) are negatively charged. Positively charged substances like Ammonium (a form of Nitrogen) are attracted to the soil. The altered pollutant by microorganisms in soil may use the pollutants and transform it to energy or food for their survival.

From the Fig. 4 it if observed that the coarser and finer solids from the wastewater are initially comes in contact with the top of soil surface as shown in part A. The coarser solids are settled and trapped near the top layer of soil as shown in part B whereas the finer solids are travelled through the pores (voids) available in next layer of soil during the transport of these solids some of them degraded and consumed by the microorganism inside the soil stratum as shown in part C. The balance or remaining solids trapped and decomposed by the pores and biota available within the layer as shown in Fig. 4 part D.

6 Conclusion

Holding and transferring of water is one of the most important roles that soils play in our ecosystem which takes place through the available pores in the soil. The pores of a soil are significant in determining if water will percolate the soil and join the subsurface water. The good structure of soil will have lots of big and small pores, even if it is clayey one [14]. Soils contain clay minerals and organic matter as a result of weathering and the addition of organic debris soilsact as buffer zone between atmosphere and ground water, and provide plants a steady supply of nutrients. The natural wastewater treatment processes are very useful for local bodies which have limited financial resources but have ample open land. Whereas if the modern treatment processes is adopted the major share of expenditures of the budget accumulated by cost of operation and maintenance of the treatment plant. Very less resource remains for other functions of local bodies.

Acknowledgement. Authors are thankful to Department of Science & Technology (Govt. of India) for funding the project DST/WTI/2K16/306 for developing a model for rejuvenation of Ausa Lake.

References

1. Bhat, T.A.: An analysis of demand and supply of water in India. J. Environ. Earth Sci. **4**(11), 69 (2014)
2. Basin Planning and Management Organisation Central Water Commission, New Delhi – 110 066, p. 3, October 2017
3. Central Pollution Control Board, Government of India, p. 1. http://cpcbenvis.nic.in/cpcb_newsletter/sewagepollution.pdf
4. Khajuria, A.: Application on reuse of wastewater to enhance irrigation purposes. Univ. J. Environ. Res. Technol. **5**(2), 72–78 (2015)
5. Food and Agriculture organization of the United Nations-AQUASTAT-India-Results
6. Siddiqui, S.A., et al.: Indian soils: identification and classification. Earth Sci. India **10**(III), 1–14 (2017)
7. Soils of India: Classification and Characteristics, Clear IAS web page
8. Water and Wastewater in India, European Business and Technology Centre, Snapshot. www.ebtc.eu
9. https://www.iss.k12.nc.us/cms/lib4/NC01000579/Centricity/Domain/1247/MS_soils-Manual.pdf
10. Mara, D.: Domestic Wastewater Treatment in Developing Countries, pp. 18–20. Earthscan, UK and USA (2004)
11. Rössle, W.H., et al.: A review of characterization requirements for in-line prefermenters. Water SA **27**(3) 405–407 (2001). ISSN 0378-4738
12. Okamoto, R.: Particle Size Analysis of Decentralized Wastewater Treatment System. California State Polytechnic University, Pomona, August 2016
13. Levine, A.D., et al.: Characterization of the size distribution of contaminates in wastewater treatment and reuse implications. Water Pollut. Control Fed. **57**, 805 (1985)
14. Soils Clean and Capture Water, April 2015, Soil Science Society of America. www.soils.org/IYS

An Experimental Examination to Achieve of High Strength Concrete Using Crushed Sand

Ajay Shelar[⊠], D. Neeraja, and Amit B. Mahindrakar

Vellore Institute of Technology, Vellore, India
ajaydidit@rediffmail.com

Abstract. Excellent cement supplanted conveyed sand is the all the more great position in the Development business. The fundamental focal point of Excellent cling to build up the compressive idea of cement by supplanting Characteristic sand into Make sand and utilizing admixture. To look at the value of conveyed sand and utilizing admixture in concrete. To examine the execution of this solid terms of its compressive quality and split inflexible nature. This paper impels the employments of beat sand as an endeavor towards supportable change in India. It will discover achievable reaction for the declining accessibility of normal sand to make eco-change. Conveyed sand is one among such materials to supplant stream sand, which can be utilized as an elective fine total in mortars and bonds. The utilization of conveyed sand in concrete is getting power nowadays. The present fundamental examinations have been made on solid utilizing made sand as fine total and watched the impacts of pummeled sand on quality properties of security.

Keywords: Compressive quality · Flexural quality · Smashed sand · Mineral admixture

1 Introduction

Before long a-days, the Legislature have put disallowance on lifting sand from Stream bed. Basic coordinating factors for HSCs are quality, entire arrangement quality, value as oversaw by break and redirection control, and moreover reaction to entire arrangement typical impacts [1]. Solid blend outline of M40 review was finished by Indian Standard code Solid 3D square; bar and round and void cases were made progress toward assessment of compressive and split adaptability freely. The solid shows unprecedented quality with 100% substitution of standard sand, so it can be utilized as a bit of concrete as sensible other contrasting option to customary sand. Strong blends with various % rates of fly fiery remains as security substitution material were examined [2]. The estimations of silica seethe were diverse % of the cementitious materials. The compressive idea of cement got at the ages of 7, 28, days.

© Springer Nature Singapore Pte Ltd. 2020
V. K. Gunjan et al. (Eds.): *ICRRM 2019 – System Reliability, Quality Control, Safety, Maintenance and Management*, pp. 28–32, 2020.
https://doi.org/10.1007/978-981-13-8507-0_5

2 Exploratory Program

2.1 Characteristic Sand Versus Made Sand

The sand from stream because of typical philosophy of crippling has a tendency to have smoother surface and better shape. It in like way passes on dampness that is gotten in the midst of the particles [3]. These characters improve solid convenience. Regardless, development and earth passed on by stream sand can be harming to the solid. Another issue related with conductor sand is that of getting required surveying with a fineness modulus of 2.4 to 2.8. It has been attested and found, at different regions transversely completed south India, that it has wound up being persistently hard to get stream sand of strong quality to the degree investigating fundamentals and obliged residue/earth content. It is in light of the way that we don't have any control over the trademark methodology. In case there should develop an occasion of made sand, the system of steadfast mishap through VSI and washing makes the pounded stone sand particles satisfactory to be mulled over shape and surface of general sand. With especially orchestrated screening framework the required investigating (Zone II) and fineness modulus (2.4 to 2.9) can in like way be master reliably by goodness of beat sand. It must be seen that fittingly organized conveyed sand can overhaul both compressive quality and flexural quality through better bond showed up distinctively in connection to course sand.

2.1.1 Fine Aggregate

River Sand - Incredible quality ordinary conduit sand is quickly open in various domains and may be easily gotten and dealt with. Moreover with the stone that they routinely run with, the sand stores won't not have been laid reliably, which implies a potential change in quality. Generally fines are gathered in perspective of size, i.e.; underneath 4.75 mm is seen as fine aggregate. The mass thickness of conduit sand was 1963 kg/m3.

Crushed Sand - Fine total utilized as a bit of this examination is Squashed sand. The totals whose size is under 4.75 mm. It was collected from RPP Arranged Mix PLANT, Break down, India. The mass thickness of made sand was 1860 kg/m3. Precisely when shake is beat and evaluated in a quarry the basic point has for the most part been to pass on coarse totals and street headway materials meeting certain unobtrusive components. For the most part, this procedure has left finished a level of overabundance fines of variable properties, by and large better than anything 5-mm measure. Made sand is utilized for mean material under 4 mm that is dealt with from squashed shake or shake and foreseen manufactured change utilize.

Sand is a material of high check, in abnormality to non-refined surplus from coarse total creation. The utilization of made totals (beat hard shake) in concrete has been known since the Roman time. In demonstrate day headway, essential totals have wound up being in a general sense gentle being used, for which reason wide utilization of made wholes has been concentrated to districts or tries where the transparency of trademark totals has been constrained.

2.2   Coarse Aggregate

Coarse aggregate of apparent size of 20 mm & 12 mm is picked and tests to choose the different physical properties as per IS 383-1970. Test results fit in with the IS 383 (Sect. 3) proposition. The mass thickness of coarse aggregate 1691 kg/m3. Coarse add up to 12 mm and 10 mm was used, which was smashed from locally open shake.

2.3   Admixture

2.3.1   Fly Fiery Debris

Fly intensely hot debris is a delayed consequence of the warm power plants and the measure of them are broadening. Clean gathering framework expels the fly intensely hot junk, as a fine particulate advancement, from begin gases before they are released into the air. The sorts and relative measures of incombustible issue in the coal utilized pick the compound structure of fly singing remains. Over 85% of most fly burning remains is consolidated blend mixes and glasses restricted from the sections silicon, aluminum, iron, calcium, and magnesium.

2.3.2   Silica Smoke

It is a thing occurring in light of diminishing of high faultlessness quarts with Coal in an electric indirect area hotter in the make of silicon or ferrosilicon composite. Silica fume ascends as an oxidized vapor. It cools, aggregates and is collected in surface packs. Silica smolder as an admixture in concrete. It is additionally orchestrated to discharge corruptions and to control molecule evaluate. Since it is an airborne material like fly singing junk, it has round shape. Silica seethe has wound up being one of the basic parts for making high gauge and transcendent cement.

2.3.3   Compound Admixture

Super-plasticizer is a principal section for brilliant concrete. Conplast SP 430 was used. Conplast SP430 is the chloride free, super-plasticizer in light of sulphonated naph-thalene polymers super plasticizer is a stand-out functionality holding unrivaled super-plasticizer with excellent properties. Super-plasticizer admixture at 1.5% by weight of cement was used to get the pined for usefulness.

3   Blend Plan

The consolidate course of action was done in light of the suggested keeps running in Indian Benchmarks. The fundamental supposition made in the Indian standard technique for blend course of action is that the compressive idea of helpful cement is everything seen as addressed by the water/security degree. In this strategy the water substance and level of fine total appearing differently in relation to a most exceptional size of total are first browsed the reference estimations of value, water-strong degree, and the inspecting of fine total. The water substance and level of fine total are then balanced for any refinement in value, water/bond degree and surveying of fine total in a specific case. Figure 1 tends to hang testing and Fig. 2 tends to hang a rousing power in fly ash and hang a helper in silica seethe.

Fig. 1. Slump testing

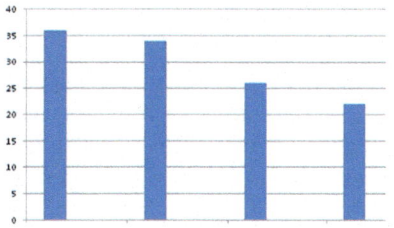

Fig. 2. Slump value on fly ash

4 Methodology

4.1 Cube Compressive Strength

For solid shape weight testing of concrete, 150 mm 3D shapes were utilized. Every last one of the solid shapes are endeavoring at 7 days years old, 28 days, 56 days of diminishing utilizing weight testing machine. Demonstrates compressive quality test. Stacking is proceeded till the dial check needles invert its course of improvement. The inversion toward improvement of the needle shows that the representation has bite the dust. The dial measure looking at immediately is note which is an entire load. A legitimate load distributed the cross sectional zone of the delineation is indistinguishable to an indisputable solid shape compressive quality.

4.2 Splitting Tensile Strength

Figure 6 addresses choose the inflexibility of barrel molded cases. part unbending nature tests were bear on barrel cases of size 100 mm width and 200 mm length at the age of 7, 28, 56 days restoring, using weight testing machine. To avoid the quick load on the illustration the round and empty cases were kept underneath the cases spilt and readings were noted. The part versatility has been test results in Table 1:

Table 1. Test results

SI. NO	Cube Id	Compressive strength of concrete (N/mm^2)		Split tensile strength of concrete (N/mm^2)	
		7 days	28 days	7 days	28 days
1	River sand	18.76	25.13	1.23	2.13
2	Creush sand	19.76	25.67	2.68	3.46
3	5.5% of silica fume	20.23	25.15	2.16	3.25
	10.5% of silica fume	21.56	25.57	2.86	3.78
	12% of silica fume	21.78	25.46	2.78	3.65
4	12% of fly ash	18.6	25.56	2.25	3.5
	22% of fly ash	18.23	25.84	2.94	3.86
	32% of fly ash	16.26	263.76	2.49	3.32

5   Conclusion

In the current circumstance reasonable enhancements there is a basic prerequisite for utilizing present day squander and other exchange materials to proximity the earth usage of pounded sand as a substitute material in the place of stream sand is portal transcendent extensively finished the globe. Use of C-sand in the amazing bond positively requires some portion of preliminary examinations. In the endeavor an undertaking has been need to consider the quality properties of m60 great bond by including differing proposal of mineral admixture like silica smoke and fly searing stays following are the hugeness conclusion landed in perspective of the examination. A Far reaching Study had been finished on various journals and books related to the top notch concrete with beat sand and diverse admixtures. The correct materials critical for the arranging of cases for experimentations have been refined. The measure of illustrations and the tests to be coordinated have been predestined. The level of substitution 0%, 5.5%, 10.5%, 12% in silica seethe. The most extraordinary quality achieves the level of 6.5.

References

1. IS 2386-1997: Techniques for Test For totals For Concrete
2. IS 1199-1999: Techniques for Examining and Examination of Cement
3. IS 516-1999: Techniques for Tests for quality of Cement

Stress and Failure Analysis of Aircraft Wing Using Glare Composite and Aluminum 7075

Piyush Sondankar$^{(\boxtimes)}$ and R. R. Arakerimath

Department of Mechanical Engineering, GHRCEM, Pune 412207, India
{sondankar_piyush.ghrcemmtechcad,
rachayya.arakerimath}@raisoni.net

Abstract. Aircraft is an advanced and fastest mode of transport in these days. Major research work for its technical advancements such as material invention, design improvement in aerospace industry is still going on. The paper will discuss the stress induced in wing made of aluminum 7075 and glare separately. The stresses induced because of lift and drag are studied. The materials in our paper consideration are glass aluminum reinforced composite and aluminum alloy 7075. The materials chosen should withstand the aerodynamic loads like drag, lift and the stresses generated inside wing. From the selection of manufacturing materials to the analysis and testing, the induced stresses in component is crucial parameter. The work presented here contains stress analysis of wing and result discussion. Two main stresses are calculated like total equivalent stress and maximum principal stress.

Keywords: Stress analysis · Lift · Wing materials

1 Introduction

In light of inaccessibility of complex machines and PCs, structure process included vast number of human power. When they finish this stage of plan effectively, another stage appears there with greater complexity than this known as analysis. If not, the procedure restarts from structure once more. As, structure of airplane is for some time run process, it is separated into some basic stages. The main names with design stage. This stage contemplates the design format of the flying machine and need of every single real trademark so as to achieve its plan objectives. If the applied structure result in progress, no major are permitted to fuse in it. The examination part here consists of static analysis of the flying machine wing. Auxiliary investigation of the air ship wing decides the anxieties and mishaps in wing. Rather than estimation of various sorts of stresses like shear, standard, coordinate and so forth. It is smarter to figure the aggregate proportionate stresses in whole structure known as Von-Mises stresses. The greatest measure of pressure prompted in any building part is known as maximum principal stresses. A wide scope of combination or composite materials might be utilized in the structure of the air ship for improvement of properties, for example, quality, versatility, explicit weight. Diverse materials can likewise be utilized in the plan air ship, as a component of the underlying necessities of the strength-to-weight proportion and the special bearings of the connected burdens. In the movement of flying machine, two noteworthy

© Springer Nature Singapore Pte Ltd. 2020
V. K. Gunjan et al. (Eds.): *ICRRM 2019 – System Reliability, Quality Control,
Safety, Maintenance and Management*, pp. 33–39, 2020.
https://doi.org/10.1007/978-981-13-8507-0_6

powers lift and drag are resolved. In this undertaking additionally, lift and drag powers are determined according to the standard formulae. The greatness of drag drive is less when we contrast it and the lift compel. This powers are then connected to the model in workbench and results are additionally examined. Stresses induced in the wing because of lift and drag powers also gives the idea of critical areas or joints where stress concentration is significant. If such critical results appear in analysis, the areas or joints are redesigned in accordance to have stress within standard permissible limits.

2 Literature Review

Sureka and Meher [1], worked on the difference between the values of equivalent stress, deformation, stress intensity, maximum principal stress and shear stress of both Al alloy and aluminum alloy 7068 are less with provision of proper validation. Aluminum alloy 7068 give more strength to the structure instead of other aluminum alloys. Pressure effect during take-off was more for Aluminum and less for Aluminum alloy 7068. Alloy 7068 being strongest and light weight, it reduces the weight of wing. Results obtained with aluminum alloy 7068 i.e. shear stress, stress intensity, equivalent stress, max principal stress are below the ultimate strength of alloy. Thus at assumed loading conditions and constraints failure of A300 flight wing will not occur suggesting replacement of aluminum alloy with aluminum 7068.

Sruthi, Lakshmana Kishore, Komaleswara Rao [2], concluded that the difference between the values of max principal stress, stress intensity, equivalent stress, deformation, and shear stress with Al alloy and aluminium + silicon carbide are minimal. Optimum results were obtained. The difference between the two result being minimal use of aluminum + Silicon carbide instead of using aluminium alloy only enhances strength of structure. Thus conclusion of research work is that at the above assumed loading conditions and constraints, flight wing structure work successfully due to material properties of aluminium + silicon carbide other than aluminium alloy. Konayapalli and Sujatha [3], watched, flying machine wing with aluminum amalgam acquired 144 MPa and for the composite material, result was 53.03 MPa. By utilizing the composite material less burdens had created on flying machine wing. An air ship wing with aluminum compound got 0.59 mm and for the composite material, it resulted 0.46 mm. By utilizing the composite material less mishapening had created in airplane wing.

Atmeh, Hasan and Darwish [4], The applied plan gives a general state of the wing. The paper at that point proceeds onward to an increasingly exhaustive plan depicting the parts of the wing and the qualities they require to perform under the determined burdens that the wing is intended to experience. The wing is then displayed utilizing Pro-E and imported into COMSOL Multi-material science to play out a limited component investigation demonstrating the wing as a decent possibility for a plane like the one presented toward the start of paper.

Soares [5], to accomplish the examination is vital to see how the heaps required for measuring an airplane structure are acquired and for achieving these heaps should experience a long procedure of enumerating and indicating a flying machine among others. The wing model made on limited component was tried for level and rising flight condition. The outcomes got are in great concurrence with the air ship working conditions and approved the numerical model made. It was discovered that basic regions of the flight condition were dissected: the cover boards in the association area to the fuselage and thusly the ribs of the fights near the root. Consequently they had embraced strengthening on the spreads, stringers on range astute heading.

3 Analytical Section

3.1 Solid Modelling of Cassena 172 Wing

The solid modelling of the airfoil was made in cad tool as shown in Fig. 1. The chord of the airfoil was taken as 1.6256 m at root 1.1303 m at tip [6]. It has span of 11 m [7]. The aerofoil shape thickness at root is 180 mm & at tip it is 122.28 mm. Figure 1 shows the software model of cassena 172 aircraft wing.

3.2 Load Calculations

The most extreme slope weight (otherwise called maximum ramp weight) is the most extreme weight approved for maneuvering an air ship on the ground as restricted via air ship quality and airworthiness prerequisites. Lift & drag coefficients, aircraft velocity are as below [8].

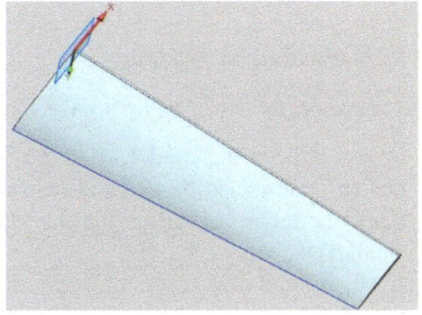

air density (ρ) = 1.225 kg/m^3
Coefficient of lift (C_l) = 1.60 Aircraft velocity (v) = 25.48 m/s
Coefficient of drag (C_d) = 0.0342

Fig. 1. NX model of aircraft wing

To calculate the lift and drag acting on wing surface, projected areas must be known. For lift, projected area is product of average chord length and half of wing span.

$$\therefore A = 1.3778 \times 5.5 = 7.578\,\text{m}^2$$
$$\text{Lift force (L)} = 1/2 \times \rho \times v^2 \times C_l \times A$$
$$= 0.5 \times 1.225 \times 25.48^2 \times 1.60 \times 7.578 \tag{1}$$
$$L = 4821.47\,\text{N}$$

For drag, projected area is product of average thickness of aerofoil and half of wing span.

$$\therefore A = 0.15144 \times 5.5 = 0.83127\,\text{m}^2$$
$$\text{Drag force (D)} = 1/2 \times \rho \times v^2 \times C_d \times A$$
$$= 0.5 \times 1.225 \times 25.48^2 \times 0.0342 \times 0.83127 \tag{2}$$
$$D = 11.30\,\text{N}$$

3.3 Material Specification

The materials utilized in aeronautic trade are for the most part aluminum, titanium, steel and their compounds. Among these, aluminum is favored over titanium for cost reason. The AA7075 is aluminum alloy with percentage concentration of materials like zinc 6.0%, copper 1.6%, magnesium 2.4%, iron 0.26%, manganese 0.21% and silicon 0.18%. A standout amongst the most alluring frameworks for new throwing combination designers was the Al–Zn–Mg–Cu, which is the essential framework for improvement of the most grounded compounds of the AA7075 type [9]. Table 1 shows properties of AA7075 materials [10]. Glare has aluminum 2024-T3 constituents with S2 glass fibers fortified together with unidirectional or biaxially strengthened cement prepreg of high-quality. Table 2 shows the properties of glare with contents of aluminum 57.9% and fiber 25.3% [11]. Glare covers offer as exceptional exhaustion obstruction, incredible effect opposition, great remaining and quality, consumption properties, and simplicity of produce and fix [12].

4 Structural Analysis in ANSYS Workbench

4.1 Mathematical Modeling

First, develop model in cad package for wing and save as this part as IGES.

- import IGES wing model in ANSYS.
- Define material properties and constants for aluminum 7075 and glare separately.
- Apply boundary condition for analysis. Here, one end is fixed. Wing is analyzed as cantilever beam.
- Apply lift and drag forces of magnitude 4821.47 N and 11.30 N respectively one by one. Lift is applied in bottom joining skin and drag is applied on top skin and leading edge.
- Insert the quantities equivalent stress and maximum principal stress which are to be found out.
- Do analysis for both AA7075 and glare composite material as static structural.
- Evaluate results.

Table 1. Properties of AA7075

Young's modulus	71.5 GPa
Density	2810 kg/m^3
Poisson's ratio	0.33

Table 2. Properties of glare

Young's modulus	54.8 GPa
Density	2446 kg/m^3
Poisson's ratio	0.25

A static analysis gives results of steady loading conditions on a structure. Static analysis done here decides the displacements, stresses, strains in structures. Here, steady loading conditions are assumed; that is, load is not going to change with time. The results of static analysis with the total equivalent stress in entire model are evaluated and maximum principal stress as well.

4.2   Structural Analysis of Wing with Material AA7075

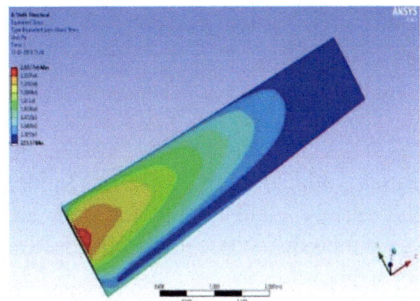

Fig. 2. Equivalent stress of AA7075 material

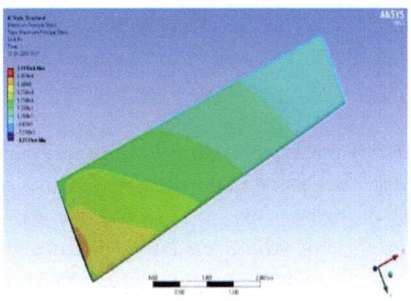

Fig. 3. Maximum principal stress of AA7075

4.3   Structural Analysis of Wing with Glare Material

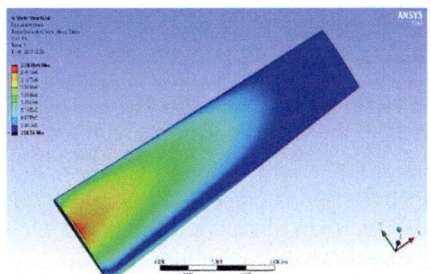

Fig. 4. Equivalent stress of wing of glare

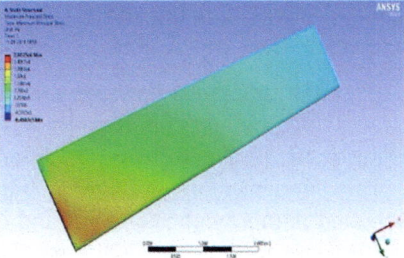

Fig. 5. Maximum principal stress of glare

5   Failure Criterion

In the work, manage of just static investigation and not the streamlined and execution is presented. For static examination, ductile materials fractures when stress generated in them exceeds their standard yield strength [13]. This is constantly maintained a strategic distance from to have fruitful plan of what we are planning. For flexible materials, they are considered as fizzled when push created inside them surpasses standard yielding quality of those materials.

Table 3. Comparison of generated and yield stress values for AA7075 and Glare

	AA7075	Glare
Max principal stress (MPa)	3.1912	2.8075
Equivalent stress (MPa)	2.6577	2.5838
Yield strength (MPa)	503	399

Both materials mainly comprise of aluminum, these materials are not expected to undergo behavior like brittle materials. Therefore their yield strength is taken as failure level of stress. Stresses created are far below the exactly yield quality for both the materials. For whatever length of time that static examination is concerned, structure is adequate to withstand the heaps. Fracture will not happen.

6 Conclusion

In this work, the flying machine wing model is demonstrated by NX programming. At that point, the model was foreign to ANSYS. We see stresses produced in wing with material AA7075 have 3.1912 MPa as greatest key stress and 2.6577 MPa as aggregate proportional stress. On the other case with composite glare, 2.8075 MPa as most extreme chief stress and 2.5838 MPa as aggregate identical stress.

(i)   Even the difference between the equivalent stresses in wing for both materials does not look large but actually it is equivalent to 73.9 kPa. Similarly, in case of maximum principal stress, the difference in generated stress is around 383.7 kPa.

(ii) This fundamentally influences material choice process. The aluminum compound is more delicate to lift and drag powers than glare composite.

(iii) Glare composite can withstand higher measure of streamlined powers than aluminum compound 7075.

References

1. Sureka, K., Meher, R.S.: Modelling and structural analysis on A300 flight wing by using ANSYS. J. Mech. Eng. Robot. Res. **4**(2), 123 (2015). ISSN 2278-0149
2. Sruthi, K., Lakshmana Kishore, T., Komaleswara Rao, M.: Design and structural analysis of an aircraft wing by using aluminum silicon carbide composite materials. IJEDR **5**(4), 949–959 (2017). ISSN 2321–9939
3. Konayapalli, S.R., Sujatha, Y.: Design and analysis of aircraft wing. IJMETMR **2**(9), 1480–1487 (2015). ISSN 2348-4845
4. Atmeh, G.M., Hasan, Z., Darwish, F.: Design and stress analysis of a general aviation aircraft wing. Jordan University of Science and Technology, Irbid, Jordan, Texas A&M University, College Station, Texas (2010)
5. Soares, L.F., Lapa, E.G., Almeida, P.A.: Structural analysis of a wing box. IJERA **5**(5), 23–31 (2015). ISSN 2248–9622
6. https://answers.yahoo.com/question/index?qid=20081018140002AAApAwsl

7. https://disciplesofflight.com/cessna-172-skyhawk/
8. McIver, J.: Cessna Skyhawk II/100 Performance Assessment. http://www.temporal.com.au/c172.pdf
9. Zolotorevsky, V.S., Belov, N.A., Glazoff, M.V.: Casting Aluminum Alloys. Elsevier Science, Amsterdam (2007)
10. Aerospace Materials Specification data sheet. http://asm.matweb.com/search/SpecificMaterial.asp?bassnum=ma7075t6
11. Botelho, E.C., Silvac, R.A.: A review on the development and properties of continuous fiber/epoxy/aluminum hybrid composites for aircraft structures. Mater. Res. 9(3), 247–256 (2006)
12. Balachandra, P., Shetty, B., Reddy, S., Mishra, R.K.: Finite element analysis of an aircraft wing leading edge made of glare material for structural integrity. J. Fail. Anal. Prev. 17(5), 948–954 (2017). https://doi.org/10.1007/s11668-017-0331-2
13. Bhandari, V.B.: Design of Machine Elements, 3rd edn. Tata McGraw Hill Education Pvt. Ltd. (2012)

Thermal Energy and Failure Study of Reheating Furnace: Model Development and Simulation

Sharad B. Masal$^{(\boxtimes)}$ and R. R. Arakarimath

Mechanical Engineering Department, GHRCEM, Pune, India
sharad.bm@gmail.com

Abstract. Reheating of steel is a continuous process. The basic purpose of the reheating furnaces is to heat the steel slabs/billets/blooms/ingots to an appropriate temperature before forging or hot rolling operation. We have noticed that the main problems in reheating furnaces are non-uniform flame distribution (Gas flow pattern and flame interaction), oxidation of metal, scale formation, carbon loss of metals and emission of pollutants. Therefore, because of these problems reheating furnaces have less productivity and high running cost. On the other hand very fast heating of slabs/billets/blooms/ingots results excessive thermal stress, cracks and distortion during forging or hot rolling operation. To avoid these operational problems reheating furnaces should be properly modeled, designed and analyzed by using heat transfer (HT) and Computational Fluid Dynamics (CFD) techniques. It is possible that numerical/mathematical heat transfer models or with special technique of CFD can improve and increase product quality with reducing high running cost. This paper presents the current modeling work demand and review on latest trends and developments available in the area of reheating furnaces for computational simulation and analysis.

Keywords: Reheating furnace · Heat transfer ·
Computational Fluid Dynamics (CFD) · Computational simulation ·
Fluid flow · Crack formation · Thermal stress · Distortion

1 Introduction

The steel and iron industry is one among the foremost industries to supply raw variety of steel and iron materials to the other one like earth moving machinery, automotive, construction, shipping and electronic one. A continuous reheating process is the major and complex process in the steel and iron industry. Chen et al. [5] and Kim [8] showed that the reheating furnace consumes substantially high amount of energy. The slabs/billets are heated nearly up to 1200 to 1250°C for the subsequent rolling or forging operation. The reheating furnace can be basically split into five zones: Non-firing, charging, preheating, heating and soaking zones. The energy for heating is supplied by roof and bottom tangential gas burners. According to Trinks [17] 'Continuous furnaces move the charged material, stock, or load while it is being heated. Material passes over a stationery hearth or the hearth itself moves. If the hearth is stationary, the material is pushed or pulled over skids or rolls, or is moved through the furnace by woven wire

© Springer Nature Singapore Pte Ltd. 2020
V. K. Gunjan et al. (Eds.): *ICRRM 2019 – System Reliability, Quality Control, Safety, Maintenance and Management*, pp. 40–45, 2020.
https://doi.org/10.1007/978-981-13-8507-0_7

belts or mechanical pushers. Except for delays, a continuous furnace operates at a constant heat input rate, burners being rarely shut off. A constantly moving (or frequently moving) conveyors or hearth eliminates the need to cool and reheat the furnace (as is the case with a batch furnace), thus saving energy'.

The combustion process from burners with direct fire is a heat source for the reheating furnaces. The hot exhaust gases raise the temperature of the billet in the preheating area or zone as the billet moves into a reheating furnace. The billet heating is mainly done by burners to achieve uniform temperature before it is sent to the rolling or forging operation. The jerky or uneven flames and non-uniform temperature of the discharged billet may cause damage to the finished product and excessive heating could lead thermal stress, cracking and distortion during final operation. This paper emphasizes on the current problems and model developments done in reheating furnaces.

2 Reheating Furnaces and Working Principle

Reheating is a process in which steel in the shape of slabs or billets is placed into the furnace and heated before metal deformation processes such as forging and hot rolling. The reheat furnaces are classified according to the method of heating: electric and combustion type. The combustion type reheat furnaces are fired by a range of hydrocarbons fuels, including natural gas, blast furnace gas and coke oven gas. The burners are installed in the roofs and walls of the furnaces which inject fuel and air mixture for combustion. There are usually three types of continuous reheating furnaces used in the rolling and forging mills [16]. These are

- Pusher type furnace
- Walking hearth furnace
- Walking beam furnace.

Pusher type furnaces are widely used for heating of rectangular slabs and billets. The billets are placed close to one another without any gap such as shown in Fig. 1. The billets are advanced by forcibly adding more billets at the entry side. The billets are advanced by the force applied at the time of new entry of billets. The advantage of this furnace is the simplest and least expensive method of transferring billet through a furnace. The billet heating occurs only from four surfaces, as billets are placed close to one another without any gap. The billets slide over the fixed skid pipe (water cooled) lay along the furnace bottom [16].

2.1 Walking Beam and Walking Hearth Furnace

Walking beam furnaces avoids the inherent disadvantages of the pusher-type furnaces. The billets are advanced by motion of walking beam and only the difference in walking hearth is that stock is placed on fixed refractory hearth which is combination of moving and steady sections. The walking beams are periodically raised to lift billets and move it to the next position in the furnaces. In the walking beam-type of furnace a gap is present between the billets, such as shown in Fig. 2. And it is also ensure that all the six faces of the billet getting exposed to heat transfer as a result, intensity of heating can be increased appreciably [16].

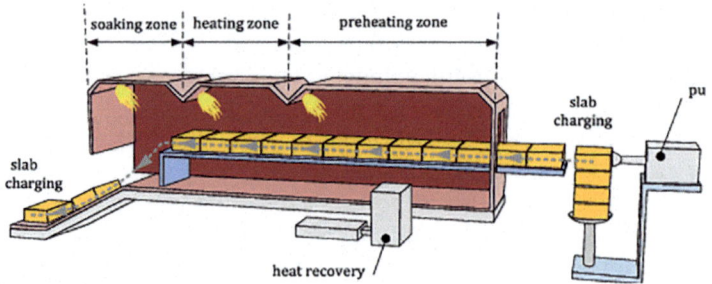

Fig. 1. Pusher type furnace [18].

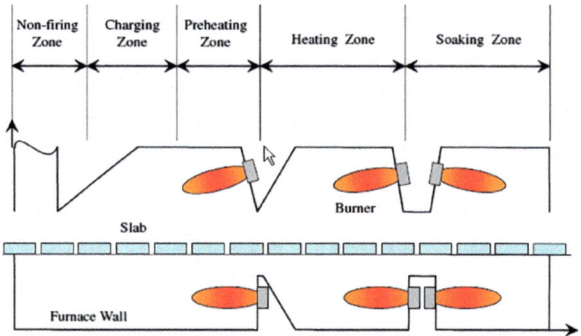

Fig. 2. Walking beam furnace [8]

3 Objectives

The aims and objectives of the present study are as follows:

- To find out numerical, mathematical and CFD model developments techniques for reheating furnaces.
- To collect data about numerical, mathematical and CFD models developments from 1990s to till date.
- Analyze reheating furnace problems: uneven or non-uniform flame distribution, oxidation of metal, carbon loss of metals, scaling or dimension loss and emission of pollutants.
- To find out possible opportunities to reduce heat losses and other problems such as crack formation, thermal stress and distortion.

4 Modeling Works- Summary and Review

The summary of work done by scientists, experts, engineers and researchers in the field of energy optimization, furnace design and model developments of continuous reheating furnaces have been described in below Table 1. In the Table 1, you can see the investigated problems in reheating furnaces with author and year wise publication. Here is the summary of the developments in reheating furnaces on literature survey.

Table 1. Modeling methods for model developments of reheating furnace

Sr. no	Author name (year)	Study/Problem type	Modeling methods
1	Chapman et al. [1]	Heat transfer modelling and parametric based study in a direct-fired reheating furnace	1-D Transient model, radiative heat exchange- Hottel's, gray gas for flux exchange
2	Venturio et al. [2]	Heat transfer and coupled fluid flow 2-D model development in a steel reheat furnace	2-D model for coupled fluid flow & radiation (FLUENT)
3	Jaklic et al. [3]	Online mathematical model of steel slab reheating in a pusher-type furnace	3-D FDM, view factor matrix, model of Heiligenstaedt
4	Tang et al. [4]	Gas flow modelling with influence on the scale accumulation in the steel slab pusher-type reheating furnace	Momentum, combustion & radiation models
5	Chen et al. [5]	Energy consumption and performance analysis of reheating furnaces in a hot strip mill	2-D transient diffusion, 2nd order implicit & central, FDM
6	Timoshpol skii et al. [6]	Conjugate heat exchange simulation in a heating furnaces with a moving bottom	Conjugate heat exchange model for temperature regimes
7	Han et al. [7]	Numerical analysis and model development of heating characteristics of a slab in a bench scale reheating furnace	FVM for radiation, k-ϵ for turbulence
8	Kim [8]	Heat transfer model for transient heating analysis of the slab in the walking beam furnace	FVM for radiation, central difference, TDMA algorithm
9	Hsieh et al. [9]	A coupled numerical study of slab and gas temperature in the walking beam reheating furnace	STAR-CD, PDF, WSGGM, radiation modeling
10	Han et al. [10]	A numerical analysis of steel slab heating in a walking beam furnace	Transient model FLUENT, UDF
11	Prieler et al. [11]	Prediction of the heating characteristics of billets in a walking hearth furnace using CFD	SIMPLE & PRISTO! algorithm, eddy-dissipation model
12	Tang et al. [12]	Model development of steel reheat process in a walking beam furnace	2-D model, FDM
13	Hosain et al. [13]	Real scale steel slab reheating furnace modelling using CFD	3-D model, k-ε, PDF methods
14	Mayr et al. [14]	Performance increase and CFD modelling of a pusher-type reheating furnace using oxy-fuel burners	RANS, UDF, realizable k-ε model
15	Pollhammer et al. [15]	CFD modelling of walking beam furnace	ANSYS, FLUENT, WSGGM, DOM

5   Discussions

From the research paper survey following points needs to be discussed:-

1. It is observed that main problems in reheating furnaces are non-uniform flame distribution, scaling (dimensional loss) or oxidation of metal, carbon loss of metals and emission of pollutants and because of these problems reheating furnaces have less production efficiency and high running cost.
2. There is a scope for reheating furnace improvements in terms of energy savings, process optimization by reducing huge amount of heat losses by using latest technique and computational models.
3. From the different papers, it is observed that conservation of waste to energy is the main and latest issue in reheating furnaces and excessive heating could lead to thermal stress, cracks and distortion.
4. It is also significantly notice that above operational problems in reheating furnace can be eliminated by proper modeling and designing techniques.
5. There are very few papers which show how CFD modeling can be used in steel reheating combustion system design optimization and burner placements.
6. Therefore, the main aim of this study is to simulate real scale reheating (walking hearth/beam type) furnace to combine with fluid dynamics, turbulent combustion and thermal radiation.

6   Conclusion

We can clearly conclude that it is possible to do reheating furnaces performance improvements through automatic control, measurements and numerical model developments by Heat Transfer and CFD techniques; this can be include direct fuel savings and indirect fuel savings by product quality improvements. Through the literature we have observed that the performance of the reheating furnace is less because of heat losses and combustion related problems. And also from the recent literature analysis it is found out that Heat Transfer and CFD analysis has huge scope in the research and development in furnaces. It is also true that insulation, excess temperature, non-uniform flame distribution and erosion rates inside the furnace are also important and challenging to understand.

References

1. Chapman, K.S., Ramadhyani, S., Viskanta, R.: Modeling and parametric studies of heat transfer in a direct-fired batch reheating furnace. J. Heat. Treat. **8**(2), 17–146 (1990)
2. Venturino, M., Rubini, P.: Coupled fluid flow and heat transfer analysis of steel reheat furnaces. In: 3rd European Conference on Industrial Furnaces and Boilers. Lisbon, Portugal, 18–21 April 1995. Eds. Leuckel, Collin Ward, Reis, INFUB, ISBN 972-8034-02-4

3. Jaklic, A., Vode, F., Robit, R., Perko, F., Strmole, B., Novak, J., Triplat, J.: the implementation of an online mathematical model of slab reheating in a pusher-type furnace. Materiali in Technologije **39**, 215 (2005)
4. Tang, Y., Laine, J., Fabritius, T., Harkki, J.: The modeling of the gas flow and its influence in the steel slab pusher-type reheating furnace. ISIJ Int. **43**(9), 1333–1341 (2003)
5. Chen, W.H., Chung, Y.C., Liu, J.L.: Analysis on energy consumption and performance of reheating furnace in a hot strip mills. Int. J. Heat Mass **32**, 695–706 (2005)
6. Timoshpol skii, V.I., German, M.L., Grinchuk, P.S., Kabishov, S.M.: Mathematical simulation of conjugate heat exchange in heating furnaces with a moving bottom. J. Eng. Phys. Thermophys. **79**, 419–428 (2006)
7. Han, S.H., Back, S.W., Kang, S.H., Kim, C.Y.: Numerical analysis of heating characteristics of a slab in a bench scale reheating furnace. Int. J. Heat Mass Transfer Sci. Direct **50**, 2019–2023 (2007)
8. Kim, M.Y.: A heat transfer model for the analysis of transient heating of the slab in a direct-fired walking beam type reheating furnace. Int. J. Heat Mass Transfer **50**, 3740–3748 (2007)
9. Hsieh, C.-T., Huang, M.-J., Lee, S.-T., Wang, C.-H.: A coupled numerical study of slab temperature and gas temperature in the walking-beam reheating furnace. In: 6th Conference on CFD in Oil & Gas, Metallurgical and Process Industries (2008)
10. Han, S.H., Back, S.W., Kang, S.H., Kim, C.Y.: Numerical analysis of slab heating characteristics in a walking beam type reheating furnace. Int. J. Heat Mass Transfer Sci. Direct **53**, 3855–3861 (2010)
11. Prieler, R., Demuth, M., Mayr, B., Holleis, B., Hochenauer, C.: Prediction of the heating characteristic of billets in a walking hearth type reheating furnace using CFD. Int. J. Heat Mass Transfer **92**, 675–688 (2016)
12. Tang, G., Okosun, T., Bai, D., Bodnar, R.: Modeling of steel slab reheating process in a walking beam reheating furnace. In: Proceeding of the ASME 2016 Heat Transfer Summer Conference HT2016. USA (2016)
13. Hosain, M.L., Bel Fdhila, R., Sand, U., Engdahl, J., Dahiqist, E., Li, H.: CFD Modeling of real scale slab reheating furnace. In: 12th International conference on Heat Transfer, Fluid Mechanics and Thermodynamics (2016)
14. Mayr, B., Prieler, R., Demuth, M., Moderer, L., Hochenauer, C.: CFD modelling and performance increase of a pusher type reheating furnace using oxy-fuel burners. Sci. Direct **120**, 462–468 (2017)
15. Pollhammer, W., Spijker, C., Six, J., Zoglauer, D., Raupenstrauch, H.: Modeling of a walking beam furnace using CFD-methods. In: INFUB-11th European Conference on Industrial Furnaces and Boilers (2017)
16. Dubey, S.K.: Computational modeling and simulation of heat transfer during steel billet reheating and transport with growth of oxide scale. Ph. D. thesis, BITS, Pilani, Rajasthan India (2014)
17. Trinks, W., Mawhiney, M.H., Shannon, R.A., Reed, R.J., Garvey, J.R.: Industrial Furnaces. Wiley, Hoboken (2004)
18. Feliu-Batlle, V., Rivas-Perez, R., Castillo-Garica, F.J.: Robust fractional – order temperature control of a steel slab reheating furnace with large time delay uncertainties. In: International Conference on Fractional Differentiation and Its Applications, ICFDA 2014 (2014)

Analysis of Steam Turbine Blade Failure Causes

Kale Dipak Rajendra$^{(\boxtimes)}$ and Rachayya Arakerimath

GHRCEM Research Center, SPPU, Pune 412207, India
drkaero@gmail.com

Abstract. The steam turbine blade is generally failed due to crack which is effect of fatigue, corrosion, pitting and creep. A various researchers have done the failure investigation of steam turbine blades and its reveal the common failure mechanisms which are revived in this paper. A various case studies are tried to summarize here which will be helpful for giving idea for life enhancement of blade. This investigation deals with the recent various failures especially in the low pressure steam turbine blades which show varying load conditions, vibrations, pitting, corrosion and fatigue are causes for crack initiation leads to failure.

Keywords: Steam turbine blade · Crack · Fatigue failure · Corrosion

1 Introduction

Steam turbine blades are critical components in steam turbine as it transforms the high temperature and pressure energy in to the rotary motion of turbine shafts. There will be various types of failure in blades like pitting, fatigue, corrosion & creep and its lead to development of crack which can be repair but will be not safe & replace the blade will be costly, so need to do the critical review of failure for life enhancement of blades. Once the crack is developed in the blade then it will grow and fail the turbine blades which cause the shutdown of plant.

The failure investigation for low pressure turbine blade of 310 MW thermal power plant is presented in [1]. It's come to know that the fracture takes place at 150 mm from root of the blade. On the edge of blades the numbers of pits are found and they were responsible for corrosion. A fossil fuel power plant [2] having capacity 110 MW had low pressure blade failure prematurely. The fracture take place at root area and causes are investigated by using various techniques which shows crack initiate due to pits. In case study of fracture analysis of steam turbine blade [3] analyzed the root of low pressure steam turbine blades at third row. To find out the crack propagating stresses the finite element analysis, fracture mechanics & experimental data are used and it come to know that the failure is done due to corrosion pits. It also shows that the bending stresses due to unsteady steam load are having role in failure of blades. The failure investigation for low pressure turbine blade of 220 MW power plant in India is done in [4]. The fracture in blade took place near the root area and don't found any evidence for degradation of blade material. The Si rich phase is detected in blade

V. K. Gunjan et al. (Eds.): *ICRRM 2019 – System Reliability, Quality Control, Safety, Maintenance and Management*, pp. 46–52, 2020.
https://doi.org/10.1007/978-981-13-8507-0_8

surface which indicate failure is due to pitting, corrosion fatigue failure. The crack is found at the root of low pressure turbine blade in 660 MW thermal power plant [5]. The investigation includes vibration analysis, bending stress analysis & metallographic analysis which conclude that the failure is due to high cycle fatigue and vibrations. In 2013 a brief review is carried out in [6] which shows a failure in steam turbine blade is mainly due to corrosion, fretting & transient load conditions. An investigation related to fracture of low pressure steam turbine blade is carried out in [7]. The failure analysis is performed in one blade at root section using various investigation processes & its shows that the failure is due to fatigue-corrosion. The cause of failure is found out for low pressure side blade in [8] which is due to faulty welding for stubs parts and vibrations in blades. The two years data of turbine is analyzed as well as metallurgical investigation is carried out. A research in [9] shows the metal magnetic memory testing for crack investigation in steam turbine blades. In [10] investigation carried out for low pressure steam turbine blade of last stage which failed after very few years of operations. A crack and several pits found on blade which conclude failure is due to pitting, corrosion and fatigue. Failure investigation [11] of first stage blade which is in service for 12 years shows crack on blade root. The test indicates that the failure was not due to material but it's happen because of corrosion fatigue. Critical review is carried out in [12] for corrosion failure in steam turbine blade, also some solution is suggested. A review carried out in [13] gives the information of steam turbine blade failure which again highlight fatigue and corrosion failure. All case study shows that the failure in steam turbine blade is majorly due to corrosion, pitting, vibration and fatigue.

2   Corrosion and Pitting Failure

The fracture surface [1] is divided in three different zones i.e. A, B & C as shown in Fig. 1. By using electronic microscope the microstructure is studied and it shows the blade leakage is spread slowly from zone A and propagates to zone C.

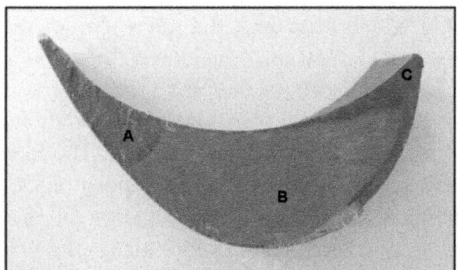

Fig. 1. Failure surface of blade left with rotor [1]

On tail of profile the pits were found and it cause due to corrosive atmosphere. In addition to this one the residual stresses leads to initiation of intergranular crack. Performing the hardness test the hardness value measured and found that it's high with standard one so its shows that some blades are not properly tempered. In case study of [2], the stereo microscopy and electron microscopy observed at pressure side the crack is originated due to pits or groove. The Fig. 2 shows result of field emission scanning electron microscopic examination. The crack is initiated at pit or groove area. The erosion pits on blade is formed due to striking of silica which was present in steam and its lead to corrosion of blade. Figure 2(a) shows the fracture blade taken for investigation and (b), (c) and (d) shows fractographs result with different magnifications which reveals crack initiation due to pitting.

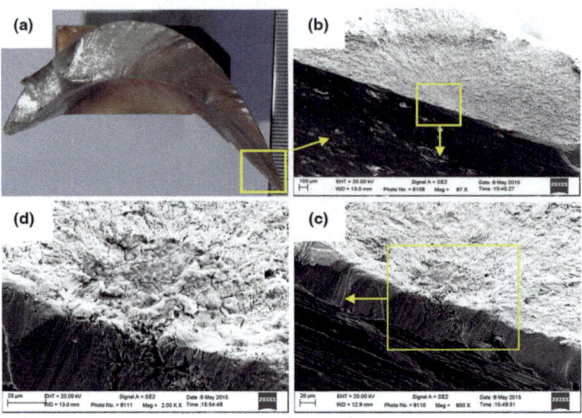

Fig. 2. Blade fracture surface (a), crack initiation area shows in fractographs with different magnifications (b), (c) & (d) [2]

When the steam expands in the turbine then nonmetallic inclusion strike on blade and get condensed. It causes the corrosion to the blade & disturbed the steam flow which leads to corrosion pits formation. The finite element analysis for notch and un-notch condition is done in [3]. In notch condition the bending stresses at notch goes beyond yield stresses and which propagate the microscopic crack to high level. In the linear elastic fracture mechanics 450 are considered the damaging loading conditions. By considering this damaging conditions the blade crack at root require 3 to 11 years to propagate from initial to failure condition. So it come to know that the turbine operator have 3 years' time span for inspecting the failure cause. The frequency of turbine blade inspection should increases and need to focus on deposition on turbine blade & avoid frequent damaging loading conditions. In investigation of blade failure in 220 MW power plant [4] reveals that the SiO2 particles are hits to the blade surface and create a small grooves which further help for corrosion. The chloride salt is found on the blade surface. Depending on load fluctuation the steam vary between dry and wet. The wet steam carries more chloride compare to dry steam. At the groove, once the crack is generated then it propagates due to corrosion & stress concentration. By considering all investigation the probability of failure was corrosion fatigue. In investigation of 110 MW power plant [7] low pressure steam turbine blade using EDS shows the silicon on fracture area. The silicon oxide present in steam hits the blade and it causes the corrosion & pitting which leads to crack generation. The corrosion on blade creates the disturbance to the steam flow which changes the velocities and accelerations of steam & its result in to the blade vibrations. So combination of the corrosion with fatigue due to vibrations is the reason of blade failure. In visual inspection of low pressure steam turbine blade [10] crack and surface damage is found out on leading edge. The chemical analysis shows the chromium is slightly lower as per slandered which is good resistance to corrosion. A surface defect like pits found on fracture surface. The hardness on leading edge area found high compare to other region & which is for resistance to corrosion and pitting in that area but on other side high hardness reduces the fatigue life of blade. Investigation for first stage blade of turbine for 55 MW power is carried out in [11]. The observation by Fractography test shows crack at tang of blade. The impurities in cracked fillet are found like S, Cl & Si which indicate corrosion in blade. The crack at the tang portion and fillet of blade is due to corrosion fatigue. When the tang portion is damaged the stresses at fillet were on higher side which led to pitting corrosion failure. The crack is also found on fillet portion and it is surface crack which propagate to blade root bottom portion. The crack is having branches which are due to stress intensity and working environment.

3 Fatigue and Vibration Failure

The characteristic of dynamic response is carried out in [5] for low pressure steam turbine blades to examine failure due to vibrations. The result of blade natural frequency analysis and blade load data shows there is failure due to resonance conditions. If blade operates continuously under resonance vibrating stress condition then it might be fail suddenly (Fig. 3).

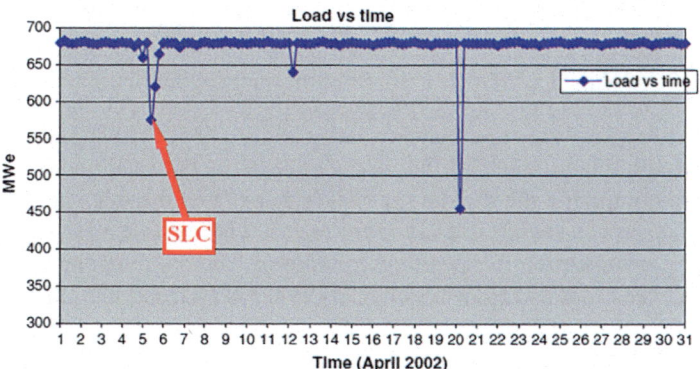

Fig. 3. Sudden load change in electric grid [5]

Due to sudden load changes (SLC) vibrations in the turbine are increases. The fatigue stresses are created in sudden load change scenario and it cause fracture initiation. The data comparison of this event and hypothetical crack initiation period are matches. The all data shows the fracture is not due to regular operating condition vibrations. It's due to sudden load change and high fatigue failure. The steam enters in stage with high fluctuation and it cause high blade vibratory stresses which lead to fatigue failure. The failure investigation for blade of 110 MW, 35 MW & 28 MW are shown in [14] which reveals the failure is due to mechanical vibrations & fatigue failure. In case study [8], of 300 MW power plant three blades were fractured which gives high vibration in turbine & force to shut down power plant. It is found that the stubs were not properly welded to the base parts and vibrations in blade initiate crack and its leads to failure. On various stubs the cracks were detected using nondestructive techniques. The stresses were developed on stubs due to high amplitude of vibrations which leads to early failure in three turbine blades. The metallurgical inspection shows erosion in blade due to moisture droplet. Firstly in stubs a fracture is occurred then a vibration in free blades increases, after this crack is initiated at the trailing edge with support of corrosion. A case study [15] shows failure investigation for low pressure steam turbine blade. The various size cracks found on quenched end of final stage blade and after various failure techniques investigation like morphology, fractography of crack region, it come to know that the failure is due to fatigue and slightly corrosion is responsible. The dye penetrant testing shows crack having length 2 to 8 mm from the edge at suction side. The crack is initiated at suction side therefore the stress concentration is the probable cause at that point. It's recommended that the suction side blade curvature radius can be enhance to reduce stress concentrations. The insufficient temper is shown in microstructure of base root part which indicates the temper temperature should be increased in modified treatment. A failure investigation [16] carried out for low pressure steam turbine blade of 26th stage shows the tempering problem in blade. In visual inspection found that the blade is fail in three regions i.e. near the boss, root and pin location which confirm by dye penetrant testing. The hardness test shows high hardness value for failed blade compare to good blades, the reason behind it is the improper tempering. The Charpy impact test shows the reduction in impact resistance value. The early failure of blade is studied in [17] for second stage of 32 MW unit. A crack on several blade roots are found out and performing various testing it is come to know that it happen due to fatigue. A case study [18] shows during maintenance of L-2 blade the cracks were observed in root portion of 350 MW power plant. Using measurement coordinate machine the root geometry of fail blade and good blade is measured. By comparing the measurement it is found that the tolerances between the blade root and rotor fastening tree are more one this is due to excessive tolerances in manufacturing or distortion in operating conditions. The cause of this is the inappropriate area between root and rotor which leads to stress concentration in root of blades.

4   Conclusion

In steam turbine the low pressure blade is a critical component. The life of steam turbine blade is averagely good however in some industries due to corrosion, vibration, fatigue and some improper manufacturing problem the blades have failed. The deposition and hitting of Si on turbine blade need to be focus which cause corrosion & pitting and disturb the steam flow. The sudden load changes should to be avoided which create vibrations & cause dynamic stresses on blades. Use of high fatigue strength & pitting resistant material, coating of anticorrosive material and some provision which used to damp out vibrations will enhance the life of blade.

Acknowledgement. The Author is grateful to Mr. Hemant Bari, member of condition monitoring society of India for giving valuable knowledge related to steam turbine blade.

References

1. Ziegler, D., Puccinelli, M., Bergallo, B., Picasso, A.: Investigation of turbine blade failure in a thermal power plant. Biochem. Pharmacol. **1**, 192–199 (2013). https://doi.org/10.1016/j.csefa.2013.07.002
2. Harison, M.C.A., Swamy, M., Pavan, A.H.V., Jayaraman, G.: Root cause analysis of steam turbine blade failure. Trans. Indian Inst. Met. **69**, 659–663 (2016)
3. Plesiutschnig, E., Fritzl, P., Enzinger, N., Sommitsch, C.: Fracture analysis of a low pressure steam turbine blade. Case Stud. Eng. Fail. Anal. **5–6**, 39–50 (2016). https://doi.org/10.1016/j.csefa.2016.02.001
4. Das, G., Chowdhury, S.G., Ray, A.K., Das, S.K., Bhattacharya, D.K.: Turbine blade failure in a thermal power plant. Eng. Fail. Anal. **10**, 85–91 (2003). https://doi.org/10.1016/s1350-6307(02)00022-5
5. Mazur, Z., Garcia-Illescas, R., Aguirre-Romano, J., Perez-Rodriguez, N.: Steam turbine blade failure analysis. Eng. Fail. Anal. **15**, 129–141 (2008). https://doi.org/10.1016/j.engfailanal.2006.11.018
6. Bhagi, K.L., Ratogi, V., Gupta, P.: A brief review on failure of turbine blades. In: Proceedings of STME-2013 Smart Technologies for Mechanical Engineering, pp. 1–8 (2013).https://doi.org/10.13140/rg.2.1.4351.3768
7. Bhagi, L.K., Gupta, P., Rastogi, V.: Fractographic investigations of the failure of L-1 low pressure steam turbine blade. Case Stud. Eng. Fail. Anal. **1**, 72–78 (2013). https://doi.org/10.1016/j.csefa.2013.04.007
8. Sz, J.K., et al.: Failure analysis of the L-0 steam turbine blade. In: Proceedings of IJPGC 2003, pp. 1–5 (2003)
9. Xing, H., Wu, D., Zhang, L., Xu, M.: Steam turbine blade MMM testing and failure analysis 141, 2561–2564 (2010). https://doi.org/10.4028/www.scientific.net/AMR.139-141.2561
10. Adnyana, D.N.: Corrosion fatigue of a low-pressure steam turbine blade. J. Fail. Anal. Prev. (2018). https://doi.org/10.1007/s11668-018-0397-5
11. Nurbanasari, M.: Carck of a first stage blade in a steam turbine. Biochem. Pharmacol. (2014). https://doi.org/10.1016/j.csefa.2014.04.002
12. Kushwaha, A.D., Soni, A., Garewal, L.: Critical review paper of steam turbine blades corrosion and its solutions. Int. J. Sci. Res. Eng. Technol. **3**, 776–784 (2014)
13. Patel, T.K., Sen, P.K.: Review on common failure of steam turbine blade. IJR **2**, 137–141 (2015)
14. Mazur, Z., García-Illescas, R., Hernández-Rossette, A.: Steam turbine low pressure blades fatigue failures caused by operational and external conditions. Struct. Dyn. Parts A B **6**, 93–106 (2009). https://doi.org/10.1115/gt2009-59040
15. Wang, W., Xuan, F., Zhu, K., Tu, S.: Failure analysis of the final stage blade in steam turbine **14**, 632–641 (2007). https://doi.org/10.1016/j.engfailanal.2006.03.004

16. Saxena, S., Pandey, J.P., Solanki, R.S., Gupta, G.K., Modi, O.P.: Coupled mechanical, metallurgical and FEM based failure investigation of steam turbine blade, Eng. Fail. Anal. (2015). https://doi.org/10.1016/j.engfailanal.2015.02.012
17. Farhangi, H., Fouladi Moghadam, A.A.: Fractographic investigation of the failure of second stage gas turbine blades. In: Proceedings of 8th International Fracture Conference, 7–9 November 2007, Istambul, Turkey (2007)
18. Sz, J.K., Segura, J.A., Gonzalez, G.R., García, J.C., Sierra, F.E., Nebradt, J.G., Rodriguez, J.A.: Failure analysis of the 350 MW steam turbine blade root. Eng. Fail. Anal. **16**, 1270–1281 (2009). https://doi.org/10.1016/j.engfailanal.2008.08.015

Design and Analysis Herringbone Gear Use in Industry

Sagar Patil$^{(\boxtimes)}$ and Prashant Ambhore

Department of Mechanical Engineering,
G. H. Raisoni College of Engineering and Management, Pune 412207, India
patilsagar475@gmail.com, prashant.ambhore@raisoni.com

Abstract. Herringbone gears, additionally called as twofold helical apparatuses are the rigging sets intended to transmit control through either parallel or less ordinarily opposite tomahawks. The one of a kind tooth structure of a herringbone equip comprises of two bordering inverse helixes that show up in the state of the letter 'V'. Herringbone adapts as a rule mate by means of the utilization of smooth, exactly fabricated V-molded teeth. Like helical riggings numerous teeth are locked in amid turn, conveying the remaining task at hand and offering a peaceful activity. Nonetheless, because of their tooth structure, herringbone gears invalidate the pivotal push not at all like helical apparatuses. The rigging set's teeth might be produced so tooth-tip lines up with the contrary tooth-tip or the contrary apparatus' tooth trough. In this undertaking 3D display is readied utilizing CATIA V5 programming, here surfacing and part modules are utilized and for investigation we utilize ANSYS programming. We change the materials of the herringbone outfit. The Material utilized in the herringbone equip are Gray Cast Iron aluminum combination and steel composite. We check the Stress Analysis and twisting of the herringbone outfit.

Keywords: Herringbone gear · CATIA · ANSYS · Fatigue analysis · Modeling

1 Introduction

Herringbone gears, additionally called as twofold helical riggings are the apparatus sets intended to transmit control through either parallel or less usually opposite tomahawks. The special tooth structure of a herringbone equip comprises of two connecting inverse helixes that show up in the state of the letter 'V'. Herringbone outfits normally mate by means of the utilization of smooth, exactly made V-formed teeth. Like helical riggings various teeth are locked in amid revolution, dispersing the remaining task at hand and offering a peaceful activity. Nonetheless, because of their tooth structure, herringbone gears invalidate the pivotal push dissimilar to helical riggings. Like helical riggings, they have the upside of exchanging power easily in light of the fact that in excess of two teeth will be in work at any minute in time. Their favorable position over the helical apparatuses is that the side-push of one half is balanced by that of the other half. This implies herringbone riggings can be utilized in torque gearboxes without requiring

© Springer Nature Singapore Pte Ltd. 2020
V. K. Gunjan et al. (Eds.): *ICRRM 2019 – System Reliability, Quality Control,
Safety, Maintenance and Management*, pp. 53–59, 2020.
https://doi.org/10.1007/978-981-13-8507-0_9

a generous push bearing. Along these lines, herringbone gears were a vital advance in the acquaintance of the steam turbine with marine drive.

1.1 Nomenclature of Gear

Rotational frequency, n
Measured in rotation over time, such as revolutions per minute (RPM or rpm).

Angular frequency, ω
Measured in radians/second. 1 RPM = 2π rad/minute = $\pi/30$ rad/second (Fig. 1).

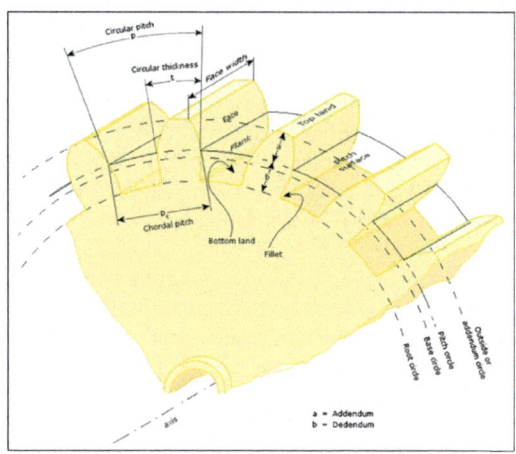

Fig. 1. Nomenclature of gear

Number of teeth, N
How many teeth a gear has, an integer. In the case of worms, it is the number of thread starts that the worm has.

Gear, wheel - The larger of two interacting gears or a gear on its own.
Pinion - The smaller of two interacting gears.
Path of contact - Path followed by the point of contact between two meshing gear teeth.

2 Motivation

1. Repeated Over Loading of herringbone gear
2. Maintenance Issue (Machinery not looked after by maintenance Engineer)
3. Life Cycle (Over Running of the machinery with respect to planned maintenance system)
4. Material used was not appropriate with respect to the load on the machinery.

3 Backgrounds and Related Work

The static examination, secluded examination and exhaustion examination was finished on 3 unmistakable materials. The examination results, i.e., ANSYS respond in due order regarding CFRP material has been presented. In the midst of the examination, it might be found that the miss-happening of the device wheel for carbon reinforced plastic material is more diminutive when diverged from that of exchange materials [1]. The investigation centers around the examination of new limited herringbone adapt and talks about its tooth confront highlights and transmission standards. The conditions of tooth face and filet are concluded and the bowing quality conditions are derived dependent on the 30° area strategy. Limitation of addendum honing is talked about and the connections between the confinement and the essential parameters are uncovered [2]. An examination a fizzled pinion adapts demonstrated that the break was caused by weariness with the weakness split started at the filet of one of the pinion teeth because of misalignment. The misalignment likewise prompted serious wear and unreasonable warmth at the mating surfaces [3]. An examination of a crown haggle disappointment presumed that the disappointment was because of a bargain made in crude material structure amid assembling [4]. There are three kinds of mistakes that can happen when center points and bores that may prompt disappointment. The primary is inaccurate bore distance across. This could result in a lot of impedance which will cause establishment issues and center point harm. The second is the drag unconventionality which could prompt similar issues. The third is the misalignment between the drag and the center point [5]. The figuring of most extreme worries in a helical apparatus at tooth root is three dimensional issues. The precise assessment of stress state is mind boggling assignment. The commitment of this postulation work can be condensed as pursues: The quality of helical rigging tooth is a urgent parameter to avoid disappointment [6]. There are two sorts of worries in rigging teeth, root twisting burdens and tooth contact stresses. These two burdens results in the disappointment of apparatus teeth, root twisting pressure results in weariness crack and contact stresses brings about setting disappointment at the contact surface [7]. The gear design the twisting pressure and surface quality of the rigging tooth are viewed as one of the primary givers for the disappointment of the rigging in a rigging set. Consequently, the examination of stresses has turned out to be prominent as a zone of research on apparatuses to limit or to decrease the disappointments and for ideal plan of riggings [8].

4 Mathematical Model

Considering the module = 3.5,
Pitch Circle Diameter = 252
Module (m):
$$m = \frac{d(\textit{pitch circle Diameter})}{t(\textit{Number of Teeth})}$$
3.5 = 252/t; t = 72
i.e. Number of teeth = **72**

Let Power $= \dfrac{\textbf{Torque}\,(\tau)}{\textbf{12 KW}}$

$\textbf{N} = \textbf{900 rpm}$

$P = \frac{2\pi NT}{60}, \quad T = \frac{P\,*\,60}{2\pi N}$

T = 127.3239 NM

Force (F_t) $F_t = \frac{1000\,*\,P}{v}$

$V = \frac{Pd p \omega}{60}; \quad V = \frac{\pi\,*\,252\,*\,900}{60}; \quad V = 11.87522 \text{ m/s}$

$F_t = \frac{1000\,*\,12}{11.87522}; \quad F_t = 1010.52 \text{ N}$

Face width (b)

$b = 4\pi m; \quad b = 4\pi 3.5$

b = 43.98 ≈ 44 mm

5 Design and Analysis

(See Figs. 2, 3, 4, 5, 6, 7, 8, and 9).

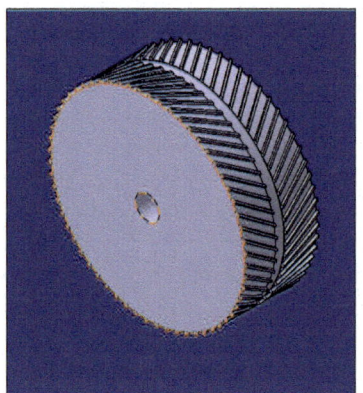

Fig. 2. Herringbone gear design.

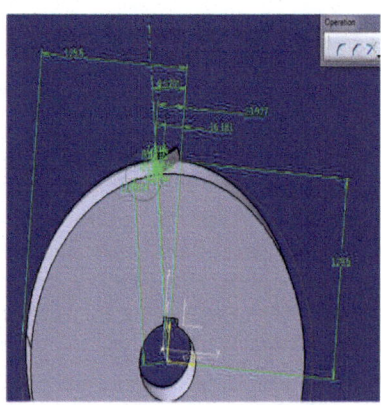

Fig. 3. Dimensions of teeth.

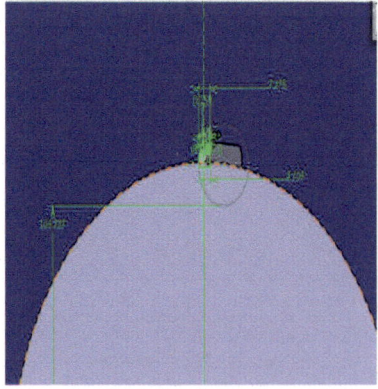

Fig. 4. Dimensions of teeth.

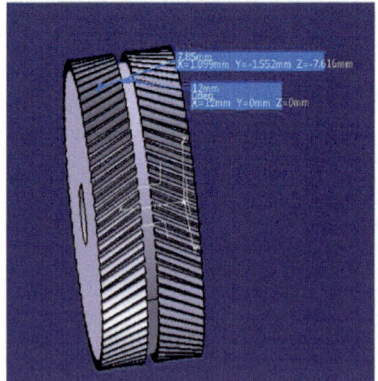

Fig. 5. Gap between teeth and height of teeth.

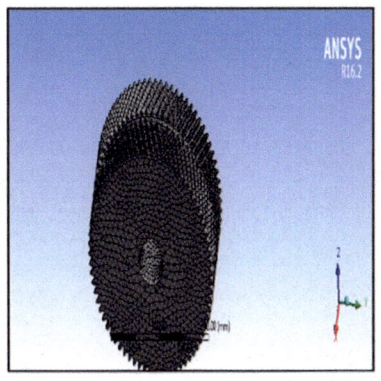

Fig. 6. Messing of herringbone gear

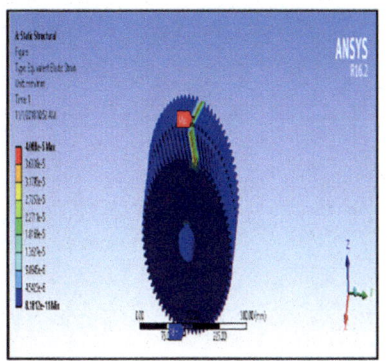

Fig. 7. Equivalent elastic strain analysis.

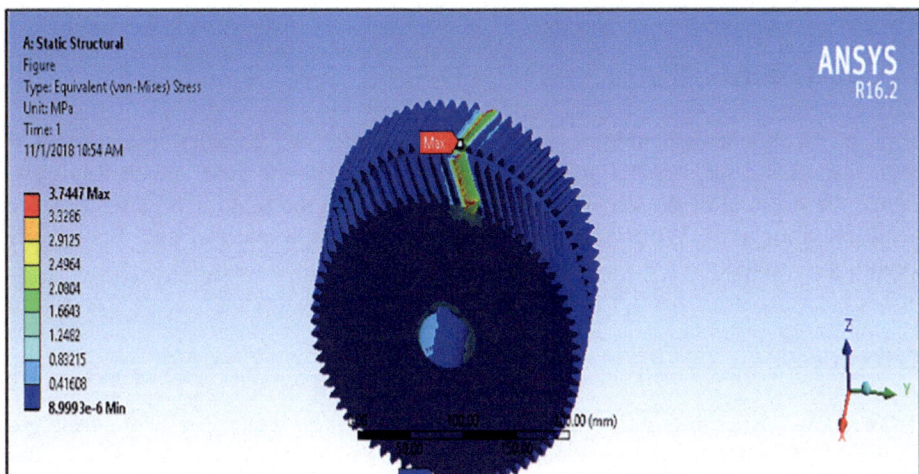

Fig. 8. Equivalent (von-mises) stress analysis.

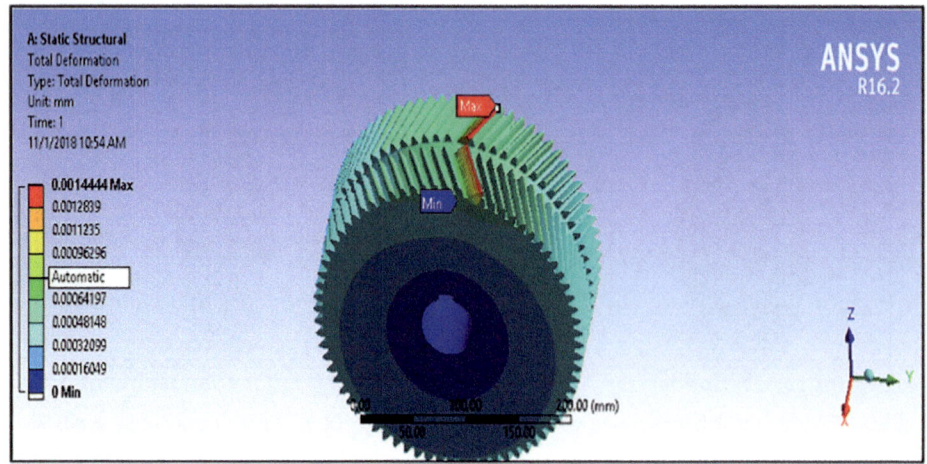

Fig. 9. Total deformation.

6 Conclusion

Proposed work obtained module, pitch circle diameter and calculate the number of teeth, torque and tangential force. Modeling the herringbone gear and in future, we change the material of the herringbone gear and analyze the best material with respect to the previous gear. We trying to reduce the cost of the material and increase the durability, hardness.

References

1. Feng, W.: Dynamic characteristics research and experimental study on herringbone gear drive system. Doctoral dissertation, Northwestern Polytechnical University, Xi'an, China (2014)
2. Xue, J.H., Li, W., Qin, C.: The scuffing load capacity of involute spur gear systems based on dynamic loads and transient thermal elastohydrodynamic lubrication. Tribol. Int. **79**, 74–83 (2014)
3. Janikow, C.Z.: A knowledge-intensive genetic algorithm for supervised learning. Mach. Learn. **13**, 189–228 (1993)
4. Hussain, S., Gabbar, H.A.: Gearbox fault detection using real coded genetic algorithm and novel shock response spectrum features extraction. J. Nondestr. Eval. **33**, 111–123 (2014)
5. Su, J., Fang, Z., Cai, X.: Design and analysis of spiral bevel gears with seventh-order function of transmission error. Chin. J. Aeronaut. **26**, 1310–1316 (2013)
6. Jiang, J., Fang, Z.: Design and analysis of modified cylindrical gears with a higher-order transmission error. Mech. Mach. Theory **88**, 141–152 (2015)
7. Bhosale, K.C.: Analysis of bending strength of helical gear by FEM. Innovative Syst. Des. Eng. **2**(4), 125–127 (2011). ISSN 2222-1727
8. Venkatesh, B., Kamala, V., Prasad, A.M.K.: Design, modeling and Manufacturing of helical gear. Int. J. Appl. Eng. Res. **1**(1), 103 (2011). ISSN 0976-4259

9. Mishra, P., Murthy, M.S.: Comparison of bending stress for different face width of helical gear obtained using MATLAB simulink with AGMA and ANSYS. IJETT **4**(7) (2013). ISSN 2231-5381
10. Patil, P., Dharashiwkar, N., Josh, K., Jadhav, M.: 3D photo elastic and finite element analysis of helical gear. Mach. Des. **3**(2) (2011). ISSN 1821-1259

Reliability Based Design Approach for Development of Friction Stir Welding Fixture

Gurunath V. Shinde[1,2(⊠)] and Rachayya Arakerimath[1]

[1] Mechanical Engineering Department, G. H. Raisoni College of Engineering
and Management, Pune, India
gurunathshinde@yahoo.com
[2] Mechanical Engineering Department,
Dr. Daulatrao Aher College of Engineering, Karad, India

Abstract. Friction Stir Welding (FSW) is a solid state welding process invented by The Welding Institute (TWI) in 1991. High speed rotating tool with small pin forced to deform work piece material using thermo mechanical action. A small pin with shoulder of tool is plunged into the work piece with axial pressure and traversed along the line of joint. During the welding process, axial force, lateral forces, traverse force and torque are applied on the workpiece which lead to distortion of workpiece materials and fixture; hence design of the fixture should be reliable for proper positioning of workpiece. In this Paper, Types of Probable failures, Cause and Effect analysis, Pre-requisites and reliability based design approach for development of FSW fixture have been discussed in brief. Further modified design was prepared in AutoCAD and FSW fixture is fabricated. Some successful trials are also carried out for testing for reliability. Results shows FSW fixture is reliable to carry out friction stir welding process.

Keywords: FSW fixture failure · Cause and effect analysis ·
Reliability based design

1 Introduction

Friction Stir Welding (FSW) is a solid state welding process invented by The Welding Institute (TWI) in 1991. High speed rotating tool with small pin forced to deform work piece material using thermo mechanical action [3, 5]. Heat generated by the thermomechanical action is used to soften the material without reaching its melting point followed by stirring mechanism to produce welds. A small pin with shoulder of tool is plunged into the work piece with axial pressure and traversed along the line of joint Fig. 1. The major functions of the rotating tool are: (i) To generate heat for softening of materials, and (ii) stirring and trailing of materials to be welded.tool rotation and translation causes material movement along the pin towards the back of the pin [1]. During FSW process, the material goes into the severe plastic distortion at elevated temperature, producing fine and equaxed recrystallized grains [2]. Generally materials used for the process are of different thickness from 1 mm to 12 mm. During the

© Springer Nature Singapore Pte Ltd. 2020
V. K. Gunjan et al. (Eds.): *ICRRM 2019 – System Reliability, Quality Control,
Safety, Maintenance and Management*, pp. 60–66, 2020.
https://doi.org/10.1007/978-981-13-8507-0_10

welding process, large axial force is applied on the workpiece hence design of the fixture for holding workpiece should be reliable for proper positioning of workpiece. Figure 2 shows types of forces acting on the workpiece during the friction stir welding.

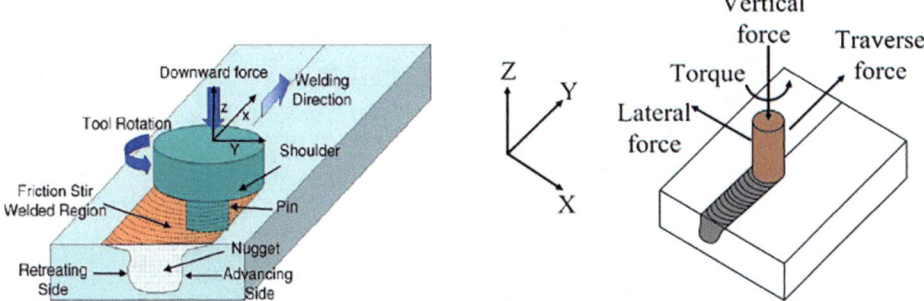

Fig. 1. Friction stir welding process   **Fig. 2.** Forces developed during FSW

2 Pre Requisite for Design of FSW Fixture

During the design of FSW fixture, A Major issues is the high temperature produced during thermo mechanical action between tool and workpiece. In this situation, workpiece material is likely to remain stuck to the baseplate deteriorating quality of the weld and the reliability of the fixture. The design of the FSW fixture should be such that it should be able to carry the forces and keep the workpiece material stable during the process [6, 7, 9]. Proper stability during the process is very important concern to avoid any distortion or sudden vibrations which affect the quality of the weld [8].

Following are the possible failures of FSW fixture during the process;

1. **Workpiece Misalignment:** Workpiece to be welded will get misaligned if not fixed properly inside the fixture.
2. **Workpiece Buckling:** Workpiece may undergo buckling due to axial, lateral forces and torque during the process
3. **Thermal Expansion:** Workpiece undergoes thermal expansion due to elevated temperature produced by thermo-mechanical action.
4. **Thermal Loss:** During the FSW, thermal must be avoided in order to keep workpiece material at elevated temperature zone.
5. **Improper Intermixing of Material:** Workpiece is required to be maintained in optimum temperature range in order to keep materials not below or above the plastic zone. Below plastic zone, materials will not intermix/stirred properly. Above plastic zone, materials may reach its melting point lead to failure of weld.
6. **Plunging:** Tool along with pin may not plunge properely inside the workpiece if sufficient axial force not applied and workpiece should be rigidly fixed inside fixture.
7. **Clamping:** Clamps may get misaligned if not fixed properly on the fixture.

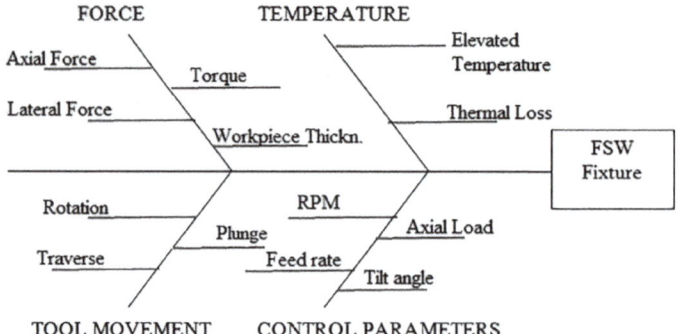

Fig. 3. Cause and effect (Fishbone diagram) diagram of FSW fixture

Figure 3 shows factors affecting weld quality and reliability of FSW fixture by the cause and effect diagram.

3   Reliability Based Design Approach of FSW Fixture

Reliability is the efficiency of the product or system that will continue to work normally over a specified interval of time, under specified conditions. The overall reliability of a system or product is described by the combination of the reliabilities of the individual components [10]. If the failure of any component will lead to the total failure of the system, then design is said to be in series reliability. In contrast better combination of components is that in which failure of all components lead to total failure of the system. Such a combination is said to be in parallel reliability. Figure 4 shows failure behavior of system or product over the time period. The highest failure rate observed during the premature phase [11]. This curve also known as "Bath Tub Curve".

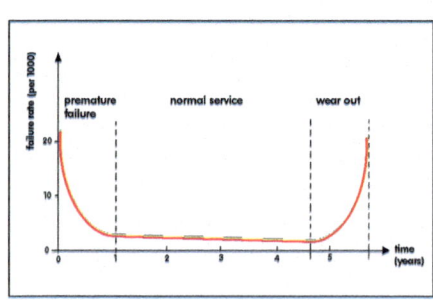

Fig. 4. Bath tub curve

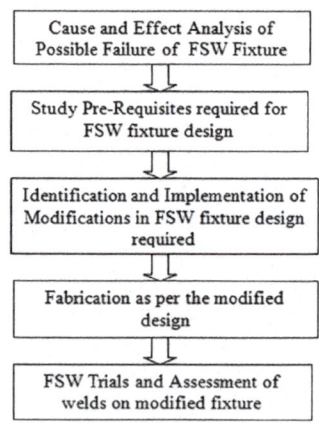

Fig. 5. Reliability based design approach

Figure 5 shows reliability based design approach for fixture required for the friction stir welding process.

Table 1 show improvements used as per the reliability based design approach for F SW fixture design.

Table 1. Improvement in FSW fixture as per the reliability based design approach

Sr No	Types of failures	Reliability based design approach for improvements in FSW fixture
1	Workpiece misalignment	Multiple clamping and dimensional accuracy has been maintained for proper alignment of workpiece
2	Workpiece buckling	Multiple and enough clamp length maintained in order to avoid buckling
3	Thermal expansion	Dimensional accuracy and high tolerance kept in order avoiding thermal expansion
4	Thermal loss	Thermal insulating material such as mica or bakelite is suggested to use as backup plate to avoid thermal loss
5	Improper material mixing	Thick fixture baseplate (40 mm) of mild steel is used to withstand against high axial load, frictional forces and elevated temperature
6	Improper plunging	High dimensional accuracy, tolerance and rigid fixture provide enough support for tool pin plunging
7	Improper clamping	Multiple and rigid clampings with proper dimensions are used to prevent misalignment from its position
8	Productivity	Arrangement has been provided to carry out two trials in single set-up
9	Fixture positioning	Universal FSW fixture design is used that can be used with any milling machine beds

4   Modelling and Fabrication of FSW Fixture

Figure 6 shows reliability based modified design of FSW fixture prepared in AutoCAD software. Figure 7 shows FSW fixture along with clamping developed using reliability design approach flexible to be used different kinds of milling machines at elevated temperature without misalignment and buckling of workpiece materials. Two trials can be carried out in single set up as shown in figure showing copper plates of 150 mm X75 mm X3 mm fixed on the fixture to increase productivity of FSW process. Base plate of large thickness of 40 mm is used in order to have rigidity to the fixture.

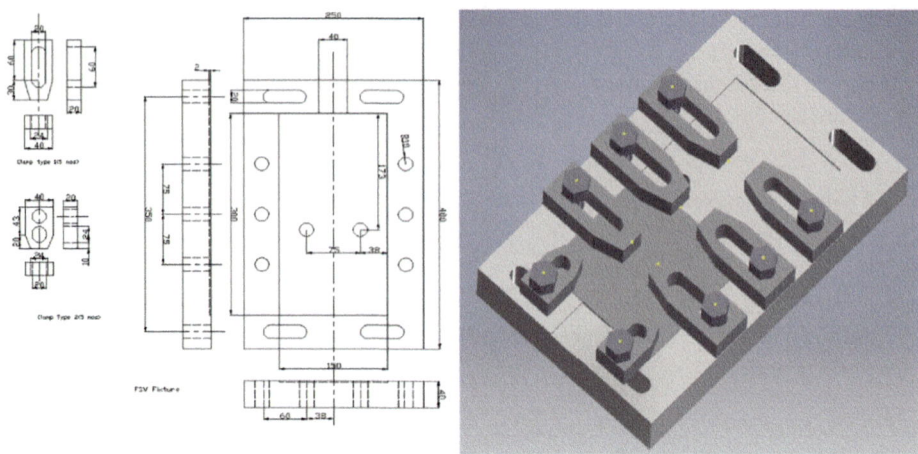

Fig. 6. Modeling of FSW fixture

Fig. 7. FSW fixture with clamping

5 FSW Trials for the Weld Quality Assessment

Various Successful FSW trials without defects have been conducted between aluminum to aluminum and aluminum to copper at different levels of RPM, Weld Speed and Tilt Angle as shown in Figs. 8(a) and (b).

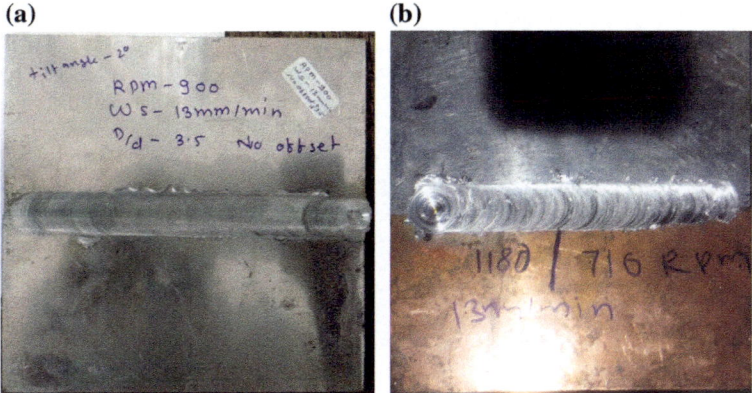

Fig. 8. (a) FSW of Al-Al (b) FSW of Al-Cu

6 Conclusions

In this paper, Reliability based design approach has been used for development of fixture for friction stir welding. Most of the design issues and probable failures are addressed from previous studies and some pre requisite are considered before designing of FSW fixture. Reliability based design approach further used considering pre-requisite to avoid possible failure of the FSW fixture. Modified design of FSW fixture has been prepared in AutoCAD and further fabricated as per design. Some successful FSW trials have been conducted which shows modified FSW fixture is reliable for carrying out friction stir welding process.

References

1. Shinde, G., Dabeer, P.: Perspectives of friction stir welding tools. Mater. Today: Proc. **5**, 13166–13176 (2018)
2. Shinde, G., Mulani, S., Gunavant, P., Suryawanshi, A.: Experimental investigation of friction welding on aluminium AA5083 Alloy. In: Springer Proceedings of First International Conference on Energy and Environment: Global Challenges, NIT Calicut, Kerala (2018)
3. Shinde, G., Dabeer, P.: Low cost friction stir welding: a review. Mater. Today: Proc. **4**, 8901–8910 (2017)
4. Shinde, G., Dabeer, P.: Review of experimental investigations in friction welding technique. In: International Conference on Science and Engineering for Sustainable Development Bankon, Thailand (2017). https://doi.org/10.21013/jte.icsesd20173
5. Shinde, G.V., Katu, P.U., Shete, H.S., Nigave, A.S., Shelke, S.S., Chougule, S.B.: Review of state of art of friction stir welding. In: National Conference on Design, Manufacturing, Energy and Thermal Engineering (2017)
6. Parida, B., Vishwakarma, S.D., Pal, S.: Design and development of fixture and force measuring system for friction stir welding process using strain gauges. J. Mech. Sci. Technol. **29**, 739–749 (2015)

7. Hasan, M.M., Ishak, M., Rejab, M.R.M.: A simplified design of clamping system and fixtures for friction stir welding of aluminium alloys. J. Mech. Eng. Sci. (JMES) **9**, 1628–1639 (2015). ISSN (Print): 2289-4659; e-ISSN: 2231-8380

8. Shinde, G., Jadhav, G., Kanunje, A.: Work assessment during assembly process by rapid analysis methods. In: International Ergonomics Conference (IIT Guwahati), pp. 420–423 (2014). ISBN(13):978-93-392-1970-3

9. Baghel, P.K., Siddiquee, A.N.: Design and development of fixture for friction stir welding. Innov. Syst. Des. Eng. **3**(12), 40–47 (2012). ISSN 2222-1727 (Paper) ISSN 2222-2871 (Online)

Book Chapter:

10. Design for Reliability: Concepts, Causes and Identifications Andrew Taylor BSc MA FRSA - Art and Engineering in Product Design

Online Document

11. Module: 5, Design for Reliability and Quality NPTEL, IIT Bombay

Reliability of Different Loads on Electro Hydrostatic Actuator

Aravindkumar D. Kotagond[1]([⊠]), Somashekhar S. Hiremath[2], and M. A. Kamoji[3]

[1] Mechanical Engineering Department,
BLDEA's V. P. Dr. P.G. Halakatti College of Engineering and Technology,
Vijayapur 586103, Karnataka, India
arvindkotgond@gmail.com
[2] Mechanical Engineering Department, IIT Madras, Chennai 600036, India
somashekhar@iitm.ac.in
[3] Mechanical Engineering Department, KLE's MSSCET,
Belagavi 590008, India
makamoji@rediffmail.com

Abstract. Electro Hydrostatic Actuators (EHA) is emerging as a viable option for industrial machine builders. The design combines the technologies of both electro-mechanical and electro-hydraulic systems. The major applications are in aerospace, robotics and automation industry etc. EHA system is an emerging technology. Its objective is to replace a centralized hydraulic system by a self-contained and localized direct drive actuator system. EHA has become an important part of modern flight control systems because of its reduced leakage, increased energy efficiency and lower overall weight compared to conventional hydraulic systems. It is treated as a mechatronics system as it consists of mechanical, electrical, hydraulic and control systems. In this work dynamic response is considered as a output parameter of EHA. The different loads like 0.5 kg, 1 kg, 1.5 kg, 2 kg and 2.5 kg are input parameters. An experimental setup of EHA system for both asymmetric and symmetric actuator is developed with measuring instruments and sensors. This forms a mechatronics system. The performance evaluation of different loads on piston displacement has been studied using both experimental setups. It is found that minimum load gives the better response. This indicates that the reliability of both actuators is high at lower loads.

Keywords: EHA · Load · Asymmetric actuator ·
Symmetric actuator and piston rod displacement

1 Introduction

The Power-By-Wire (PBW) systems is the recent form of flight surface actuation. It requires only the attachment of power and control wires. It is also lower in weight, modular and fault tolerant actuators embedded in the flight surfaces. PBW technology has several advantages and potential for energy saving and weight reduction compared to mechanical linkages and Fly -By-Wire (FBW). The basic structure of a modular PBW flight surface actuation system consists of control system, velocity feedback and actuator. This structure is used in the more electric aircraft concept. The PBW actuation

© Springer Nature Singapore Pte Ltd. 2020
V. K. Gunjan et al. (Eds.): *ICRRM 2019 – System Reliability, Quality Control, Safety, Maintenance and Management*, pp. 67–72, 2020.
https://doi.org/10.1007/978-981-13-8507-0_11

system is classified into three categories. Those are Electro Mechanical Actuation (EMA), Integrated Actuator Package (IAP) and Electro Hydrostatic Actuation (EHA). EMA sometime leads to excessive heating and even jamming due to big stresses developed in mechanical transmission devices. IAP uses a variable displacement pump and fixed speed motor. Here the flow is varied by changing the swash plate angle.

There are number of advantages in choosing Electro Hydrostatic Actuator system like increased energy efficiency, reduced overall weight, reduced noise and reduced leakage etc. Electro Hydrostatic Actuation (EHA) replaces EMA and IAP by modular, compact and electric signal wired subsystems. The main parts of the EHA are electric servo motor, bi-directional pump, hydraulic actuator, control elements, accumulator and controller. These are built on a single manifold which eliminates a network of pipes. According to the input instruction from sensors, the controller sends the electric control signal via DAQ card to the servo motor. The servo motor takes this electrical signal and gives the corresponding speed and rotational direction to the pump. The pump in turn sends the hydraulic flow to the piston end and creates a pressure difference on both sides of the piston. The pressure difference across the piston makes the piston to move along with the rod which is attached to it. The sensor present in the path of piston rod sends the information of displacement back to the controller. This forms the closed loop control system. Hu et al. (2014) discussed the effects of flight control in both dynamic and energy consumption by replacing the traditional hydraulic actuator with EHA.

Application of EHA is found to be vast and even small improvement in the configuration of the system is found to be reflecting significantly in the performance characteristics. As EHA has its main application in aerospace industry, weight of the unit should be kept as minimum as possible. However recently they are actively used as industrial actuators. EHA has advantages like increased energy efficiency, reduced pressure losses, reduced noise, reduced leakage, reduced maintenance and failsafe functionality. A new EHA system with axial piston pump instead of gear pump will improve the performance and helps to achieve precise positioning. EHA covers benefits of both Electro-mechanical actuation (EMA) and Electro-hydraulic servo systems (EHSS). Those are high energy efficiency, environmental cleanliness, low noise emission, high forces and no backlash. The general configuration of EHA consists of mechanical, electrical, hydraulic and control domains. Again each domain consists of various components.

Different configurations of EHA uses different types of pumps like gear pumps, vane pumps and piston pumps. Again there are variety pumps are available in each case. Gear pumps are essentially of the fixed displacement type whereas vane pumps and piston pump can be either of the variable displacement type or of the fixed displacement type. Big displacement pumps are used in heavy load with low speed applications, whereas small displacement pumps are used in light weight with high speed applications. McCullough (2011) discussed two main forms of hydrostatic actuation systems. One that uses a fixed-displacement external gear pump and another that uses a variable-displacement piston pump. Because of their low cost, higher reliability and simple structure, gear pumps are suitable for high temperature applications (Xu et al. 2010).

Gear pumps are specially suitable for EHA (Gendrin and Dessaint 2012) because of their high speed rotation. According to Ivantysyn and Ivantysynova (2003), distinctly non-uniformity grade of the flow rate can be achieved by the internal gear pumps. Lubrication is provided by meshing topology but it possesses backlash effect. Manring and Kasaragadda (2003) concluded that, it is advantageous to design an external gear pump with a fewer number of teeth on the driven gear and a large number of teeth on the driving gear. It helps in reducing the physical size of the pump without affecting the volumetric displacement. It also reduces the amplitude of the flow pulsation, at the same time increasing the natural frequencies of the machine. These points are useful for an EHA in flight control actuation system, while designing an external gear pump.

The use of symmetrical actuator (double rod end) keeps the input flow of the pump equal to the output flow during the movement of the load. The aerodynamic load is approximately symmetrical around zero as explained by Gendrin and Dessaint (2012). The characteristics of the single-rod cylinder is investigated by Oh et al. (2012). They explained the characterization using mathematical models along with the other sub-systems of EHA. They also explained the performance of a closed-loop hydraulic control system incorporating the EHA with load and controller. Hydraulic reservoir is built inside the piston rod of the EHA (Takahashi et al. 2008) instead of having a full-volume reservoir outside the cylinder. This type of arrangement leads to weight and volume reduction.

Fluid power is controlled primarily through the use of control devices called valves. But these valves are avoided in EHA. This itself is the important point in EHA. The EHA system has non-linear effects like friction in cylinder and dead band observed in pump. These non-linear effects play an important role in developing a controller. At the same time it increases the complexity of the control method. A mathematical model is developed by Zhang and Li (2011) and used a feedback linearization control strategy for position control of cylinder. The simulation results show the appropriate response with no overshoot and better rapidity. But there is lack of experimental investigations. Sliding Mode Controller (SMC) is robust controller (Wang et al. 2008) for linear and non-linear systems.

2 Experimental Setup

The circuit for an experimental setup of the EHA system with asymmetric actuator is shown in Fig. 1. In case of conventional hydraulic system, servo valves are used for position control of actuator. But in case of Electro Hydrostatic Actuator the position of the actuator can be controlled by varying the motor speed itself. The circuit design includes controller, electrical, hydraulic and mechanical subsystems. Load is suspended over the arm as shown in the figures. The experimental setup of the EHA system with symmetric actuator is shown in Fig. 2. The load on the actuator ranges from 0.5 kg to 2.5 kg. The experiment is conducted initially with no load followed by increase of 0.5 kg after each turn. The same circuit and experimental setup is used for symmetric actuator only by replacing asymmetric actuator with symmetric actuator. In the present work, experimental investigations of loads is carried out on both asymmetric (single rod end) and symmetric (double rod end) actuators. It is important to note that the asymmetric cylinder has volume difference in the forward and return direction. But symmetric cylinder has same volume in both directions

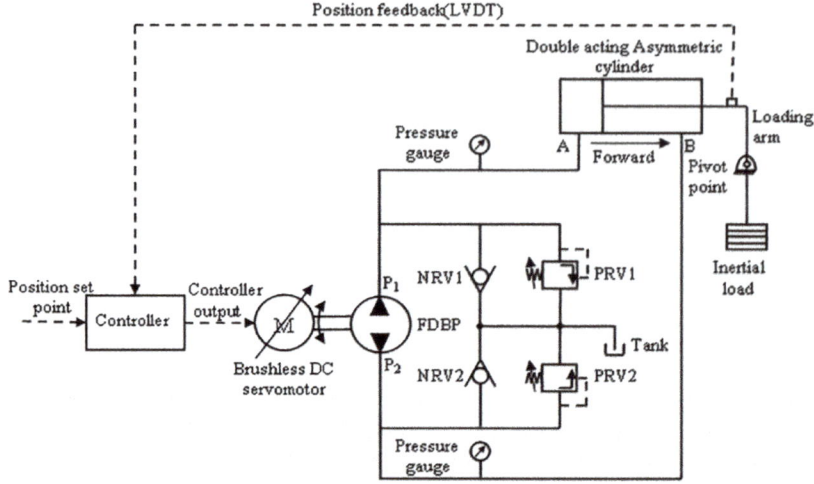

Fig. 1. Circuit for experimental setup of EHA System

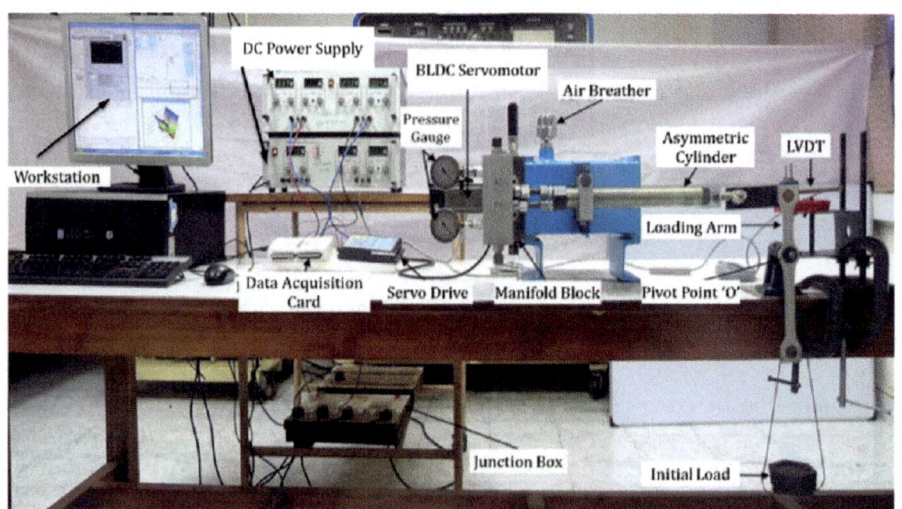

Fig. 2. Asymmetric actuator setup

3   Experimental Results and Discussion

The experiment is carried out on both asymmetric (single rod end) and symmetric (double rod end) actuators by varying the loads from no load to 2.5 kg. The results are elucidated in the following graphs shown in Figs. 3 and 4.

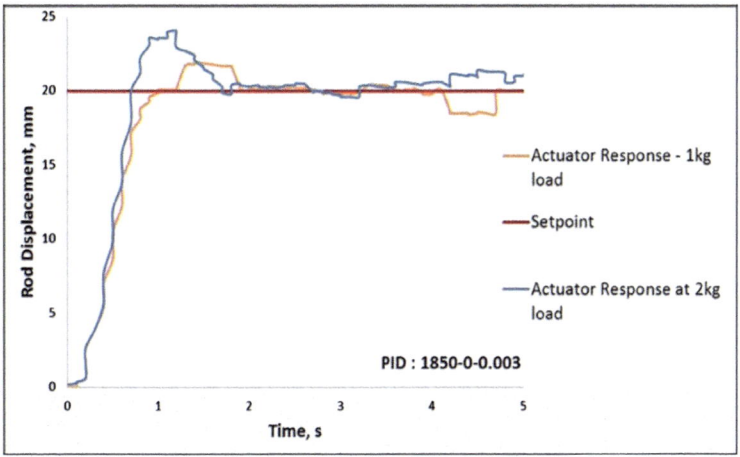

Fig. 3. Asymmetric actuator response

The graph shown in Fig. 3 gives a comparison of the EHA system subjected to one and two kg loads respectively. The overshoot of the asymmetric actuator when subjected to two kg load is more when compared to a lesser load of one kg. The actuator tends to settle earlier (1.1 s) when subjected to one kg load, where it takes more time (1.9 s) when subjected to two kg load. This is in agreement with the literature.

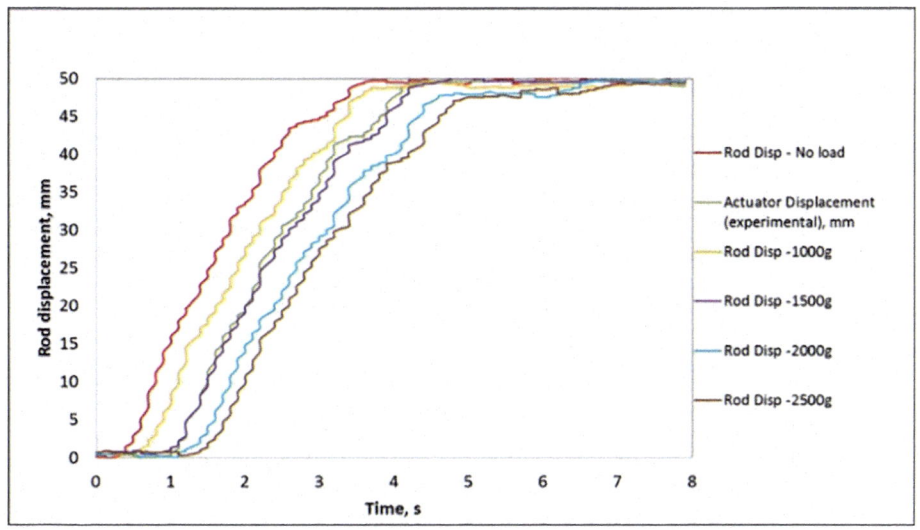

Fig. 4. Symmetric actuator response

It is inferred from the Fig. 4 that the settling time tends to increase as the load on the symmetric actuator is added. At no load the symmetric actuator settles at 3 s, however at 2.5 kg load the actuator settles at 8 s. The results obtained in the above section are in cognizance with the data available in the literature for load analysis being carried out on symmetric actuator. As the load on the actuator increases, a corresponding force needs to be exerted by the system to overcome the load factor. It leads to consumption of more time.

4 Conclusion

From the results it is clear that the response of the actuator is fast for lower loads. The overshoot also increases with increase in load. The settling time of the actuators is also less for no load and lower loads. For higher loads i.e. more than 2.5 kg, the EHA does not show any response. As the load increases, the resistance offered against the actuator movement increases. This leads to the hydraulic oil present in the actuator to compress because of high resistance and takes more time to move the actuator. It is concluded that EHA is reliable for lower loads.

References

Hu, X., Mao, Z., Jiao, Z., Wu, S., Yu, X., Li, F.: Analysis of the characteristics by modeling and simulation of actuator in flight control system. In: Proceedings of 2014 IEEE Chinese Guidance, Navigation and Control Conference, pp. 2618–2623. IEEE (2014)

Xu, S., Zhou, M., Dong, H.: High temperature hydraulic pump and flow control based on frequency conversion technology. In: International Conference on Mechatronics and Automation (ICMA), pp. 1203–1207. Xian, 4–7 August 2010

McCullough, K.R.: Design and characterization of a dual electro-hydrostatic actuator. Thesis (Master). McMaster University, Hamilton, Canada (2011)

Gendrin, M., Dessaint, L.: Multi domain high detailed modeling of an electro-hydrostatic actuator and advanced position control. In: 38th Annual Conference on IEEE Industrial Electronics Society, IECON 2012, Montreal, 25–28 October 2012, pp. 5463–5470. IEEE (2012)

Ivantysyn, J., Ivantysynovs, M.: Hydrostatic Pumps and Motors; Principles, Design, Performance, Modeling, Analysis, Control and Testing. Tech Books International, New Delhi (2003)

Manring, N.D., and Kasaragadda, S.B.: The theoretical flow ripple of an external gear pump. J. Dyn. Syst. Meas. Contr. **125**(3), 396–404 (2003)

Takahashi, N., Kondo, T., Takada, M., Masutani, K., Okano, S., Tsujita, M.: Development of prototype electro-hydrostatic actuator for landing gear extension and retraction system. In: Proceedings of the JFPS International Symposium on Fluid Power, vol. 2008, no. 7–1, pp. 165–168 (2008)

Zhang, Q., Li, B.: Feedback linearization PID control for electro-hydrostatic actuator. In: 2nd international Conference on Artificial Intelligence, Management Science and Electronic Commerce, pp. 358–361. Deng Leng. IEEE, 8–10 August 2011

Wang, S., Habibi, S., Burton, R.: Sliding mode control for an electro hydraulic actuator system with discontinuous non-linear friction. In: Proceedings of the Institution of Mechanical Engineers, Part I: Journal of Systems and Control Engineering, vol. 222, no. 8, pp. 799–815 (2008)

Application of MOORA to Optimize WEDM Process Parameters: A Multi-criteria Decision Making Approach

Himadri Majumder[1], Srimant Kumar Mishra[1($\boxtimes$)], Anil R. Sahu[1],
Anil Laxmanrao Bavche[1], Mahadev Valekar[1],
and Bijaya Kumar Padaseti[2]

[1] Department of Mechanical Engineering, G. H. Raisoni College of Engineering
and Management, Pune 412207, India
`srimantnitrkl@gmail.com`
[2] Department of Mechanical Engineering, National Institute of Technology,
Rourkela 769008, India

Abstract. Wire electrical discharge machining (WEDM) is a versatile, most useful non-conventional machining process used to machine conductive materials with high precision for cutting intricate shapes, complex profiles etc. This paper exemplifies a multi-criteria decision making (MCDM) approach, multi-objective optimization on the basis of ratio analysis (MOORA) to optimize major process parameters for several contradictory responses during WEDM of inconel 718. The selected WEDM responses are cutting rate (CR), arithmetic mean roughness (Ra) and machining time (MT). All of these have been examined by varying pulse-on time (T_{ON}), pulse-off time (T_{OFF}), pulsed current (I) and servo voltage (SV). The optimal machining condition for multi-performance features has been found as T_{ON} = 120 µs., T_{OFF} = 46 µs., I = 230 A. and SV = 20 V. Hence, current research concentrated on the utilization of MCDM method MOORA as a critical selection approach to when deal with multi criteria optimization environment.

Keywords: Inconel 718 · MCDM · MOORA, optimization · WEDM

1 Introduction

The hypothesis of the electrical discharge machining (EDM) was established by Lazarenko [1]. Wire electrical discharge machining (WEDM), a variant form of EDM, turn out to be one of the most useful non-conventional machining process in the modern era of machining [2]. WEDM becomes the best alternative to produce miniaturized scale products with the high accuracy in dimensional precision as well as high degree of surface. By adopting this thermo-electrical approach, almost any conductive material can be machined regardless of their hardness and mechanical properties [3]. Combination of pulse generator, wire electrode and de-ionized medium are the main feature behind a simple WEDM. Electrical spark is generated by pulse generator between workpiece and wire electrode when immersed into de-ionized medium. Such repeated tiny spark causes electro-erosion to erode material from

© Springer Nature Singapore Pte Ltd. 2020
V. K. Gunjan et al. (Eds.): *ICRRM 2019 – System Reliability, Quality Control,
Safety, Maintenance and Management*, pp. 73–78, 2020.
https://doi.org/10.1007/978-981-13-8507-0_12

workpiece which also increases the machine zone temperature near about 10,000 °C [4]. Higher and lower flushing arrangement flushes away these eroded materials from the machining zone. These constant sparking, erosion and flushing makes uneven, hard and brittle surface [5].

However, complex mechanism, ample of conflicting process parameters, instability in the manufacturing process makes it awfully difficult to set ideal process parameter setting to get most preferred responses. Here comes the necessity of multi-criteria decision making (MCDM) approach. This operation research technique exceptionally appraises various contradictory criteria to ease in decision making. As a potential tool, MCDM approach widely used to analyze complex real problems to get possible collection of the best choices which will be cost effective for all the manufacturing processes [6]. Though there are substantial MCDM approach are available for selection of ideal setting [7–13], multi-objective optimization on the basis of ratio analysis (MOORA) is detected to be simple and computationally easy.

Madic et al. [14] showed application potential of MOORA to solve problem related to non-conventional machining methods. Comparison with technique for order preference by similarity to ideal solution (TOPSIS) indicated perfect correlation to solve this kind of problems. Gadakh et al. [15] applied MOORA to solve different multiple objective optimization difficulty in welding which proved MOORA as a potential, flexible technique. Patel et al. [16] successfully applied MOORA method in connection with AHP to select optimum value of WEDM process parameters when machining EN31 alloys steel using brass wire. Achebo et al. [17] successfully applied MOORA coupled with standard deviation (SDV) to find out the ideal welding process parameters.

Inconel 718, a nickel based super alloy, used in many fields like spacecraft, gas turbines, nuclear reactors, rocket motors, etc. where the material can withstand at very high temperature [18]. High toughness, strength and work-hardening characteristic holds a great challenge to cut this difficult to machine material by conventional machining techniques [19]. Thus, the use of WEDM is very much prominent to machine this type of super alloys. Due to the plenty option of process parameters it is very much necessary to select optimum condition in WEDM. In this work, to regulate WEDM process parameters namely pulse-on time (T_{ON}), pulse-off time (T_{OFF}), pulsed current (I) and servo voltage (SV) MOORA was used as an application potential of MCDM approach.

2 Materials and Methods

2.1 Materials, Experimental Setup and Data Collection

Following Taguchi's L_9 orthogonal array, 9 experiments was accomplished on 4-axis AGIECUT CNC WEDM (maker: AGIE). Four important input variables namely pulse-on time (T_{ON}), pulse-off time (T_{OFF}), pulsed current (I) and servo voltage (SV) were considered to detect significant machinability features like cutting rate (CR), arithmetic mean roughness (Ra) and machining time (MT). De-ionized water and brass wire having diameter 0.25 mm was utilized as dielectric medium wire electrode and respectively. 5 mm length was cut by respective input setting from 5 mm thickness plate of inconel 718. Chemical composition of inconel 718 is shown in Table 1. CR was taken from WEDM machine monitor during respective setting machining and average value from five measurements was booked for final CR value. Using stopwatch MT was taken for every experiment. After machining, Ra values of the machined surface were measured utilizing Taylor Hobson 3D profilometer and average value from five measurements were taken as final Ra value. All the experiment was accomplished in a single loading to diminish loading-unloading time.

Table 1. Chemical composition of inconel 718 [20].

Element	Ni	Cr	Cb	Mo	Ti	Al	Co	Si	Cu	Mn	C	P	Fe
Content (%)	53.4	18.8	5.27	2.99	1.02	0.5	0.17	0.12	0.07	0.07	0.03	0.01	Bal

2.2   Methodology

In this study, a MCDM model MOORA has been practiced to optimize different WEDM process parameter for inconel 718.

2.2.1   Multi-objective Optimization on the Basis of Ratio Analysis (MOORA)

To simultaneously optimize different conflicting attributes MOORA, which was first introduced by Brauers [21], is used subject to some definite limitations [22]. The associated steps were involved:

1st step: After defining the objective, decision matrix was formed to represent different performance features with respect to diverse variables.

$$Q = \begin{bmatrix} q_{11} & q_{12} & \cdots & \cdots & q_{1n} \\ q_{21} & q_{22} & \cdots & \cdots & q_{2n} \\ \cdots & \cdots & \cdots & \cdots & \cdots \\ \cdots & \cdots & \cdots & \cdots & \cdots \\ q_{m1} & q_{m2} & \cdots & \cdots & q_{mn} \end{bmatrix} \tag{1}$$

where, q_{mn} = Performance measure of the mth alternative on n^{th} response.

m = Number of variables.

n = Number of performance features.

2nd step: Decision matrix was normalized to turn it dimensionless quantity to compare all components. Normalization was done following Eq. 2.

$$q_{mn}^* = \frac{q_{mn}}{\sqrt{\sum_{i=1}^{r} q_{mn}^2}} \tag{2}$$

where, q_{ij}^* display the normalized value m^{th} alternative on n^{th} response $(0 < q_{ij}^* < 1)$.

3rd step: In the next step, normalized values were added together for beneficial condition and subtracted for non-beneficial condition to find out overall assessment of the performance measures.

$$y_i = \sum_{n=1}^{n} q^*_{mn} - \sum_{n=g+1}^{n} q^*_{mn} \qquad (3)$$

Usually few responses are more dominant than others. Corresponding weights were multiplied with the specific objective to give it more preference [23]. Overall assessment value (Yi) calculated as follows:

$$y_i = \sum_{n=1}^{n} w_n q^*_{mn} - \sum_{n=g+1}^{n} w_n q^*_{mn} \qquad (4)$$

where, w_n known as the weight of n-th response.

4th step: In the last step, overall assessment values were arranged in descending order. The highest value of Yi signifies the best optimized setting while lowest value of Yi signifies the least preferred setting.

3 Results and Discussion

The responses measured in the current exploration are cutting rate (CR), arithmetic mean roughness (Ra) and machining time (MT) while varying T_{ON}, T_{OFF}, I and SV. Among these responses, Ra and MT are non-beneficial condition whereas CR is beneficial condition. Using Eq. 2, normalized values for each response were calculated. Relative weights for each responses were given as CR = 0.3; Ra = 0.3 and MT = 0.4. Overall assessment value for each experimental setting was calculated following Eq. 4 and tabulated in Table 2. After that, respective overall assessment value prepared in descending order to find out the best setting. It was calculated that, experiment no. 7 has the highest Yi value. So, according to MOORA the optimized process parameter setting like T_{ON} = 120 μs., T_{OFF} = 46 μs., I = 230 A. and SV = 20 V.

Table 2. Calculation using MOORA.

Exp. no.	Normalized values			Yi	Rank
	CR	R_a	MT		
1	0.204797	0.304231	0.406804	−0.192552	9
2	0.388262	0.351459	0.299488	−0.108754	3
3	0.319996	0.327296	0.315294	−0.128308	5
4	0.221864	0.303133	0.381847	−0.17712	8
5	0.281597	0.29215	0.371032	−0.151579	7
6	0.366929	0.366835	0.30032	−0.1201	4
7	0.430928	0.374523	0.26538	−0.08923	1
8	0.405328	0.347065	0.298656	−0.101984	2
9	0.30293	0.322903	0.334844	−0.13993	6

4 Conclusions

The workpiece inconel 718 was machined using WEDM practice and the outcomes were optimized using MCDM approach MOORA. Following major conclusions might be drawn:

- MCDM approach MOORA found fairly easier supportive strategy which involves less mathematical calculations.
- However applying MCDM approach MOORA gave optimized setting as T_{ON} = 120 μs., T_{OFF} = 46 μs., I = 230 A. and SV = 20 V which yield the preferred results.
- The capacity to illuminate process fluctuation makes MOORA more useful for those conditions where numerous responses are need to be improved all along.

References

1. Lazarenko, B.: To invert the effect of wear on electric power contacts. Dissertation of the All-Union Institute for Electro Technique in Moscow/CCCP, Russian (1943)
2. Majumder, H., Maity, K.: Predictive analysis on responses in WEDM of titanium grade 6 using General Regression Neural Network (GRNN) and Multiple Regression Analysis (MRA). Silicon **10**, 1–14 (2018)
3. Majumder, H., Maity, K.: Multi-Response Optimization of WEDM Process Parameters Using Taguchi Based Desirability Function Analysis. In: IOP Conference Series: Materials Science and Engineering. IOP Publishing (2018)
4. Majumder, H., et al.: Use of PCA-grey analysis and RSM to model cutting time and surface finish of Inconel 800 during wire electro discharge cutting. Measurement **107**, 19–30 (2017)
5. Çaydaş, U., Hasçalık, A., Ekici, S.: An adaptive neuro-fuzzy inference system (ANFIS) model for wire-EDM. Expert Syst. Appl. **36**(3), 6135–6139 (2009)
6. Majumder, H., Saha, A.: Application of MCDM based hybrid optimization tool during turning of ASTM A588. Decis. Sci. Lett. **7**(2), 143–156 (2018)
7. Naik, D.K., et al.: Experimental investigation of the PMEDM of nickel free austenitic stainless steel: a promising coronary stent material. Silicon **11**, 1–9 (2018)
8. Saha, A., Majumder, H.: Multi criteria selection of optimal machining parameter in turning operation using comprehensive grey complex proportional assessment method for ASTM A36. Int. J. Eng. Res. Afr. **23**, 24–32 (2016)
9. Kumar, A., et al.: NSGA-II approach for multi-objective optimization of wire electrical discharge machining process parameter on inconel 718. Mater. Today: Proc. **4**(2), 2194–2202 (2017)
10. Saha, A., Mondal, S.C.: Multi-objective optimization in WEDM process of nanostructured hardfacing materials through hybrid techniques. Measurement **94**, 46–59 (2016)
11. Saha, A., Mondal, S.C.: Multi-objective optimization of manual metal arc welding process parameters for nano-structured hardfacing material using hybrid approach. Measurement **102**, 80–89 (2017)
12. kumar Naik, D., Maity, K.: Optimization of Dimensional accuracy in plasma arc cutting process employing parametric modelling approach. In: IOP Conference Series: Materials Science and Engineering. IOP Publishing (2018)

13. Majumder, H., Maity, K.: Application of GRNN and multivariate hybrid approach to predict and optimize WEDM responses for Ni-Ti shape memory alloy. Appl. Soft Comput. **70**, 665–679 (2018)
14. Madić, M., Radovanović, M., Petković, D.: Non-conventional machining processes selection using multi-objective optimization on the basis of ratio analysis method. J. Eng. Sci. Technol. **10**(11), 1441–1452 (2015)
15. Gadakh, V., Shinde, V.B., Khemnar, N.: Optimization of welding process parameters using MOORA method. Int. J. Adv. Manuf. Technol. **69**(9–12), 2031–2039 (2013)
16. Patel, J.D., Maniya, K.D.: Application of AHP/MOORA method to select wire cut electrical discharge machining process parameter to cut EN31 alloys steel with brasswire. Mater. Today Proc. **2**(4–5), 2496–2503 (2015)
17. Achebo, J., Odinikuku, W.E.: Optimization of gas metal arc welding process parameters using standard deviation (SDV) and multi-objective optimization on the basis of ratio analysis (MOORA). J. Miner. Mater. Charact. Eng. **3**(04), 298 (2015)
18. Kumar, A., Bagal, D.K., Maity, K.: Numerical modeling of wire electrical discharge machining of super alloy Inconel 718. Proc. Eng. **97**, 1512–1523 (2014)
19. Kumar, A., et al.: Multi-objective optimization of wire electrical discharge machining process parameterson Inconel 718. Mater. Today Proc. **4**(2), 2137–2146 (2017)
20. Rahman, M., Seah, W., Teo, T.: The machinability of Inconel 718. J. Mater. Process. Technol. **63**(1–3), 199–204 (1997)
21. Brauers, W.K., Zavadskas, E.K.: The MOORA method and its application to privatization in a transition economy. Control Cybern. **35**, 445–469 (2006)
22. Khan, A., Maity, K.: Selection of optimal machining parameters in turning of CP-Ti Grade 2 using a hybrid optimization technique (2017)
23. Majumder, H., Maity, K.: Optimization of machining condition in WEDM for titanium grade 6 using MOORA coupled with PCA—a multivariate hybrid approach. J. Adv. Manuf. Syst. **16**(02), 81–99 (2017)

Productivity Improvement in a Manufacturing Industry Using Value Stream Mapping Technique

Vahid M. Jamadar[1](✉), Gurunath V. Shinde[1], Sandip S. Kanase[2], Ganesh S. Jadhav[1], and Anant D. Awasare[1]

[1] Department of Mechanical Engineering, AGTI's DACOE, Karad, India
vmjamadar.mech@dacoe.ac.in
[2] Department of Mechanical Engineering, BVCOE, Kharghar, India

Abstract. The purpose of this paper is to develop a value stream map for a compressor assembly company in Emerson Climate Technologies, India Ltd. Atit. The main aim is to identify, analyze and eliminate waste in the shop floor which is any activity that does not add value to their final product, in the production and assembly process. Value Stream Mapping has the one of the too which status of finding waste in manufacturing, production, assembly and business processes by identifying, control and removing or streamlining non-value-adding steps and processes. The tool helps to identify the current flow of material and information in processes for a family of products, highlighting the opportunities for improvement that will most significantly impact the overall production and assembly system. VSM application will help to identify a current situation diagnostic of enterprise logistic and supply chain, so detect the problems and linked to the process and finally chose the wastes lean thinking in the production techniques to eliminate and control those wastes. VSM ends with the future state showing the proposed improvements in the process by reducing wastage the proportion of value adding time in the whole process rises and the process throughput speed is increased.

Keywords: Lean manufacturing · VSM · Current state map · Future state map

1 Introduction

The aim of this study was to develop suggestions to the company on how production waste can be reduced on the shop floor. The study used concepts of Lean thinking on the Value Stream Mapping tool to analyze, identify and examine value and non-value adding activities during assembly of compressor, to reduce wasteful operations while flow of compressor, to improve quality within processes to standardize working processes and methods. Value Stream mapping of line is essentially concerned with finding better ways of doing operations. Value Stream Mapping is one of the tool used in identifying, analysis and minimizing major wastes in manufacturing of compressor assembly line of CR series [1]. The objective is to reduce the wastes and non value added activity.

© Springer Nature Singapore Pte Ltd. 2020
V. K. Gunjan et al. (Eds.): *ICRRM 2019 – System Reliability, Quality Control, Safety, Maintenance and Management*, pp. 79–84, 2020.
https://doi.org/10.1007/978-981-13-8507-0_13

Value Stream Mapping (VSM) is one of the lean manufacturing tools. Lean manufacturing is a systematic approach for process improvement. It is based on identifying and reducing waste together with continuous improvement [2]. Value stream mapping is a powerful yet simple tool which allows the user to see the waste throughout the stream. It consists of sketching the current and future state map. The current state map charts the present flow of material and information as a product goes through the manufacturing process [3, 4]. It is a simplified visual waste all over the system and encourages systematic move towards eliminating waste. The future state map is a chart that shows how to create a lean flow. It adopts lean manufacturing techniques to eliminate waste and reduce non-value added activities to the minimum. Earlier studies reveal study of productivity improvement by ergonomic evaluation approach using REBA-RULA, Motion study, Why-Why Analysis etc. [5–8].

2 Methodology

Methodology provides an overview of the approach, the chosen study approach and data collection methods. Before going on actual line complete safety training in the industry is essential. Generally VSM has four major steps as follows

(1) Selection of Product
(2) Drawing mapping of recent
(3) Drawing mapping for expectations
(4) To develop planning as per new line

2.1 Selection of Product

A particular produce should be defined as the target for improvement time to time [2]. This study focuses on CR series type compressor designed by Emerson Climate technologies manufacturing industry as it comprises 65.5% of the customer demand. There are several other product families but as "CR" series compressor's product family contributes to a major portion of the customer demand, it is a good product family to focus the study on.

2.2 Current State Description

According to volume and standardization, the manufacturing system of industry is mass production type. Product layout is combination of inverted L type and line type layout which is prepared by utilizing the automated conveyor and roller conveyor respectively. This is a combination or hybrid type of layout. This layout has separate stations provided for each operation. It is automated by using the jigs provided to each assembly of the compressor. The raw materials come from different suppliers as per the orders. Ideally, these would have had due communication of production requirements from production control. On manual assembly line, workstation is designed along the work flow path so that one or more workers can perform the task. The work elements represent small portion of work that must be accomplished to assemble product.

Workstations designed should include productivity, operator comfort, operator variety and safety. The number of operators may be different and one operator might monitor several workstations. Information flow in Assembly Line is done manually and electronically too. The pattern of material flow is an important consideration in the plant layout decision because good layout aims at minimizing flow of material. The flow pattern of materials helps in eliminating bottle-necks, rushing and backtracking and ensures goods supervision and control.

Value Stream Analysis

- **Data Collection**
  1. Personal observations
  2. Interactions
  3. Customer Requirement
- **Current State Map**
  The current state map was drawn with the help of information collected from assembly line. With the information collected, a clear understanding of the whole manufacturing process is depicted for the CR series product family.
- **Wastes Analysis**
  By visiting the assembly line sufficient data collection regarding the activities, processes, and mudas, ergonomics of workers and aesthetical analysis of the work place, value added and non-value added activities, implementation of the lean tools on the line was done. All the data collected from the line was tabulated and analyzed as follows (Figs. 1 and 2), Table 1:

Table 1. Assembly line process flow

Activity	Head counting	Non-value added activity	MUDA	Rejection reason
Body part loading	4	Yes	WIP	
Piston insertion	1	Yes	WIP DR	Transportation
Crankshaft insertion	2		W WIP	Oversize, undersize of shaft
Gasket selection	2		DR W WIP	Improper straightness
Cylinder head assly Plcmnt	6		W	Bending of circlip
Stator Plcmnt	3		W WIP DR	Stucking of winding with mandrel
Rotor Plcmnt	1		DR	
Upper & lower housing Plcmnt	3		W	Transportation
Run test	1	Yes	WIP DR DS	Low volt, noise. valve plate leak,

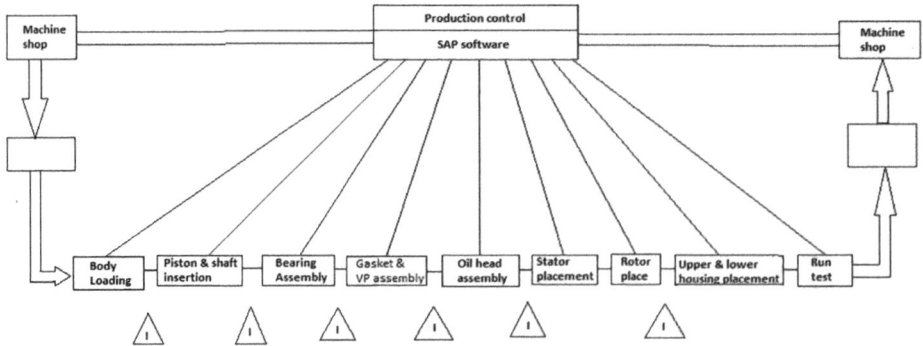

Fig. 1. Current state map

Suggestions

- **Inventory:** Overall system can be upgraded by using E-Kanban system. A traditional card based system has some limitations that can be improved with an electronic kanban. The number of card movement increases as well. The cards may be lost or misplaced sometimes, which nearly cause immediate problems. Electronic Kanban (e-Kanban) is a system of signaling that makes use of information technology (IT) to trigger the movement of raw materials to allow for a real time view of inventory throughout an assembly line [3].
- **Ergonomics of the workers:** The ergonomics of the workers on the production line is maintained by some measures that should be enhanced at the critical processes where the fatigue level is seen higher such as the run test process and rotor placement area.

  1. **Run test**
     At this process the twisting and back motion of a worker causes the excessive stress which could be reduced by improving the design of conveyor and display and control panel. Conveyor system requires the worker to pull the compressor by using a long hook which can be avoided by adding pushing arm operated automatically. Control panel height is more so the sitting arrangement should be such that worker can easily read the display at upper portion.
  2. **Rotor placement area**
     At this process the worker needs to lift the rotor with the help of hand lifter which is heated initially. So there is a danger of dropping down or slipping of the rotor in the case of the heavy rotors. Pneumatic lifter can do it without danger and with more speed and fatigue reduction of the worker.
  3. **Upper placement section**
     In this area the upper heads are stocked on a table, then these are stamped with the spring and one socket is hammered to the assembly. This process includes excessive movement and twisting of the workers. This can be reduced by installing the roller conveyor. This will also increase the speed of the process and simultaneously the production rate.

Visual Management

There is a less evidence of visual control and visual management throughout the whole system. As per the E kanban system, the development of ICT especially RFID and wireless can used for more opportunities to develop mechanical visual management system [4]. 5S is the basic step of making system more visual so the proper maintenance of the 5S is required at some places, which are described as per each S below:

1. **Seiri-sort out:**
   Storage areas should be observed for unnecessary things to be sorted out at the initial body loading section.
2. **Seiton-set in order:**
   All the materials are arranged in a proper way in the storage and material handling devices, but the devices are not properly positioned on the shop floor. Proper positions should be decided by allowing the easy travelling without destructions. For example, the trolleys carrying the storage of subassembly parts are making barriers to materials & people travelling on the shop floor can be set in order by allotting a place marked by the lines on floor.
3. **Seiso-shine & clean up:**
   Daily cleaning process is maintained at the line. Each process and work station is cleaned properly after work is finished. Shop floor shoes, goggles and ear plugs are provided to each person on the shop floor which ensures the safety measures.
4. **Shitsuke- standardize:**
   As per the company standards the audits and the periodic reviews are taken and the quantitative and qualitative goals are set according to the observations and standards.
5. **Shiketsu-sustain:**
   The maintenance of the attained cleanliness and systematic work environment is more important, which is responsibility of each personnel.

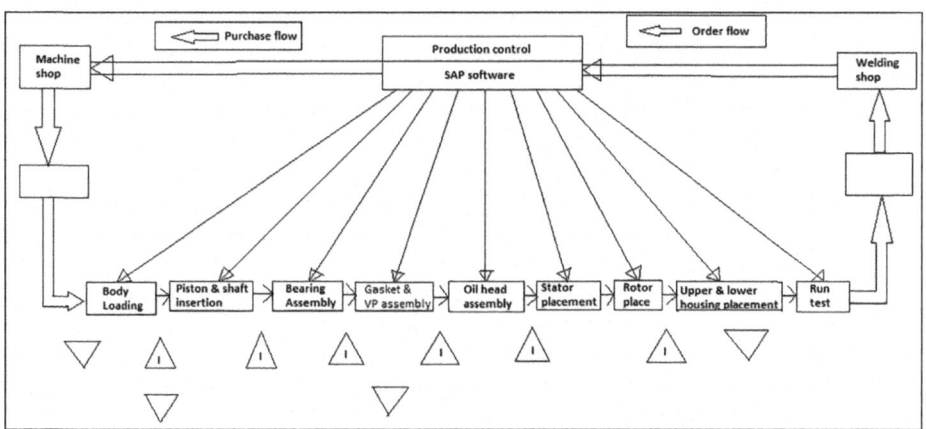

Fig. 2. Future state map

3 Conclusion

Value stream mapping has proven to be an effective tool and different ways to analyze a company's current production rate in the state and point out different problem areas. The visual nature of value stream mapping, by combining information and material flow on one map, depicts how the two relate to waste. First a current VSM was created, providing an overview of the material flow and transportation between the factories. Then future state map by using the lean principles was designed. These suggestions will help to increase the productivity in future development, reduce fatigue of workers and also reduce wastes on assembly line. With these objectives in consideration, recommended future state maps have been created to improve the current situation with the assistance of supervisors at Emerson Climate Technologies India, Ltd., Atit.

Acknowledgment. The authors would like to express sincere thanks to Emerson Climate Technologies Ltd., Atit India and AGTI'S Dr. Daulatrao Aher College Engineering Banawadi Karad, for their guidance and support.

References

1. Sheth, P.P., Deshpande, V.A., Kardani, H.R.: Value stream mapping: a case study of automotive industry. Int. J. Res. Eng. Technol. **3**, 310–314 (2014)
2. Yang, M.G.M., Hong, P., Modi, S.B.: Production flow analysis through value stream mapping: a lean manufacturing process case study (2011)
3. Rahani, A.R., Al-Ashraf, M.: Production flow analysis through value stream mapping: a lean manufacturing process case study (2012)
4. Yadav, R., Shastri, A., Rathore, M.: Increasing productivity by reducing manufacturing lead time through value stream mapping (2012)
5. Shinde, G., Jadhav, G., Kanunje, A.: Work assessment during assembly process by rapid analysis methods. In: International Ergonomics Conference (IIT Guwahati), pp. 420–423 (2014). ISBN(13) 978-93-392-1970-3
6. Shinde, G., Jadhav, G., Sawant, S.M.: Ergonomic evaluation tools RULA and REBA: a case study. In: National Conference on Industrial Engg. And Technology Management (NITIE Mumbai), pp. 31–36 (2014). ISBN 978-93-83842-90-2
7. Shinde, G.V., Jadhav, V.S.: Ergonomic analysis of an assembly workstation to identify time consuming and fatigue causing factors using application of motion study. Int. J. Eng. Technol. **4**(4), 220–227 (2012). ISSN 0975–4024, 220-227
8. Shinde, G.V., Jadhav, V.S.: A Computer based novel approach of ergonomic study and analysis of a workstation in a manual process. Int. J. Eng. Res. Technol. **1**(6), 1–5 (2012). ISSN 2278- 0181

Determination of Specific Heat of Nagpur Orange Fruit (Citrus-Sinesis L) as a Function of Temperature and Moisture Content

Prashant M. Rewatkar[1](✉) and M. Basavaraj[2]

[1] Department of Mechanical Engineering, SSPACE, Wardha, MS, India
revatkar.prashant@gmail.com
[2] BIT, Ballarpur, Chandrapur, MS, India
drmbraja@gmail.com

Abstract. Nagpur Orange fruit (Citrus - Sinesis L) is an important source of vitamins, fibers and poly nutrients. They are seasonal in nature. Due to perishable nature and high moisture content its shelf life is insignificant. As such we do not have well equipped processing units with storage facilities, most of the Nagpur Oranges are getting spoiled. To avoid putrefaction of oranges we require processing and preservation equipment. Specific heat is an important thermal property of food, which is necessary for analyzing thermal behavior and designing of food processing equipment involved in heat transfer. Specific heat of Nagpur Orange fruit (citrus sinesis L) can be evaluated experimentally at different moisture contents from 40% to 90% (wb) for different densities, $\rho 1 = 1006$ kg/m^3 and $\rho 2 = 1062$ kg/m^3. It is noticed that the specific heat (Cp) of Nagpur Orange fruit increases with increase in moisture content and temperature in the range of 2.126 to 6.50 kJ/Kg°C for above said densities. The experimental values are analyzed and compared with standards Dickerson and Siebel's model. The experimental values are coordinated with predicted models.

Keywords: Nagpur Orange fruit · Moisture content · Density · Specific heat (Cp)

1 Introduction

India is categorized as a developing country with majority of the population currently in the agricultural field, its development is very important. Production is one side of the coin while the marketing is another side. Orange fruit has great economic importance. Fruits are nature; s wonderful gifts to mankind, indeed; they are like medicines packed with vitamins, minerals, antioxidants and many poly-nutrients. They not only look pleasing to eyes, but also protect the human body from diseases and keep our body healthy. This is because of their unique nutritional values. It is a rich source of vitamin AB and phosphorus. Commercially oranges are available for consumption directly and in the form of jam, juice, squash and medicines. It is main source of citric acid and other cosmetics which have high international market value. Citrus is one of the most important crops among fruits in India, which covers about 0.62 million hectares with the total production of 4.72 million tons. Maharashtra ranks first in production of

© Springer Nature Singapore Pte Ltd. 2020
V. K. Gunjan et al. (Eds.): *ICRRM 2019 – System Reliability, Quality Control, Safety, Maintenance and Management*, pp. 85–91, 2020.
https://doi.org/10.1007/978-981-13-8507-0_14

Oranges followed by Andhra Pradesh, West Bengal, Assam, Punjab, Karnataka and Tamil Nadu. As per agricultural statistical information of Maharashtra state in the year 2008-09, the area of production and productivity of Oranges in Maharashtra state 123495 hectares with a total production of 721217 tons, 9.8 mt./ha respectively [1].

Nagpur district has the largest area of production and productivity, that's why this city is known as "The Orange City". Orange crops in Nagpur Dist occupies 15,205 haters with a production of 1,40,613 tons of Oranges per year [1, 2]. Oranges have limited shelf life therefore it is necessary to utilize the fruits for making different products to increase its accessibility for a long period of time. This would definitely help to control the price during off seasons. Oranges can be consumed in fresh or into the forms juice, syrup, orange bar etc. These orange products have economical value. The primary goal of the food processing industry is to supply the consumers with safe processed products [2].

To keep up the demand, oranges are preserved by means of thermal processing. The application of heat can induce favorable product change, deactivate enzymes and kills microorganisms. All this is done to ensure that the oranges are hopefully appetizing and have increased shelf-life and safe for consumers. Heating and cooling of food is one of the earliest methods of applying science to food. In thermal processes, heat energy is transferred from the food products. It has ability to conduct, store and lose heat. Thermo-mechanical properties are indispensable in process calculation and quality optimization of foods that includes heating, blanching, freezing and drying [3]. It has been known from literature survey that biological materials exhibit Thermo-Mechanical properties. The food stuffs are dependent on temperature, moisture content and composition of food. Compositions of the fruit depend on the physical characteristics from variations in soil, irrigation system, climate condition and fertilizers used. They would manifest themselves in major Thermo- physical properties [4, 9].

Specific heat is the property needed in estimation of the amount of heat required to change the properties of the product. The value of Specific heat (Cp) should be known for modeling, simulating and optimizing of process operation which involves heat transfer [5].

2   Materials and Methods

2.1   Sample Preparation and Moisture Content

Freshly harvested Nagpur Orange of uniform shape size and colors are selected from Nagpur Santra Market (Karanja Region Area, Vidarbha India). They are washed in clean water and rinds and fibers of the sample removed manually. The moisture content of the sample orange are measured by using standard AOAC (Association of Official Analysts and Chemists) [6] in oven at 70 °C for 24 h with 05 replicates. The moisture content range in obtain samples are found to be 90.0% to 93.0% (Wb), the samples were dried for various periods in designed hot air cum solar dryer manufactured in MGIRI, Ministry of MSME Govt. of India Wardha, at 55 °C, 65 °C and 75 °C. The partly dried sample are sealed in polyethylene bag and stored at constant temperature of 29 °C for 24 h to ensure uniform moisture content throughout the sample.

2.2   Experimental Set up for Determination of Specific Heat (Cp) of Nagpur Orange Fruit

Laboratory apparatus used for measurement of Specific Heat (Cp) is shown in Fig. 1. Constant temperature hot water is circulated over the tube throughout experimentation. It consists of an Aluminum cylinder of 29.5 mm diameter, 104 mm length, a chromium wire of 33 gauge is used as heating element of 200 mm. It is placed co-axially into test cylinder. Both ends of tube are covered by Teflon and are used to minimize axial conduction of heat. An insulated iron constantan thermocouple is inserted through the Teflon cover on the top to measure the temperature rise. The DC power supply is connected to heater wire to supply constant power to it. The outer surface of the test cylinder is encircled by a network of tubes to maintain constant temperature totally insulated to act as an adiabatic container [8].

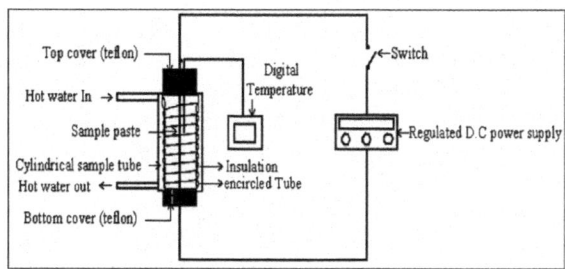

Fig. 1. Schematic diagram of experimental set up

2.3   Methodology

A known mass of desired moisture content is filled in the test cylinder. Prepared sample was placed in open end of cylinder. It was sealed off with polyethylene, to prevent the evaporation of the moisture content. The temperature of sample is equilibrated with the ambient temperature, When the sample reaches uniform targeted temperature by circulating controlled water bath through the encircled tube network, Stop watch and power supply are switched on simultaneously. The time required for every degree rise in temperature at constant power is determined. The experiment is repeated for various MC samples and Cp is calculated by heat energy balance equation [7].

$$Cp = V \times I \times t/m \times \Delta T = I^2 R/m \times \Delta t. \tag{1}$$

Where,

V   =Volts,
I   = Current (Amp),
R   = Resistance of heater (Ω/m),
T   = Time (Sec),
M   = Mass of orange fruit sample,
ΔT = T_f–T_i, T_f = Final temperature °C,
T_i = Initial temperature °C

Predicated Model for Specific Heat: - Dickerson proposed model for Specific heat for calculating the Specific heat (Cp) for unknown fruit product for different moisture content [8]

$$Cp = 1.675 + 0.025M. \tag{2}$$

Where, M = Moisture content of fruit in % (wb).

Siebel proposed the below equation to know Specific heat of unknown fruit product.

$$Cp = 0.8374 + 0.0335M. \tag{3}$$

Where M = Moisture content of fruit % (wb).

3   Results and Discussion

3.1   Assay of Specific Heat at Density ρ_1 = 1006 Kg/m^3

Figure 1 shows the effect of moisture content (40%–90%), temperature difference (5 °C–15 °C) on Specific Heat of Nagpur Orange fruit at 1006 kg/m^{3}density. The Specific Heat of Nagpur Oranges increases with increase in moisture content. This is because of higher thermal content present between molecules of sample and also some bound molecules in the sample. The average Specific Heat increases from 2.26 kJ/Kg °C to 5.78 kJ/Kg °C from moisture content 40% to 90% (wb). It's observed that graph is not linear this is may be due to the increase in porosity formation [8] (Fig. 2).

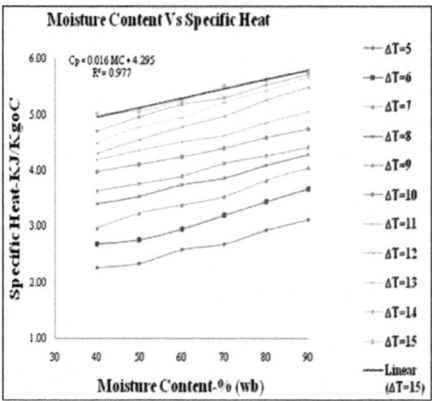

Fig. 2. Variation of specific heat with moisture content

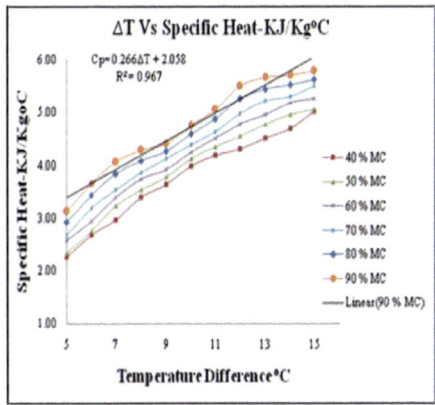

Fig. 3. Variation of specific heat with temperature difference

It has been observed that the specific heat increases with increase in moisture content due to the presence of bounded water molecules in the sample. The specific heat also increases with moisture content due to the higher thermal contact between the particles of the sample but this increase may not be linear. For particular moisture content the specific heat increases with increase in temperature difference as the sample was have a tendency of increased rate of heat transfer because of bound water present in the sample. It is observed that specific heat increases with increase in temperature difference for the moisture range of 40% to 80% (wb). This result shows that, higher moisture levels of 70% and 80% (wb) specific heat increases with increase in temperature differences but this increase may not be linear. Temperature mainly influences the reduction in specific heat values due to the formation of porosity [9, 10] (Fig. 3).

3.2 Assay of Specific Heat at Density ρ_2 = 1062 Kg/m^3

The Fig. 4 indicates the effects of moisture content and temperature difference on Specific Heat of Nagpur Orange fruit at density ρ_2 = 1062 kg/m^3, moisture content varies from 40% to 90%(Wb). The specific heat increases with increase in moisture content that varies from $Cp_{min.}$ = 2.31 kJ/Kg °C to $Cp_{max.}$ 6.50 kJ/kg °C for moisture range 40% to 90% (wb) for temperature difference of 5 °C to 15 °C (Fig. 5).

3.3 Assay of Specific Heat at Density ρ_1 and ρ_2

For different bulk densities ρ_1 = 1006 kg/m^3 and ρ_2 = 1062 kg/m^3 at moisture content 80% Wb the Specific heat increases gradually with increase in temperature difference as shown in the Fig. 6.

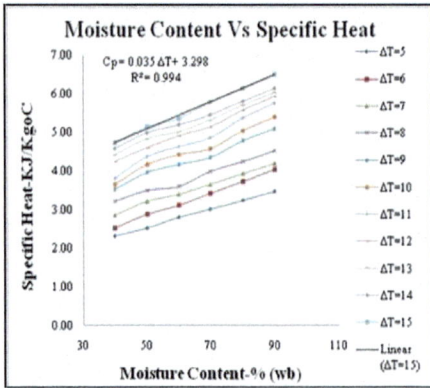

Fig. 4. Variation of specific heat with moisture

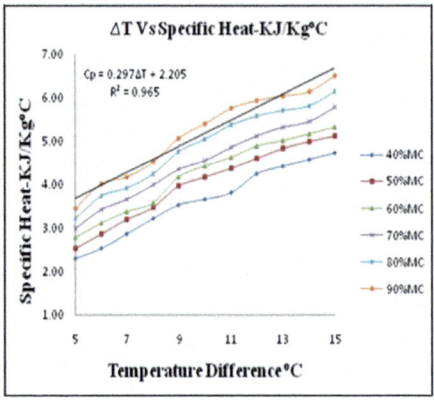

Fig. 5. Variation of specific heat with temperature difference

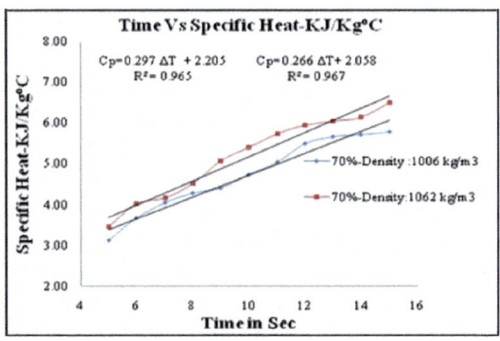

Fig. 6. Variation of specific heat vs time for different densities at 70% moisture content

4 Conclusion

The present work evaluates the Specific heat (Cp) of Nagpur Oranges fruit in the temperatures difference ΔT (5 °C–15 °C) experimentally. It is observed that specific heat increase with increase in moisture content. It is because of the good thermal contact between particles of dried orange with decreased porosity. It is found that specific heat does not increase linearly, because of non-homogeneity particle size, temperature distribution and surface resistance during experimentation in the sample. The results obtained in this experiment will be useful in the thermal analysis of orange fruit.

References

1. Bante, R., Pallewar, S.: Economics of orange production in Nagpur district of Maharashtra. IRJACS **6**(1), 136–139 (2015)
2. Kurozawa, E., El-Aouar, A.A., Simoes, M.R., Murr, F.E.X.: Determination of thermal properties of (carica papaya L) as a function of temperature. In: 4th Mercosur Congress on Process Engineering (2005)
3. Saenmuang, S., Sirijariyawat, A., Aunsri, N.: The effect of moisture content, temperature and variety on specific heat of edible- wild mushrooms: model construction and analysis. Eng. Lett. **25**(4). EL_25_4_12
4. Azadbakht, M., Khoshtaghaza, M.H., Ghobadian, B.: Thermal properties of soyabean pod as a function of moisture content. Am. J. Food Sci. Technol. **1**(2) 9–13 (2013)
5. Ekpunobi, U.E., Ukatu, S.C.: Investigation of thermal properties of selected fruits and vegetables. Am. J. Sci. Technol. **1**, 293–297 (2014)
6. Mohsenin, N.N.: Thermal Properties of Food and Agricultural Materials. Garden and Breach, New York (1980)
7. Kaleemullah, S., Kailappan, R.: Thermal properties of chillies. J. Food Sci. Technol. **41**(3), 259–263 (2004)
8. Modi, S.K., Durga Prasad, B., Basavaraj, M.: Experimental investigation on specific heat of psidium guajava L (guava fruit) as a function of moisture content and temperature. IJMET **4**, 180–185 (2013)
9. Polley, S.L., Snyder, O.P., Kotnour, P.: A compilation of thermal properties of foods. Food Technol. **34**, 76–82 (1980)
10. Gupta, T.R.: Specific heat of Indian unleavened flat bread (chapati) at various stages of cooking. J. Food Process Eng. **13**, 217–227 (1990)

FEM for Modeling and Analysis of Retreaded Tire with Abaqus Software

Uday Gudsoorkar[⊠] and Rupa Bindu

Department of Mechanical Engineering,
DY Patil Institute of Engineering and Technology, Pimpri, Pune, India
udaygudsoorkar@gmail.com

Abstract. FEM is used for modeling and analysis of retreaded tire in various working conditions. In this paper 2D axisymmetric geometric tire model is developed with all its components, carcass, plies, belts, bead, cushion gum, tread in Auto cad. These sketches are imported in ABAQUS (sat file). Separate part for cushion gum (cured) is made. Neo Hoookean model is used for Hyper elastic rubber. Rim is modeled as analytical rigid surface. First 2D axisymmetric tire model and then full 2D model is developed. Finally 3D model is developed by symmetric model generation and symmetric result transfer option. Analysis of rim mounting, tire inflation and static loading of retreaded tire model is studied.

Comparison is done between experimental and simulated deformation up to 12 mm at inflation pressure of 0.241 MPa on horizontal surface. It is found that simulated tire is stiffer than actual tire.

Keywords: Axisymmetric · Neo Hookean · Rim mounting · Inflation · Cushion gum

1 Introduction

Tire durability and reliability are dominating parameters in the performance of tire. Due to fast development in the field of computers, virtual tire modeling and analysis has become reality now. It plays vital role in reducing the cost related to, repeated testing of product and its complexity. Simulation of tire has many advantages over old methods of tire design. This research paper attempts to predict the behavior of tire in mounting of bead on rim, inflation in carcass, cyclic loading during static and stead state rolling.

2 Literature Review

Literature review on retreaded tire specially in the context of model is sparse. This research paper delves:

Gall et al. (1995) and Mischke et al. (1997): Application of global-local analysis is developed to reduce number of degrees of freedom. Schmeitz's model (2004), It can be used to simulate for multiple obstacles case but it is valid at 2D road surface only. Mf-swift (2005), It is relatively simple with high accuracy but still undergoing development. It is not suitable to analysis the influence of the parameter change. CD-Tire

© Springer Nature Singapore Pte Ltd. 2020
V. K. Gunjan et al. (Eds.): *ICRRM 2019 – System Reliability, Quality Control, Safety, Maintenance and Management*, pp. 92–97, 2020.
https://doi.org/10.1007/978-981-13-8507-0_15

(2007), It can be used for ride comfort analysis and durability analysis at both 2Dand 3D cases. It includes different tire models for different applications. Baecker's Tire Model (2010), It is valid for low aspect ratio tires but only 2D simulation results are given. ABAQUS-Based surrogate tire model (2011), It is valid for 2D cases only. The tire cord reinforcements are modeled as rebar elements. This is also incorporated in ABAQUS. Mechanical behavior of rubber (hyper elasticity) is having different models. Neo-Hookean is the simplest to use and good approximation for small strains. In this paper same hyper elastic model is used [8]. It is found that development in FE Tire Modeling and analysis is done for new tires. To summarize literature review of this paper, same method is used for retreaded tires [9].

3 Methodology

3.1 Tire Modeling

A retreaded tire of size 165/80/R14 is taken for modeling. The construction of this tire model is done in Auto cad. All components like carcass, plies, belts, bead, cushion gum and rim are sketched separately. These sketches are imported in ABAQUS/CAE software in .sat file. Since this is a retreaded tire, cushion gum is made as separate part. It is assembled between tread rubber and carcass. Once all sketches are converted in parts then assembly of that is now ready for material modeling. Rubber is nonlinear material and it is modeled in Neo-Hookean Model [2]. Properties of rubber, belts, carcass, and bead are modeled by the effective moduli of the individual material properties of cord and rubber. Material properties are as below [6].

3.1.1 Material Properties
See Table 1.

Table 1.

Sr. No.	Tire parameters	Values
1	Neo Hookean hyper elastic constant Rubber C10 Rubber D1	0.60 0.03
2	Young Modulus-rebar	$2 * e^5$ MPa
3	Poisson's ratio	0.3
4	Density-rebar	$7.5e^{-9}$ MPa

3.1.2 Assignment of Material Property
For nonlinear elastomeric material Neo-Hookean Model is used. Material density is considered because it needs at the time of steady state transport analysis [1]. Nylon and steel are commonly used reinforcement material in car tire and they are modeled as linear elastic or hyperplastic material model. Rebar is a reinforcement material embedded in rubber material, which can be oriented in any angle with respect to Y-Z plane (meridional) [3]. Angle 0 corresponds to meridional plane and angle 90

corresponds to circumferential rebar. All material properties are assigned to respective part section.

3.1.3   Boundary Conditions

After material properties, contact load and boundary condition are given for rim mounting at inflation pressure (0.241 MPa). Boundary condition-Symmetry, Step-initial, Region selection-carcass -1 Select U2 and UR2 in the edit boundary condition box to create boundary condition. The setting will restrain the displacement in Y direction and rotation about the Y-axis. Displacement boundary condition for rim: Step-Mounting Category-mechanical, Type-Displacement/Rotation, Choose –rim-1 in region selection dialog box.

3.1.4   2D Axisymmetric-Tire Geometry Modeling

To reduce the work of analysis axisymmetric model is used. Rim mounting, inflation, can be successfully carried out by 2d axisymmetric model. Creation of rim, carcass, ply and belt geometries are sketched initially or designers provide that. We have drawn sketches of all components of tire in AutoCAD (.sat file).

All sketches are having only one common point for accurate positioning. All sketches are imported in Abaqus software and created separate part for each sketch. Finally, all parts including rim are assembled as one core part (Figs. 1, 2, 3, 4, 6 and 7).

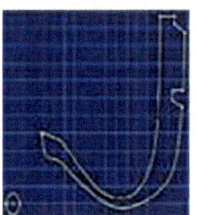
Fig. 1. Carcass

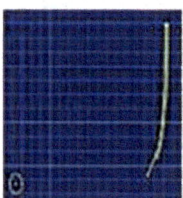
Fig. 2. Belt

Fig. 3. Bead

Fig. 4. Plies

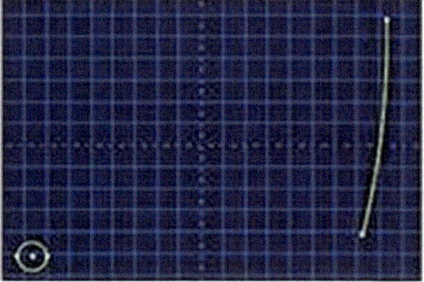
Fig. 5. Cushion gum cured

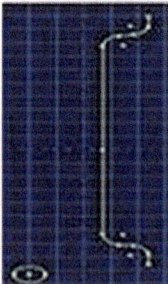
Fig. 6. Rim

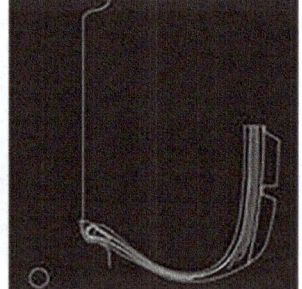
Fig. 7. Assembly

4   2D Rim Mounting and Inflation Analysis

For contact between rim surface and bead, automatic shrink fit and axial displacement option is used. Symmetry boundary condition and rim mounting are shown in Fig. 7. Now we use symmetry boundary condition because only half of the tire is modeled. After mounting of bead and air inflation, full tire model is created by rotation of axisymmetric model around tire axis.

4.1   3D Model Generation and Loading Analysis

Each part of tire is meshed separately to ease the meshing algorithm produce a better quality mesh. Partitions are made in each part to make it close to rectangle shape with four non split edges. Partitions are necessary to have control over mesh distortion. Without these partitions, we cannot have control on mesh density and quality of mesh especially at apex area. In meshing we assigned hybrid type of element and full quad shape Seed sizes are used to produce mesh sufficient enough to predict forces and moments. It also gives rough estimate of contact pressure, contact patch and shape [5]. Carcass geometry is meshed with triangular and quadrilateral, axisymmetric, fully integrated, linear elements with twist (CGAX3 and CGAX4). Belt element is meshed using axisymmetric membrane element with twist (MGAXI). For meshing plies global mesh size of 4 mm is applied for reinforcement component (Fig. 8). Axisymmetric membrane element with twist (MGAXI) is used, same as used in belts. After successful completion of all steps. Job is submitted for analysis and results are visualized [6]. Symmetric model generation option is used for getting 3D model by revolving 2D half model after rim mounting and inflation analysis. Separate surface is created as road model (Fig. 10). Foot print analysis with rated load and rated inflation pressure. In this analysis, results of 2D model are transferred to 3D model. First surface comes in contact and load is applied until equilibrium comes (Fig. 8).

Fig. 8. 3D half tire   **Fig. 9.** 3D full tire   **Fig. 10.** UTM

5   Results and Discussion

Geometric tire model, material model, assembly and air pressure is successfully developed which can be used for predicting tire characteristics. Finally 3D model is developed by symmetric model generation and symmetric result transfer option. Analysis of rim mounting, tire inflation of tire model is studied Finite element method is only reliable tool for modeling of tire. From the load–deformation curve it shows that simulated model is stiffer than the actual. This is because linear material like steel is used in model. Simulated results are in good correlation with the experimental one so FEA helps to accelerate the process of tire design.

Experimental setup, Fig. 10 Tire size-165/80/R14 is inflated with 241 kPa. Apply load slowly to axle, we get deflection in carcass. Plot a graph between load and deflection.

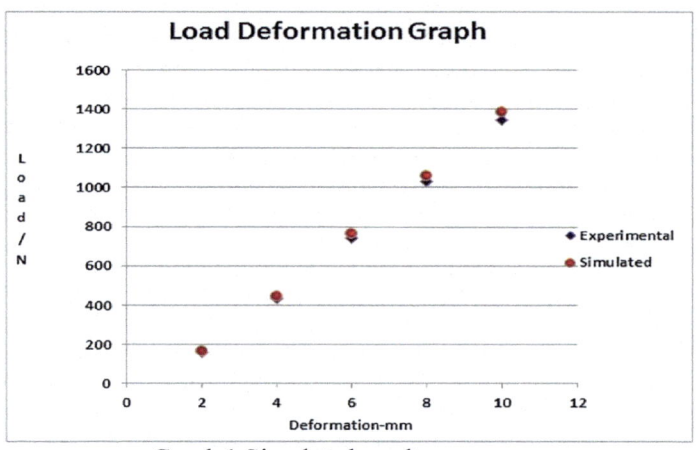

Graph 1 Simulated results
● Simulated results ◆ Experimented result5

6   Research Future Scope

There is a scope for rolling tire analysis of retreaded tire at different load and air pressure.

7   Conclusion

The vertical stiffness test and simulation is carried out under different load condition. Graph between load and deformation is plotted (0.241 MPa inflation pressure) for experimental and simulated results. It is found that vertical load and displacement show approximately linear relationship. We can very easily calculate the difference between contact pressure of deformed and un deformed tire, at the same time. It can be predicted the possible deflection for certain load on it. Size and shape of footprint gives the grip of tire with the road, which helps to avoid slip and contact between rim, and bead helps to avoid slip between to surfaces.

Acknowledgement. Research project is assisted by IRMRA (Indian Rubber Manufacturers Association) Thane Mumbai.

References

1. Kennedy, R.H., Terziyski, J.M.: Experiences with Embedded Elements in Tire Modeling. Hankook Tire Co. Akron Technical Center (2004)
2. Moisescu, A.R.: Investigation of the influence of vertical force on the contact between truck tire and road using finite element analyses (2017)
3. Kim, S., Kim, K.: Nonlinear material modeling of a truck tire, pavement, and its effect on contact stresses. In: Proceedings of the ASME 2012 International Engineering Technical Conference and Computers and Information in Engineering (2012)
4. Gent, A.N.: Engineering with Rubber How to Design Rubber Components. A455.R8E54 2012,620.1'94–dc23 (2012)
5. Kim, K.: Nonlinear material modeling of truck tire pavement and its effect on contact stresses. Conference Paper (2012). https://doi.org/10.115/detc2012-70664. Published
6. Tire analysis with Abaqus-fundamental, by Dassault Systems. India Private Limited
7. Moisescu, R., Fratila, G.: Finite element model of radial truck tire for analysis of tire-road contact stress. UPB Sci. Bull. Ser. D **73**(3) (2011). ISSN 1454-2358
8. Modeling of rubber and viscoelasticity with Abaqus-2017 by Dassault Systems India Private Limited
9. Li, B., Yang, X., Yang, J.: Tire model application and parameter identification-a literature review. SAE Int. J. Passeng. Cars - Mech. Syst. **7**(1), 231–243 (2014)

Application of Linear Programming - Genetic Algorithm Combination for Urmodi Reservoir Operation

A. S. Parlikar[✉] and P. D. Dahe

Department of Civil Engineering, S.G.G.S.I.E. & T., Nanded, Maharashtra, India
asparlikar2007@rediffmail.com

Abstract. The reservoir operation is a challenging task to researchers and water managers. The tempo-spatial variability of rainfall in an area causes uneven distribution of water supply for various requirements. Therefore, researchers are in search of new improved technique that satisfies the requirements to the maximum extent. Linear Programming (LP) is a widely adopted technique for reservoir operation. The evolutionary algorithms (EA) have a specific status in the study of reservoir operation. The attempts are being made to make the reservoir operation technique easy. In view of this, the reduction in time of computation for the determination of reservoir yield is studied in this paper. Linear Programming and Genetic Algorithm (LP-GA) combination are used to compute reservoir yield. Its computation time is compared with the time required for simple GA. It is observed that the LP-GA combination is faster and produces nearly equal results as produced by simple GA. From the present study, it is concluded that the LP-GA can reduce the time of computation in reservoir operation studies. The maximum over year yield from Urmodi Reservoir in Maharashtra, India using LP-GA approach is found to be 187.965MCM at the cost of comparatively lesser duration as compared to simple GA technique.

Keywords: Linear Programming · Genetic Algorithm · Urmodi Reservoir · LP-GA · Time of computation

1 Introduction

The reservoir operation involves a large number of hydrologic variables which include rainfall, inflow, demand, evaporation, releases, etc. The demand for irrigation in the command area depends on the cropping pattern. The municipal water supply and hydropower generation generally require a constant volume to be supplied each month. The limited availability of water resources and increasing demand for water creates a huge gap in the supply of water to the beneficiaries. Hence, a system analysis technique is useful in managing the scarce resource of water. The reservoir operation is done using conventional techniques like Linear Programming, Nonlinear Programming, and Dynamic Programming. In recent years Evolutionary Algorithms are applied for reservoir operations. The successful application of Artificial Neural Network (ANN), Fuzzy Logic, genetic algorithms (GA), genetic programming (GP), Ant Colony

© Springer Nature Singapore Pte Ltd. 2020
V. K. Gunjan et al. (Eds.): *ICRRM 2019 – System Reliability, Quality Control, Safety, Maintenance and Management*, pp. 98–103, 2020.
https://doi.org/10.1007/978-981-13-8507-0_16

Optimization (ACO), Particle Swarm Optimization (PSO); Simulated Annealing (SA), Tabu Search (TS), etc. is also done by researchers and water managers.

The reservoir system analysis was studied in depth by Loucks [1]. The Linear Programming technique was successfully applied to the Narmada reservoir system to optimize the reservoir yield [2]. A comprehensive study of simulation–optimization modeling for reservoir operations was done by Rani et al. [3]. The Genetic algorithms (GAs) are a particular class of EA based on mechanics of natural selection and natural genetics [4]. It is based on Darwin's Principle of "Survival of the fittest". Esat and Hall [5] applied GA to a four reservoir problem with the objective of maximizing the power generation from power production and irrigation. Successful application of GA was also reported through the works by various researchers [6–9]. Genetic algorithms in combination with other algorithms were implemented for reservoir operation studies by many researchers [10, 11]. The present study deals with the use of a combination of Linear Programming (LP) and Genetic Algorithm (GA) in context with the single reservoir operation of Urmodi Reservoir in Maharashtra, India. The objective function for the study is to maximize the over year yield from the reservoir for the 33 years historic inflow.

2 Study Area

The system considered for the study is Urmodi reservoir in Krishna Basin in Maharashtra, India. It is located between $17°40'0''$ N Latitude and $73°54'40''$ Longitude; in District Satara of Maharashtra State in India. For the historical inflow data of 33 years from 1975 to 2007, it is observed that the average inflow into the reservoir is 349.32 MCM out of which about 77.5% inflow occurs in the months of July and August. The evaporation rate in the months of March, April and May have almost 48.5% share in the total annual evaporation of 1727 mm (Table 1).

Table 1. Salient features of Urmodi reservoir

Type of dam: earthen dam with gated ogee spillway on the left bank	
Maximum height of dam	50.10 m
Gross catchment area	116.86 sq km
Gross capacity of reservoir	282.14 MCM*
Dead storage capacity of reservoir	8.867 MCM
Live storage capacity of reservoir	273.273 MCM

(*MCM – Million Cubic Meter)

3 Methodology

The reservoir operation study is done by the approximate yield model [1]. In the linear programming model, the objective functions, as well as the constraints, are linear in nature. Generally for the reservoir operation studies LP is preferred due to its computational simplicity, easy availability of computer software and handling of a large number of variables. Often it is observed that the objective function(s), as well as some of the constraints, are nonlinear, however, the piecewise linearization technique can be successfully used in such situations for the solution. The Genetic Algorithm (GA) is based on the Darwinian Principle of survival of the fittest. The GA technique starts from a randomly generated solution space. It progresses towards the optimal solution through the steps of selection, crossover, and mutation.

The yield model is an approximation of the full optimization model [1]. For a hydrologic record of n years, each having T periods, the number of constraint equations is reduced from $2nT$ to $2(n + T)$. The number of variables is reduced from $(2nT + T + 2)$ to $(2n + 2T + 3)$. The yield model consists of a set of annual constraints and an additional set of within a year or monthly constraints based on a critical year.

The objective function of approximate model or yield model is to maximize the over year yield i.e. R for the following constraints.

(1) **Over year storage continuity equation:** The relationship between initial storage of a year and the final storage in that year is given by storage continuity equation. It is given by

$$S_j + I_j - \theta_{p,j} * R - Sp_j - E_j = S_{j+1} \qquad \forall j \qquad j = 1...33$$

Where, S_j = initial storage in the beginning of year 'j'; I_j = inflow during the year 'j'; E_j = Evaporation loss from the reservoir during the year 'j'; Sp_j = Surplus from the reservoir during year 'j'; $\theta_{p,j}$ = failure fraction ($0 < \theta_{p,j} < 1$); $\theta_{p,j} = 1$, for successful year and 0, for failure year.

(2) **Within year storage continuity:**

$$S_{t-1}^w + \beta_t (R + \sum_t El_t) - Wyt - El_t = S_t^w \qquad \forall t$$

Where S_{t-1}^w = Within year storage at the beginning of a period in a year; S_t^w = Within year storage at the end of a period in a year; β_t = Relative proportion of the critical year's inflow that is likely to occur in period t; El_t = Within year evaporation loss in each period t of the critical year; Wyt = Within year yield in a period.

(3) **Total reservoir capacity**

$$Max\ (S_j) + S_{t-1}^w \le Y_a \qquad \forall t$$

Where, Y_a = Total active storage capacity of the reservoir.

The Linear Programming model is solved using software LINDO-6.1 (Linear, INteractive, and Discrete Optimizer). In the LP approach, the evaporation loss is computed by approximating the relationship between storage and water spread area to be linear above the dead storage level.

In Genetic Algorithm (GA) the population consists of randomly generated chromosomes of decision variable i.e. over year yield from the reservoir. The number of generations is the stopping criteria used for the present GA model. As an initial search, the yield is computed by considering a population of 10 chromosomes, 60% crossover probability, 0.1% mutation probability and 30 numbers of generations. After performing the sensitivity analysis, the best parameters of GA are fixed. For sensitivity analysis, the population size varied by 10 numbers, crossover probability 0.2 and mutation probability by 0.001 in every run. The optimal genetic parameters obtained in simple GA are: over year yield at 100% reliability-187.965 MCM; Population-70; Crossover Probability (Pc)-0.8; Mutation Probability (Pm) – 0.008; No. of Generations – 30.

4   Results and Discussion

The simple GA when used for optimization, it takes much time to compute the decision variables as it starts from a random search space. The size of the search space i.e. population also affects the computation processing time. The GA reaches the optimal solution after passing through the steps of selection, crossover, and mutation. However, if the search space is bound by the lower and upper limits, the initial random population is generated in a limited space of search.

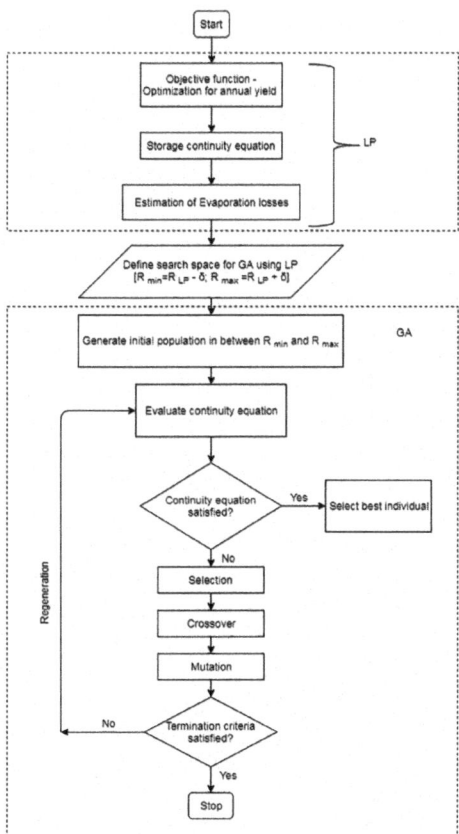

Fig. 1. Flowchart of combined LP–GA approach

Table 2. Yield computation

Population	Generations	Max yield (MCM)		Computation time (seconds)	
		GA	LP + GA	GA	LP + GA
10	30	185.368	**185.825**	196	**181**
	40	186.222	**187.185**	111	**92**
	50	186.859	**187.825**	130	**105**
	60	**186.186**	186.124	**150**	170
20	30	186.255	**187.771**	70	**54**
	40	186.295	**186.699**	110	**92**
	50	187.129	**187.931**	205	**175**
	60	**186.221**	186.035	**241**	217
50	30	186.047	**187.965**	216	**185**
	40	186.791	**186.342**	290	**264**
	50	**185.564**	185.259	**382**	350
	60	185.905	**186.063**	461	**453**
100	30	185.483	**185.972**	340	**336**
	40	**186.321**	185.575	**423**	420
	50	**185.195**	185.173	**559**	565
	60	185.213	**185.678**	720	**720**

The number of generations is used as the termination criteria in the present study. The approximate yield model is used to find the maximum annual yield from the Urmodi reservoir. The study of GA and LP-GA combination is done for the reservoir operation and determined the objective function and genetic parameters (Fig. 1). The range for the decision variable (i.e. annual yield) for LP-GA was defined based on its separate computation using LP only. The results from LP model are used as input for the GA model with the addition and subtraction of the discrete value 'δ'. The range of 'δ' is varied from 5% to 15% to bind the search space [11]. GA gives a near optimal solution if implemented in combination with LP at the cost of less time as compared to simple GA. Optimizations herein were carried out in a computer with an Intel-Core i3-3217U CPU @ 1.80 GHz and 4.0 GB RAM capacity.

The GA model used binary coding for representing the decision variable. After the evaluation of the fitness function, the Roulette Wheel Method is applied to form the next generation [4]. The uniform crossover is applied to the selected chromosomes. The last genetic operator 'mutation' injects new genetic material in the chromosome thereby changing the character of the chromosome. In the initial search, the population is varied as 10, 20, 50 and 100. For every population size, the number of generations varied is as 30, 40, 50, 60. Keeping crossover probability and mutation probability constants, the number of generations is changed for various populations. It is observed that at lower population size, the maximum yield values are obtained for the combination of LP-GA for (almost) every changed number of generations. But, as the population size is increased, the optimal values using both the approaches viz. GA and LP-GA are obtained at nearly equal durations (Table 2).

5 Conclusion

A Genetic Algorithm (GA) is a robust search technique used for many optimization studies in water resources engineering in recent years. The time of computation in GA is a matter of concern as it searches the optimal solution from a large randomly generated search space. To reduce the computation time in GA, it is combined with LP which is an optimization technique. The combined LP-GA and simple GA are applied for reservoir operation of Urmodi Reservoir in Maharashtra, India. The convergence in LP-GA is observed to be faster as compared to simple GA. For the same population size and number of generations, the convergence in LP-GA is about 20%–25% faster than that in simple GA. The reservoir over year yield for 100% reliability is found to be 187.965 MCM at the cost of 185 s by using LP-GA combination. The objective functions achieved for various populations and number of generations are also nearly equal by both simple GA and LP-GA, but at the cost of less time of computations in LP-GA combination. The present method is quite applicable where the decision space is large and the optimal results are required in less duration. The population size when increased, the results by both the approaches viz. GA and LP-GA are obtained nearly equal. Hence, here it can be concluded that there is no need to increase the population size beyond a certain value. This threshold population size will help reduce the time of computation.

References

1. Loucks, D.P., Stedinger, J.R., Haith, D.A.: Water Resources Systems Planning and Analysis. Prentice Hall, Englewood Cliffs (1981)
2. Dahe, P.D., Srivastava, D.K.: Multireservoir multiyield model with allowable deficit in annual yield. J. Water Resour. Plann. Manag. 128(6), 406–414 (2002)
3. Rani, D., Moreira, M.M.: Simulation–optimization modeling: a survey and potential application in reservoir systems operation. Water Resour. Manag. 24(6), 1107–1138 (2010). https://doi.org/10.1007/s11269-009-9488-0
4. Goldberg, D.E.: Genetic Algorithms in Search, Optimization and Machine Learning. Wiley, New York (1989). https://doi.org/10.1023/A:1022602019183
5. East, V., Hall, M.J.: Water resources system optimization using genetic algorithms. In: Proceeding 1st International Conference on Hydroinformatics, Baikema, Rotteram, The Netherlands, pp. 225–231 (1994)
6. Kumar, N., Raju, S., Ashok, B.: Optimal reservoir operation for irrigation of multiple crops using genetic algorithms. J. Irrigat. Drain. Eng. 132(2), 123–129 (2006)
7. Jothiprakash, V., Shanthi, G.: Single reservoir operating policies using genetic algorithm. Water Resour. Manag. 20, 917–929 (2006). https://doi.org/10.1007/s11269-005-9014-y
8. Chen, L., Chang, F.J.: Applying a real-coded multi-population genetic algorithm to multi-reservoir operation. Hydrol. Process. 21(5), 688–698 (2007)
9. Mathur, Y.P., Nikam, S.J.: Optimal reservoir operation policies using genetic algorithm. Int. J. Eng. Technol. 1(2), 184–187 (2009)
10. Shokri, A., Haddad, O.B., Mariño, M.A.: Algorithm for increasing the speed of evolutionary optimization and its accuracy in multi-objective problems. Water Resour. Manag. 27(7), 2231–2249 (2013). https://doi.org/10.1007/s11269-013-0285-4
11. Rani, D., Srivastava, D.K.: Optimal operation of mula reservoir with combined use of dynamic programming and genetic algorithm. Sustain. Water Resour. Manag. 2, 1–12 (2016). https://doi.org/10.1007/s40899-015-0036-1

Designing of Shell and Tube Heat Exchanger for Methylene Di Chloride (MDC)

Nilima Khare[1(✉)], Shilpa Dhopavkar[1], Suhas Chavan[1], and Gurunath Gore[2]

[1] Department of Mechanical Engineering, GHRCEM, Pune 412207, India
nilimakhare1994@gmail.com,
Shilpa.dhopavkar@raisoni.net,
suhaschavan5015@gmail.com
[2] Consulting Engineers PVT LTD., Pune, India
guru.gore88@gmail.com

Abstract. Heat exchanger exchanges the thermal energy between two or some times more than two fluids between or it can be used for transfer the heat between a solid surface and flowing fluid. There are some heat exchangers which transfers heat between fluid and solid particulates at different temperatures with maximum heat transfer rate and minimum investment is known as Heat exchanger. The shell and tube heat exchanger with straight tubes and double pass is selected to study among different types of heat exchanger. In different industries Heat exchangers haves many application. In present study the shell and tube heat exchanger with straight tubes and double pass is under study. Task is to design the heat exchanger to exchange the heat between Methylene Di Chloride (hot fluid) and Water (cold fluid). The temperature of the hot fluid at inlet i.e. Methylene Di Chloride is considered 40 °C our aim is to reduce the temperature up to 34° and the inlet temperature of the water i.e. cold fluid is considered as 26 °C inlet. and outlet temperature 34 °C. The process is carried out under vacuum. Only the design calculation is shown in this paper. The heat exchanger is designed using kerns Method.

Keywords: Section of area · Heat exchanger parameters · Modelling

1 Introduction

Heat exchanger transfers heat from one fluid to another fluid. Heat exchangers are used in chemical industries and process industries. Heat exchangers that are observed by us in our day to day life are evaporators and condensers used in refrigeration and air conditioning units. Condensers and Boilers are the examples of large industrial heat exchangers. Heat exchangers are termed as per their applications. Heat exchangers used for condensing the fluid are named as condensers. Also heat exchangers used for vaporizing liquids are kwon as evaporators. [6] Heat exchanger performance is calculated on the basis of minimum area of heat transfer and pressure drop through heat exchanger. A presentation of efficiency is done by computing over all heat transfer coefficient for the heat exchanger. Area required for a definite amount of heat transfer

© Springer Nature Singapore Pte Ltd. 2020
V. K. Gunjan et al. (Eds.): *ICRRM 2019 – System Reliability, Quality Control, Safety, Maintenance and Management*, pp. 104–111, 2020.
https://doi.org/10.1007/978-981-13-8507-0_17

and pressure drop provides an idea about the total cost of a heat exchanger and power requirements. For designing the heat exchanger there is a lots of literature is obtainable since lot of work is done on the heat exchanger previously. For a better performance the heat exchanger design must considered smallest amount of probable pressure drop and heat transfer areas to accomplish the requirements of the heat transfer [9].

2 Literature Review

Aik et al. [7] the paper is based on as the length of the shell and tube heat exchanger decrease then effect of it on the cost, manufacturing time and important factor is space can be reduced. Patel et al. [3] In this paper the tube length, tube layout Tube pitch ratios, as well as baffle spacing ratio were found to be important design parameters which has a direct effect on pressure drop and causes a conflict between the effect uniformity and total cost. Shinde et al. [4] In this paper baffle inclination angle will provide an optimal performance of heat exchangers. Gowthaman et al. [2] the pressure drop in segmental baffle is more as compare helical baffle.

3 Methodology (Analytical Calculation)

3.1 Shell and Tube Type Heat Exchanger Design Using Kern's Technique [1]

Kern's technique is formed on work on heat exchangers which are used in commercial industries.

3.2 Exchanger Type and Dimensions

In order to start the calculations four tube pass and two shell pass and four tube passes will be used to start with. The LMTD-Method will be used to evaluate the required equipment's area. Analysis will start by first calculating the mean temperature. Both fluids will be used in counter-flow to maximize the log mean temperature difference (Table 1).

$$\Rightarrow Q = UA\Delta T_m \tag{1}$$

Where

ΔTm = Mean temperature difference in °C
U = Overall heat transfer coefficient in W/m^2 °C,
A = Area of heat transfer in m^2,
Q = Heat transfer per unit time in W,

Table 1. Initial condition

Parameters	
Cold fluid mass flow rate, m_c	0.01945 kg/sec
Hot fluid mass flow rate, m_h	0.06722 kg/sec
Heat transfer rate	0.65068 kw
ΔTm	6.95 °C

3.3 Physical Properties (MDC and Water)

See Table 2.

Table 2. Area calculation

Physical properties	MDC Hot fluid	Water Cold fluid
Specific heat of fluid Cp (J/kg K)	1210	4182
Density (kg/m^3)	$1.32 * 10^3$	998.2

3.4 Layout of Tube

For the better surface area per unit length and heat transfer the triangular pitch (30° layout) is used. For the flow rate and identical tube pitch, the tube layout is in declining order of shell-side heat pressure drop and heat transfer coefficient are: 90°, 60°, 45°, 30°. The 90° layout will have the low pressure drop and the low heat transfer coefficient (Table 3).

Table 3. Tube pattern

Parameters	
Cold fluid inlet temperature, t_1	26 °C
Hot fluid inlet temperature, T_1	40 °C
Cold fluid outlet temperature t_2 outlet temp. of cold fluid	34 °C
T2 outlet temp. of hot fluid	34 °C
U	40 W/m^2 °C
A	3 m^2
L	1.250 m
d_i	22.48 mm
d_o	26.7 mm
N_t No. of tubes	30

Area, $A = \pi DL$, L = length of tube

$$\text{No. of tube} = \frac{Area\ required}{Area} \tag{2}$$

3.5 Tube Pattern

See Table 4.

Table 4. Calculation of heat transfer coefficient

K_1	0.319
n_1	2.149
D_b	234.09 mm
D_s	250 mm

$$D_b = d_o (N_t / K_1)^{1/n1} \tag{3}$$

$$\text{Shell diameter} - D_s = D_b + \text{clearance} \tag{4}$$

3.6 Tube Side Calculation

$$\text{Tube side mean temp Tm} = \frac{t1 + t2}{2} \tag{5}$$

$$\text{Tube cross section area } A_t = \frac{\pi}{4} * di^2 \tag{6}$$

Tube per pass $= \frac{Nt}{2}$

Calculate the linear velocity us and the tube-side mass velocity Gs (Table 5).

$$Gs = \frac{Ws}{A_t}$$

$$us = Gs/\rho$$

$$Re = \frac{\rho \mu d_i}{\mu}$$

$$Pr = \frac{\mu Cp}{K}$$

$$\frac{hidi}{k} = j_h RePr^{0.33} \left[\frac{\mu}{\mu\omega}\right]^{0.14} \tag{7}$$

Table 5. Pitch of tube calculation

Tm	30 °C
A_t	396.90 mm^2
Tube per pass	15
Gs	3.2689 kg/s
us	0.00328 m/s
Re	91.7071
Pr	5.7
hi	50.14 w/m^2 °C

3.7   Shell Side Calculation

1. Cross flow area calculation

$$\Rightarrow As = \frac{(Pt - do)Dsls}{Pt} \tag{8}$$

Where,

s = Shell inside diameter in m,
Pt = Pitch of tube,
ls = spacing between baffles, m.
do = tube outside diameter,

The ratio of the clearance between tubes and the total distance between tube centers is $\frac{(Pt-do)Dsls}{Pt}$ is

$$\text{Mean shell side temperature } Tm = \frac{T1 + T2}{2} \tag{9}$$

2. The linear velocity us and the shell side mass velocity Gs is given by:

$$Gs = \frac{Ws}{As} \tag{10}$$

$$us = Gs/\rho \tag{11}$$

Where,

ρ = fluid density at shell side, kg/m^3.
Ws = shell side fluid flowrate, kg/sec,

Choose baffle spacing = $\frac{D_s}{5} = \frac{250}{5} = 50$ mm baffles are placed to the shell side to direct the fluid flow and create the turbulence [5] (Table 6).

$$\text{Tube pitch} = P_t = 1.25 * d_0 \tag{12}$$

Table 6. Shell side calculation

As	$2.68 * 10^{-6}$ m^2
Gs	25.07 kg/sm^2
Tm	37 °C
P_t	34 mm

3. For a triangular pitch arrangement shell side equivalent diameter is calculated.

$$d_e = \frac{1.1}{do}\left(p_t^2 - 0.785d_0^2\right) \tag{13}$$

4. Reynolds number at shell side can be calculated by

$$\text{Re} = \frac{Gsde}{\mu}$$

5. For the computed Reynolds number read the value of j_h from Figure for the selected tube arrangement and baffle cut [10] (Table 7).

$$Nu = \frac{h_s d_e}{k_f} = j_h \text{RePr}^{1/3}\left(\frac{\mu}{\mu w}\right)^{0.14} \tag{14}$$

Where,

Pr = Prandtl number
N_u = Nusselt number $\frac{h_s d_e}{k_f}$
Re = Reynolds number
de = equivalent or hydraulic mean diameter in m
hi = Inside coefficient in W/m^2 °C,
k_f = Thermal conductivity of the fluid W/m °C,
Ut = fluid velocity in m/s,
Gs = mass velocity or mass flow per unit area in kg/m^{2}s,
Cp = heat capacity of fluid, J/kg °C
μ = viscosity of the fluid at wall
μw = fluid viscosity in Ns/m^2,

Table 7. Shell side heat transfer coefficient calculation

d_e	20.69 mm
Re	1556.67
Pr	3.7
h_s	650.09 w/m² °C

3.8 Estimate Wall Temperature

Thermal conductivity of SS316 = 16.3 W/m² °C.

Consider the coefficients of fouling from Table (MDC light organic) 5000 W/m² °C for water take as highest value, 3000 W/m² °C [11].

Mean Temperature 37 − 30 = 7 °C

All resistance

Across MDC film $= \frac{U}{h_0} * \Delta T == \frac{40}{50.14} * 7 = 5.58$ °C We have study the

$$\frac{1}{U_0} = \frac{1}{h_0} + \frac{1}{h_{od}} + \frac{d_o \ln \frac{d_o}{d_i}}{2k_w} + \frac{d_o}{d_i} * \frac{1}{h_{id}} + \frac{d_0}{d_i} * \frac{1}{h_i} \tag{15}$$

Where,

- k_w = Tube wall material thermal conductivity W/m °C
- h_i = heat transfer coefficient for inside film in W/m² °C
- U_0 = based on the outside area of the tube overall heat transfer coefficient in W/m² °C
- h_0 = heat transfer coefficient for outside film in W/m² °C
- h_{id} = fouling factor for inside in W/m² °C
- h_{od} = fouling factor for outside in W/m² °C
- U_0 = 38.38 W/m² °C
- Well above assume value of 40 W/m °C

4 Conclusion

In this study industrial problem is considered for cooling methylene di chloride by using shell and tube heat exchanger with straight tubes and double pass. The basic dimensions of the single shell and double tube heat exchanger are calculated. The present study focuses on thermal design only have shown. The method is used for the calculation is kern's method. The process involved is the iterative process. Only the final iteration of the calculation is represented. As coolant is assigned to the tube-side because it is corrosive and The MDC is low temperature and Low pressure gas assign it to the shell side. Selecting the segmental baffle for increase the turbulence. Selecting the baffle cut as 25% it gives the best result. Selecting the triangular pitch arrangement As per calculation the number of tube is 30 and the required area for the heat exchanger is 3 m².

References

1. Raj, T.R., Ganne, S.: Shell side numerical analysis of a shell and tube heat exchanger considering the effects of baffle inclination angle on fluid flow using CFD. Therm. Sci. **16**(4), 1165–1174 (2012)
2. Gowthaman, P.S.: Analysis of segmental and helical baffle in shell and tube heat exchanger. Int. J. Curr. Eng. Technol. **2**, 625–628 (2014)
3. Patel, S.K., Mavani, A.M.: Shell and tube heat exchanger thermal design with optimization of mass flow rate and baffle spacing. J. Int. J. Adv. Eng. Res. Stud. **2**(1), 130–135 (2012)
4. Shinde, S.S., Samir, S.J., Pavithran, S.: Performance improvement in single phase tubular heat exchanger using continuous helical baffles. J. Int. J. Eng. Res. (IJERA) **2**, 1141–1149 (2012)
5. Abd, A.A., Naji, S.Z.: Analysis study of shell and tube heat exchanger for clough company with reselect different parameters to improve the design. Case Stud. Therm. Eng. **10**(2017), 455–467 (2017)
6. Heat exchanger applications (2010). http://www.wcr-regasketing.com. http://www.wcr-regasketing.com/heatexchanger-applications.htm
7. Aik, L.E.: Computational Fluid Dynamics Analysis of Shell and-double Concentric-tube Heat Exchanger. Science Direct
8. Rehman, U.U.: Heat transfer optimization of shell-and-tube heat exchanger through CFD studies (2012)
9. Arjun, K.S., Gopu, K.B.: Design of shell and tube heat exchanger using computational fluid dynamics tools. Res. J. Eng. Sci.
10. Sinnott, R.K.: Coulson & Richardson's Chemical Engineering. Chemical Engineering Design, 4th edn, vol. 6

Experimental Study on Ready Mix Concrete Plant Waste Concrete as a Aggregate for Structural Concrete

Abhay Shelar[(✉)], D. Neeraja, and Amit B. Mahindrakar

Vellore Institute of Technology, Vellore, India
abhayshelar@yahoo.com

Abstract. The use of Arranged blend solid plant abuse concrete is broadening very much arranged, paying little mind to how infers were taken to diminish its utilization. This makes liberal waste each day which is much troublesome. A sound and supportable reuse of Arranged blend solid plant abuse solid offers a huge get-together of reasons for interest [1]. The reasonableness of reused abuse concrete as coarse total in bond and its focal concentrations are talked about here. The essential demand creating of the bond quality and the sparkle of hydration concerning Arranged blend solid plant squander solid total were dealt with. Tests were coordinated to pick the properties of Arranged blend solid plant squander solid total, for example, thickness, particular gravity and total smashing worth. As 100% substitution of steady coarse total [2] (NCA) with Arranged blend solid plant squander bond coarse total (RMCW) is possible, deficient substitution at different rate were analyzed [3]. The rate substitution that gave higher compressive quality was utilized for picking trade properties, for example, modulus of flexibility, split rigidity and flexural quality. Higher compressive quality with increment bond was found with 100% NCA supplanted concrete.

Keywords: Basic concrete · Coarse total ·
Prepared blend solid plant squander solid waste

1 Introduction

1.1 A Subsection Sample

Concrete is the most extensively utilized man made change material on the planet and its second just to water as the most used substance in the planet. Hunting down totals for concrete and to discard the disaster from different items is the present bond. Today sensibility has got top need being created industry. In the present examination the Reused Arranged blend solid plant squander were utilized to set up the coarse totals along these lines furnishing a supportable choice to manage the Readied blend solid plant abuse. There are different reusing plants over the world, yet as Readied blend solid plant abuse is reused they lose their quality with the measure of reusing. So these Readied blend solid plant squanders will wind up as earth fill. In this condition as opposed to reusing it endlessly, on the off chance that it is used to configuration totals

© Springer Nature Singapore Pte Ltd. 2020
V. K. Gunjan et al. (Eds.): *ICRRM 2019 – System Reliability, Quality Control,
Safety, Maintenance and Management*, pp. 112–118, 2020.
https://doi.org/10.1007/978-981-13-8507-0_18

for solid, it will be a manual for the headway business. The majority of the misstep in solid structures occur because of the disappointment of bond by beating of entireties. Orchestrated blend solid plant abuse solid Waste, Coarse Sums which have low beating sees won't be walloped as effectively as the stone aggregates. These entireties are additionally lighter in weight wandered from stone wholes. Since an entire substitution for NCA was not found rational, an incomplete substitution with different level of RMCWCAC was finished. Both volumetric and grade substitution was utilized as a part of this examination Time of Arranged blend solid plant abuse is one of the snappiest making areas. Dependably in excess of 500 billion Arranged blend solid plant squander. In a landfill or in condition, Arranged blend solid plant abuse not to deteriorate. Different examines were coordinated to utilize industry by things, for example, fly scorching debris, silica of cement. Flume, solid plant squander in attach to enhance the properties. (17%) is higher than for the Readied blend solid plant squander industry somewhere else on the planet. India has a people of more than 1 billion and a Readied blend solid plant abuse utilization of 4 million tons. 33% of the majority is discouraged and won't not have the extra exchange to gobble up much out the methodology for Arranged blend solid plant squanders or assorted things. The virgin business does not revolve around this masses to expand its business zones. Regardless, 33% of the general population is the administrative class whose needs could be shaped to develop utilize. Orchestrated blend solid plant abuse makers make necessities for this piece of individuals. The rising needs of the consistent laborers, and breaking points of Arranged blend solid plant abuses to fulfill them at a more reasonable cost when veered from different materials like glass and metal, has added to an augmentation in the utilization of Arranged blend solid plant squanders over the most recent couple of years. In this venture Prepared blend solid plant squander materials were used to deliver Basic cement. The alpha work to start with, the fabric, combine degrees, gathering and relieving of the illustrations square measure elucidated this can be then trailed by depiction of forms of illustrations used, take a look at parameters, and take a look at procedures. Amendment of the trail toward poignant ready combine to sturdy plant waste sturdy combination.

.

2 Mixing, Casting and Solidification

The ready combine sturdy plant misuse concrete is factory-made by as just like the created bond initially the dry materials Concrete, Totals and Sand square measure mixed. The liquid piece of the combination was then supplemental to the dry materials and also the admixture continuing for urge around four minutes to make the new concrete. The new concrete was tossed into the molds efficiently within the wake of blending, in 3 layers for 3D form illustrations. For compaction of the illustrations, every layer was offered sixty to eighty manual strokes employing a rodding bar, and afterward vibrated for twelve to fifteen seconds on a moving table. Before the new concrete was tossed into the molds, the droop estimation of the recent bond was measured (Figs. 1, 2 and 3).

Fig. 1. Casting of concrete

Fig. 2. Mixing of concrete

Fig. 3. Curing of concert

The ready combine sturdy plant misuse Squander sturdy cases ought to be wrapped within the thick of relieving at brought temperature up in an exceedingly dry circumstance to deflect unconscionable dissemination. Wide trails discovered wrapping of sturdy cases by victimisation vacuum stowage film is possible for temperature up to 100c for some of long stretches of relieving. To settle the film to the sturdy form, a fast jolt seal or a twist tie wire was used. The later was used as a chunk of all more alpha work as a results of its ease and fund views. Preliminary check what is more discovered that ready combine sturdy plant misuse Squander - primarily based concrete failed to harden in an exceedingly New York minute at temperature. Precisely once the space temperature was underneath 30c, the Chief Executive failed to occur (rather than setting time used as a chunk of the case of OPC concrete) for ready combine sturdy plant misuse Squander - primarily based Cement.

2.1   General Procedure

Inside the check examine program regarding the modification of mechanical properties of a ready combine sturdy plant waste bond of audit M thirty was thought of with the going with association, similarly. The w/c-extent is zero. 50 Coarse aggregates were picked, having associate degree atom live primarily varied between twelve millimeter and twenty millimeter. Basic cognitive process the last word objective to direct this and to visualize early-age breaking, extra inward restoring water are going to be given by techniques for SAP. A heightened preliminary program is performed to think about the result of inward restoring on different types of sturdy properties: (I) contemporary properties (hang and thickness); (ii) mechanical properties (compressive quality, flexural quality, half versatility).

2.2 Compressive Strength Check

At the season of testing, each example should confine compressive testing machine. The foremost extreme load at the breakage of solid sq. are going to be noted. From the distinguished esteems, the compressive quality might patterned by utilizing beneath formula (Fig. 4).

Compressive Strength $1/4$ Load/area

Size of the test specimen $1/4$ 150 mm $\times$ 150 mm $\times$ 150 mm

Fig. 4. Compression test

Fig. 5. Split tensile test

Fig. 6. Flexural strength test

2.3 Split Tensile Test

The extent of chambers 300 mm length and 150 mm breadth are set in the machine with the goal that heap is connected on the contrary side of the blocks are threw. Adjust precisely and stack is connected, till the example breaks. The recipe utilized for calculation (Fig. 5).

$$Split\ tensile\ strength 1/4\ 2P=ndl$$

2.4 Flexural Strength Test

In the midst for the testing, the pole cases of size 700 m × 150 m × 150 mm were used. Illustrations were dried in outside after 7 long stretch of restoring and subjected to flexural quality test under flexural testing get together. Apply the load at a rate that consistently manufactures the most outrageous stress until the point when the moment that burst happens. The split shows in the strain surface inside the middle third of find the way length (Fig. 6).

3 Ratio for Special Concrete (Extra Ingredients)

RATIO –I:
Ready mix concrete plant waste Waste with – 0% by replacement of Aggregate
RATIO – II:
Ready mix concrete plant waste Waste with – 10% by replacement of Aggregate
RATIO – III:
Ready mix concrete plant waste Waste with – 20% by replacement of Aggregate.
RATIO – IV:
Ready mix concrete plant waste Waste with – 30% by replacement of Aggregate.
RATIO – V:
Ready mix concrete plant waste Waste with – 40% by replacement of Aggregate.
RATIO – VI:
Ready mix concrete plant waste Waste with – 50% by replacement of Aggregate.
RATIO – VII:
Ready mix concrete plant waste Waste with – 60% by replacement of Aggregate.
RATIO – VIII:
Ready mix concrete plant waste Waste with – 70% by replacement of Aggregate.
RATIO – IX:
Ready mix concrete plant waste Waste with – 80% by replacement of Aggregate.
RATIO – X:
Ready mix concrete plant waste Waste with – 90% by replacement of Aggregate.
RATIO – XI:
Ready mix concrete plant waste Waste with –100% by replacement of Aggregate.

Above all ingredients are added by weight of Aggregate (Figs. 7, 8, 9, 10, 11, 12, 13 and 14) shows results obtained from expriment.

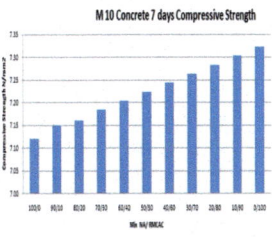

Fig. 7. Compressive strength of 7 days

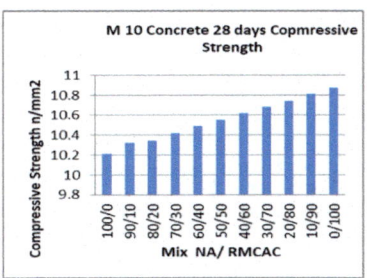

Fig. 8. Compressive strength of 28 days

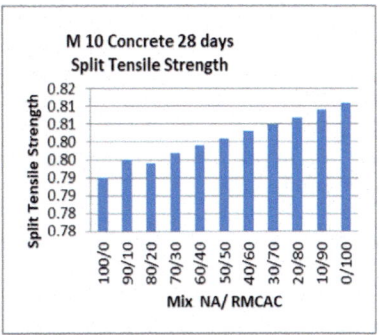

Fig. 9. Split tensile strength

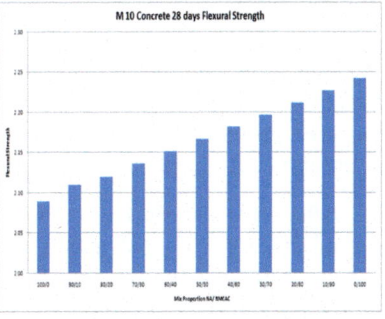

Fig. 10. Flexural strength

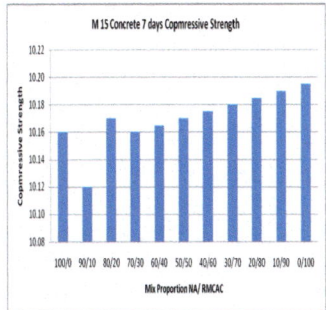

Fig. 11. Compressive strength at 7 days

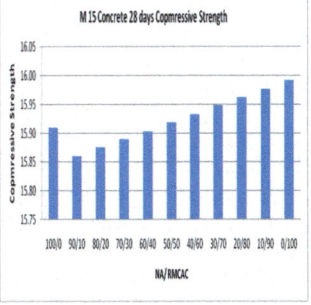

Fig. 12. Compressive strength at 28 days

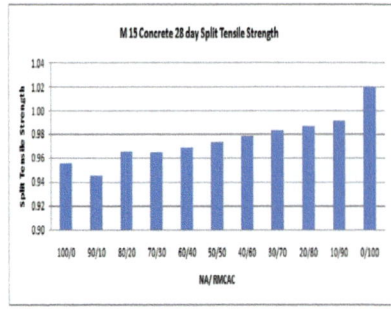

Fig. 13. Split tensile strength

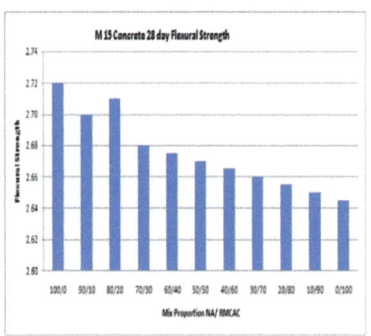

Fig. 14. Flexural strength

4   Conclusion

This examination anticipated that will realize the convincing ways in which to contend with reutilize the arduous ready combine sturdy plant waste particles as sturdy combination. Examination of the standard properties of bond containing reused misuse ready combine sturdy plant waste have the going with results.

- It's perceived that ready combine sturdy plant waste may be masterminded by victimisation them as advancement materials.
- The compressive quality and split skillfulness of concrete containing ready combine sturdy plant misuse add up to is command much in examination with controlled sturdy cases. Anyway quality detectably extended once the ready combine sturdy plant misuse content was over 2 hundredth.
- Has been wrapped 2 hundredth of ready combine sturdy plant waste combination may be melded as coarse combination substitution in concrete with no whole deal interference impacts and with satisfactory quality amendment properties.

References

1. Martinelli, E., Koenders, E.A.B., Caggiano, A.: A numerical recipe for modelling hydration and heat flow in hardening concrete. Cem. Concr. Compos. **40**, 48–58 (2013)
2. Koenders, E.A.B.: Simulation of volume changes in hardening cement-based materials. Ph.D. thesis, Delft University of Technology, Delft (Netherland), 171 p. (1997)
3. Lokhorst, S.J.: Deformation behavior of concrete influenced by hydration related changes of the microstructure. Internal Report Nr. 5-99-05, Delft University of Technology, Delft (Netherland), 178 p. (1999)

Life and Risk Assessment of Semiactive Suspension System Using Ride Comfort Advisory System (RCAS)

R. N. Yerrawar[1(✉)], M. A. Joshi[1], and R. R. Arakerimath[2]

[1] MES College of Engineering, Wadia College Campus, Pune 411001, India
rahul.yerrawar@gmail.com
[2] G. H. Raisoni COE and M, Wagholi, Pune 412207, India

Abstract. In this paper, the design of experiment approach is used for analysis and optimization of ride comfort of the vehicle. Also, optimized suspension parameters combination was presented through Taguchi DOE method. From Taguchi L16 array and SN ratio analysis, it is observed that the Cylinder Material with Al and CS for damper cylinder is a key parameter for performance measure of Semi active suspension system. From regression analysis, linear mathematical model is developed. The optimized Ride Comfort observed from the DOE table is 0.99 m/s^2 which is in the range of Fairly Uncomfortable to Uncomfortable as per IS 2631. The advisory system, known as Ride Comfort Advisory System (RCAS) is developed. The optimized model of Magnetorheological (MR) damper will have widely helpful to Indian Auto Industry. From the measured acceleration and RCAS acceleration results, it is observed that there is error between ±8 to ±10% and it is under limit as per comfort level suggested by IS 2631. The error arise due to the linear model equations and actual readings.

Keywords: Semiactive suspension · Ride comfort ·
Magnetorheological damper

1 Introduction

The ride comfort and road holding have compromise in a vehicle. The relation between that is shown in Fig. 1. There is a need of today's vehicle manufacturer to combine both in a single vehicle. The good comfort level is difficult to resolve with low body resistance and steering feel in contact with road surface which makes car handling better. Newly vehicle uses electronic system to improve vehicle handling stability with passenger safety for comfortable driving. Ride quality is measured in terms of the level of isolation from road inputs the suspension transfers to the vehicle without compromising vehicle control. The performance of a suspension depends on suspension elements characteristics i.e. spring stiffness and damping rate.

The good ride comfort and good handling are conflict to each other and require spring and damper characteristic differently. So, it is necessary to select such a damping coefficient which gives the better ride comfort with vehicle handling.

© Springer Nature Singapore Pte Ltd. 2020
V. K. Gunjan et al. (Eds.): *ICRRM 2019 – System Reliability, Quality Control,
Safety, Maintenance and Management*, pp. 119–126, 2020.
https://doi.org/10.1007/978-981-13-8507-0_19

1.1   Ride Comfort

The road condition has an important influence on ride comfort. The proper design of vehicle components is the only way to improve it. Therefore, it is essential to describe the various facets of position and shape of the vehicle body and suspension may be done in a general way applicable to all forms of suspensions, in terms of suspension bump position, body ride heights etc. Ride comfort is a measure of how good a vehicle feels to drive while it depends on the degree to which the irregular shape of the road surface get transferred to the vehicle body and therefore to the passengers [1–3]. Ride comfort is defined by the IS 2631-1-1997 standards state that the capability of suspension system to provide a range of comfort levels for a seated human in a vehicle to the vertical acceleration that are transferred from a road. The IS 2631 standard recommended the Acceleration as ride comfort measure for vehicle which is represented in Table 1.

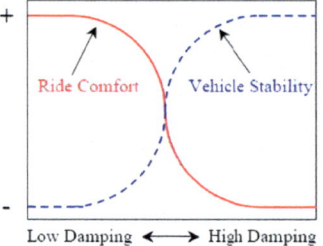

Fig. 1. Compromise between ride comfort and vehicle stability (road holding)

Table 1. Comfort reactions to vibration environment [IS 2631]

Sr No.	Acceleration (m/s^2)	Level
1	Less than 0.315	Not uncomfortable
2	0.315 to 0.63	A little uncomfortable
3	0.5 to 1	Fairly uncomfortable
4	0.8 to 1.6	Uncomfortable
5	1.25 to 2.5	Very uncomfortable
6	Greater than 2	Extremely uncomfortable

Every human has a certain tolerance level that can sustain these accelerations at various frequencies that vary as the road profile varies. This comfort levels are inversely proportional to the vertical acceleration experienced by the passengers.

1.2   Ride Position

The simplest quarter car model or heave model has a body mass and wheel mass associated with suspension, tyre stiffness and damping. For racing cars, the suspension structures compliance may be significant and passenger on a seat may be added. The Quarter Car Model terminology is considered because it has one suspension only. However, the quarter car model has one drawback regarding to the mass of the passenger (quarter of the passenger). It is better to consider it to be a heave model with a mass equal to the whole vehicle body, suspension compromising the total damping coefficients and one 'complete' passenger on one seat cushion.

2 Parameters Affecting Ride Comfort and Road Holding

According to literature survey, it is observed that there are number of factors which affect ride comfort and road holding [4, 5] but for the design of experiment following are the factors which are more significant.

The factors available are:

- Camber
- Frequency
- Current
- Sprung mass
- Unsprang mass
- Voltage
- Spring stiffness
- Damping coefficient
- Tyre pressure
- Speed
- Toe
- Castor

3 Design of Experiments

In the field of product or process development, the DOE that is Design of Experiment method is used as most global approach. It is based on statistical equations that try to provide a predictive knowledge of a complex, multi-variable process with few trials.

3.1 Taguchi Method

Taguchi has projected a new method of conducting the design of experiments which are based on well-defined guidelines. This method uses a special set of arrays called orthogonal arrays. These standard arrays stipulate the way of conducting the minimal number of experiments which could give the full information of all the factors that affect the performance parameter. Parameter design is related to finding the appropriate design factor levels to make the system less sensitive to variations in uncontrollable noise factors, i.e. to make the system robust.

3.2 Selection of Control Factors

An optimization of vehicle performance can be obtained by choosing values of stiffness and current under certain range of vehicle load and road conditions. The system will be highly stable with extremely stiff suspension but acceleration of sprung mass will be high and passenger comfort will be low. In case of less stiff suspension, passenger comfort will increase but vehicle become unstable.

Table 2. Taguchi parameter table

Parameter	Notation	Unit	Levels	Level 1	Level 2
Tyre pressure	P	Kgf/cm^2 (PSI)	2	34	40
Spring stiffness	K	N/mm	2	14000	8000
Damper cylinder material	CM	–	2	Aluminum	Structural carbon steel
Sprung mass	M	Kg	2	72	97
Current	I	(A)	2	0.5	0.7

Table 3. Orthogonal array as per Taguchi DOE approach with acceleration and SN ratio

Run	Air pressure (P)	Spring stiffness (K)	Cyl. mtrl. (CM)	Sprung mass (M)	Current (I)	Accl^n (RC)	SNRA1
1	34	14000	1	72	0.5	0.7250	2.79324
2	34	14000	1	97	0.7	0.6820	3.32431
3	34	14000	−1	72	0.7	1.1150	-0.94550
4	34	14000	−1	97	0.5	0.8852	1.05917
5	34	8000	1	72	0.7	0.9980	0.01739
6	34	8000	1	97	0.5	0.9250	0.67717
7	34	8000	−1	72	0.5	1.2500	-1.93820
8	34	8000	−1	97	0.7	0.9960	0.03481
9	40	14000	1	72	0.7	0.6520	3.71505
10	40	14000	1	97	0.5	0.8250	1.67092
11	40	14000	−1	72	0.5	0.8524	1.38713
12	40	14000	−1	97	0.7	0.7920	2.02550
13	40	8000	1	72	0.5	0.5240	5.61337
14	40	8000	1	97	0.7	0.9152	0.76968
15	40	8000	CS	72	0.7	1.2100	-1.65571
16	40	8000	CS	97	0.5	1.1600	-1.28916

To study the large number of variables with a small number of experiments, Taguchi method utilizes orthogonal arrays from design of experiments theory. Using orthogonal arrays significantly reduces the number of experimental configurations to be studied. Furthermore, the conclusions drawn from small scale experiments are valid over the entire experimental region spanned by the control factors and their settings. In present work, the five parameters with two level is used presented in Table 2 and as per availability L16 array is selected for experimentation.

- The number following the "L" indicates the number of runs in the design.
- The L16 array is denoted as L16 (2^5).
- L16 means the array requires 16 runs.

3.3   Analysis of SN Ratio

The performance characteristics that are deviating from the desired target value is known as loss function and Taguchi recommends the use of this loss function. The value of this loss function is further transformed into signal-to-noise (SN) ratio. There are three standard types of SN ratios depending on the desired performance response. Usually, there are three categories of the performance characteristics to analyze the SN ratio. They are: nominal-the-best, larger-the-better, and smaller-the-better. In present work, SN ratio with lower is better characteristics is selected as lower value of acceleration which is a measurement parameter for ride comfort of passenger. The Table 3 represents the L16 array table as per Taguchi DOE approach with acceleration measure and respective SN Ratio.

From Table 4 and Fig. 2, it is observed that Cylinder Material performs first rank and from main effect plot of SN Ratios run number L8 shows optimum level of parameter. The maximum Ride Comfort can obtain for P34 K8000 CM-1 M97 I0.7.

Table 4. Signal to noise ratio response table

Level	Air pressure	Spring stiffness	Cylinder material	Sprung mass	Current
1	0.6278	0.2787	−0.1652	1.1233	1.2467
2	1.5296	1.8787	2.3226	1.0341	0.9107
Delta	0.9018	1.6001	2.4879	0.0893	0.3360
Rank	3	2	1	5	4

4   Ride Comfort Advisory System (RCAS)

The expert advisory system "RCAS" is developed for inexperienced vehicle operators those are unable to understand exact comfortness in the vehicle. In some region of India due to poor road condition and bad quality of road, it is need of vehicle system to define the advisory system to the vehicle operator so that it will be beneficial to him with good comfort. As suggested by IS 2631 which define the Ride Comfort Levels need to be maintained in vehicle so that it will not damage to the vehicle operator. In expert advisory system the operator makes decisions based on information provided by advisory system [9].

According to literature work it is observed that there are different parameters responsible for Ride Comfort and total five parameters are considered for Ride Comfort Measurement. Hence same parameters are used to define RCAS (Ride Comfort Advisory System) represented by Fig. 3. The Visual Basic 6 software is used to define the Graphical Interfacing to vehicle operator. The GUI (Graphical User Interfacing) is represented in Fig. 4. As per sample Ride Comfort Calculation shown in Fig. 5, the Ride Comfort measured is 0.6654 which shows the status of Fairly Uncomfortable as per IS 2631 Standard.

The Table 5 represents the measured accelerations at the time of experimentation and RCAS Acceleration suggested by advisory system, from this it is observed that there is error between ±8 to ±10% between the measured and RCAS values and it is under limit as per comfort level suggested by IS 2631. As advisory system suggested the acceleration which is linear based equation the error arise with respect to actual readings.

The expert system identifies the quality of road condition and trying to suggest the required comfortness in the vehicle. The system developed is user friendly and the operator does not require any expertise in the computer system. Apart from the above advantages, the system is economical. The RCAS developed for semi active suspension system suggested the Ride Comfort with damper material as Al and CS. This system will further develop by considering more numbers of materials for cylinder and a greater number of parameters for RC measurements.

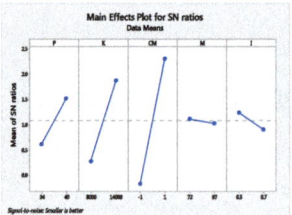      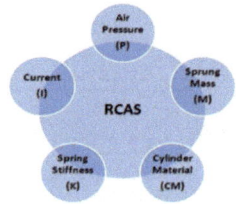

Fig. 2. Main effect plot for SN ratio   **Fig. 3.** Ride comfort parameters   **Fig. 4.** Ride comfort advisory system parameters

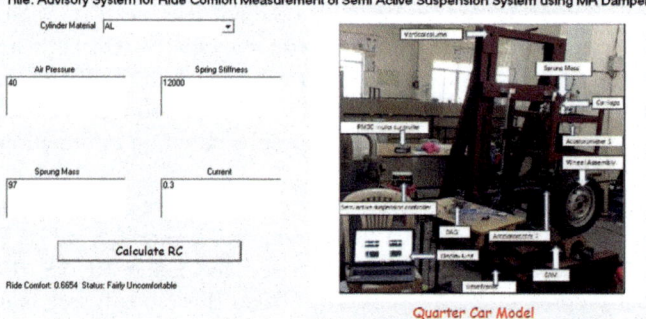

Fig. 5. GUI of RCAS

Table 5. Measured acceleration and RCAS acceleration

Run	Air pressure	Spring stiffness	Cyl. mtrl	Sprung mass	Current	Measure accln (Rc)	RCAS Rc
1	34	14000	AL	72	0.5	0.7250	0.7308
2	34	14000	AL	97	0.7	0.6820	0.7394
3	34	14000	CS	72	0.7	1.1150	1.0096
4	34	14000	CS	97	0.5	0.8852	0.9646
5	34	8000	AL	72	0.7	0.9980	0.9376
6	34	8000	AL	97	0.5	0.9250	0.8926
7	34	8000	CS	72	0.5	1.2500	1.1628
8	34	8000	CS	97	0.7	0.9960	1.1714
9	40	14000	AL	72	0.7	0.6520	0.6772
10	40	14000	AL	97	0.5	0.8250	0.6322
11	40	14000	CS	72	0.5	0.8524	0.9024
12	40	14000	CS	97	0.7	0.7920	0.9110
13	40	8000	AL	72	0.5	0.5240	0.8304
14	40	8000	AL	97	0.7	0.9152	0.8390
15	40	8000	CS	72	0.7	1.2100	1.1092
16	40	8000	CS	97	0.5	1.1600	1.0642

5 Conclusion

The optimized Ride Comfort observed from the DOE table is 0.99 m/s^2 which is in the range of Fairly Uncomfortable to Uncomfortable as per IS 2631. The advisory system, known as Ride Comfort Advisory System is developed. From the measured acceleration and RCAS acceleration results, it is observed that there is error between ±8 to ±10% and it is under limit as per comfort level suggested by IS 2631. The error arise due to the linear model equations and actual readings.

References

1. Lam, H.F., Lai, C.Y., Liao W.H.: Automobile suspension systems with MR fluid dampers, smart materials and structures laboratory, Department of Mechanical and Automation Engineering, The Chinese University of Hong Kong, pp. 1–18 (2002)
2. ISO 2631-1:1997: Mechanical vibration and shock, evaluation of human exposure to whole-body vibration, Part 1: General requirements, 2,1–31 (1997)
3. Mathews P.G.: Design of experiments with MINITAB, Classroom Exercises and Labs (2007)
4. Roy R.: Design of Experiments Using Taguchi Approach. John Wiley Inc (2001)
5. Devdutt and Aggarwal M. L, Simultaneous optimization of semi active quarter car suspension parameters using taguchi method and grey relational analysis. Int. J. Recent Adv. Mech. Eng. **4**(1) (2015)
6. Argade, P.V., Arakerimath, R.R.: Parametric investigations on CO2 laser cutting of AISI 409 to optimize process parameters by Taguchi method. Int. J. Eng. Trends Technol. (IJETT) **37** (6), 1–6 (2016)

7. Nemani, R., Arakerimath, R.R.: Taguchi based design optimization of side impact beam for energy absorption. Int. J. Adv. Res. Eng. Technol. **3**, 9 (2015)
8. Liao, H.T., Enke, D., Wiebe, H.: An expert advisory system for the ISO 9001 quality system. Expert Syst. Appl. **27**, 313–322 (2004)
9. Mansyur, R., Atiq, R., Rahmat, O.K., Ismail, A., Kabit, M.R.: Knowledge based expert advisory system for transport demand management. In: Proceeding of the International Conference on Advanced Science, Engineering and Information Technology, pp. 652–657 (2011)

Prediction of MRR for VMC Five Axis Machining of D3 Steel Using Desirability Function Approach

Arun Patil[1(✉)] and Ramesh Rudrapati[2]

[1] Department of Mechanical Engineering,
JJT University, Churela, Rajastan, India
patil.mailme@gmail.com
[2] Department of Mechanical Engineering, Hawassa University,
Hawassa, Ethiopia
rameshrudrapati@gmail.com

Abstract. The optimum selection of machining parameters plays a important role to achieve the desired performance of the machining process at a low cost within short time. Aim of the present research investigation is to develop a mathematical model for predicting material removal rate (MRR) in terms of speed, feed and depth of cut. Experiment are conducted on D3 tool steel material in VMC five axis operation based on the full factorial design. A second order mathematical model in terms of machining parameters and MRR is developed by response surface methodology (RSM). Analysis of variance (ANOVA) technique is utilized to determine the significant process parameters which expected to influence MRR. Contour plots are developed by RSM approach from the mathematical model to study the interaction effects of machining parameters on MRR. Desirability function approach has been applied to solve the mathematical model of MRR to maximize it.

Keywords: Five axis VMC · RSM · DFA · D3 steel · MRR

1 Introduction

The selection of optimum input process parameters is of great concern for metal cutting industries to produce high quality products to meet the current growing trend of advanced industrial applications. Identification of optimal cutting parameters like number of passes, feed rate, depth of cut and cutting speed is considered as a crucial for obtaining better qualities on machined part in turning, drilling, milling, etc. Vertical milling center (VMC) with five-axis is one of the important advanced manufacturing processes which used to create complex features on parts with one setup by the simultaneous interpolation of a machine with five degrees of freedom [1]. Conducting VMC five axis machining and producing dimensionally accurate milled parts are difficult task. To conduct VMC five axis machining process economically high material removal rate (MRR) and predictably to produce of parts with high surface quality is important area of research.

© Springer Nature Singapore Pte Ltd. 2020
V. K. Gunjan et al. (Eds.): *ICRRM 2019 – System Reliability, Quality Control, Safety, Maintenance and Management*, pp. 127–132, 2020.
https://doi.org/10.1007/978-981-13-8507-0_20

2   Literature Survey

Steel and its alloys are important industrial materials, which are treated as difficult to machine materials. The machining of advanced materials like steel-based alloys need advanced machining processes. VMC five axis is one of the advanced metals cutting process which is used to cut intricate shapes on parts. Here, literature survey has been made to study the reported articles on analysis and optimization of milling using RSM and optimization techniques like GA, DFA, etc.

Rishi et al. [2] had been optimized milling parameters for minimizing surface roughness in CNC end milling. Mathematical relations had been developed between the input parameters and output response for predicting surface roughness. Model adequacy tests were conducted using ANOVA table and the effects of various parameters were investigated. Zhang et al. [3] had been optimized the MRR along with Ra and carbon emission using regression analysis and genetic algorithm. Rajeswari and Amirtha gadeswaran [4] had made an experimental investigation to maximize the MRR of aluminum composites in end milling operation using RSM and grey relational analysis. Gopikrishnan et al. [5] optimized the cutting conditions of peripheral milling of AISI 4340 steel using RSM technique. Riadh et al. [6] analyzed, modeled and optimized the input parameters in turning operation using integrated RSM and DFA approach. Again, Muhammad and Khaled [7] had investigated the effects of turning parameters on output responses: MRR and surface roughness in single point diamond turning operation using RSM and DFA.

3   Materials and Methods

D3 steel: Tool steel refers to a various carbon and alloy steels that are particularly suited to made into good tools. one of the advanced materials which used for high demand tooling applications, this material is selected for the present research work. It originates from their high hardness, deformation, good resistance to abrasion and their ability to hold a cutting edge at high temperatures.

Full factorial design is used to plan the experiments. full factorial design is very useful to study and analyze the entire process by considering all possible combinations while planning and conducting the experiments. Analysis of variance is a statistical technique which used to identify the significant parameters which has detrimental effects on output responses. Response surface methodology (RSM) is a collection of mathematical and statistical techniques for empirical model building. By careful design of experiments, the objective is to optimize a response (output variable) which is influenced by several independent variables (input variables). RSM is the most effective method for modeling manufacturing processes to make relationships between the input and output responses. Desirability function approach (DFA) is important optimization tool used to predict the output quality characteristics by optimizing the input parameters. In the present work, DFA is used to maximize the MRR in VMC five axis when machining of D3 steel.

4 Experimental Procedure

The full factorial design has been used to design the experiments by considering depth of cut (A), cutting speed (B) and feed (C) as input control parameters for VMC-5-axis machining of D3 steel material. The selected cutting parameters and their varying levels are: depth of cut (A) = 0.1, 0.15 and 0.2 mm; cutting speed (B) = 3000, 3500 and 4000 rpm; feed (C) = 1500, 2000 and 2500 Rev/min. the full factorial design is shown in Table 1. The output response is measured and given in Table 1.

5 Results and Analysis

As mentioned earlier, the present work is planned to study the significance of VMC process parameters on MRR in VMC five axis machining of D3 steel material. The data shown in Table 1 has been used for analyzing and interpreting the MRR with use of statistical analysis of variance (ANOVA) and RSM technique. Finally, DFA has been utilized to solve the mathematical model of MRR.

5.1 Analysis of Variance for MRR of D3 Steel Material

Analysis of variance from MINITAB 16.2 software is applied on the experimental data of MRR of D3 steel material to identify the significant control parameters which has detrimental effects on MRR. The results of ANOVA are given in Table 2. ANOVA test is performed at 95% confidence level i.e. 5% significant level. If the value of probability (P) is less than 0.05, it indicates that the effects of process parameters on corresponding responses are significant.

From the ANOVA table shown in Table 2, it is notified that direct effects of depth of cut (A) and feed (C) most significant for MRR of D3 steel in VMC five axis machining as its P values are zero. The effects of cutting speed (B) and interaction effects of depth of cut (A) – cutting speed (B), depth of cut (A) – feed (C) and cutting speed (B) - feed (C) are insignificant as its P values are more than 0.05.

5.2 Mathematical Modeling of MRR

Mathematical modeling is made by using RSM application from MINITAB 16.2 software to postulate the relationships between the milling input parameter and output response, MRR and given in Eq. 1.

$$\begin{aligned} Y_{MRR} = {} & 2.82 + 23.67 * A - 0.0013 * B - 2.40E - 04 * C + 41.6 * A * A + 1.91E \\ & - 07 * B * B + 4.27E - 08 * C * C - 6.30E - 04 * A * B + 0.0021 * A \\ & * C + 5.83E - 08 * B * C \end{aligned}$$

(1)

Contour plots are drawn from the developed mathematical model as given in Eq. 1 and shown in Fig. 1 almost straight lines in the contour plot indicating that interaction effects of process parameters are negligible.

Table 1. Full factorial design and output response (MRR)

Sr. No	Input parameters			Output response
	Depth of cut (A)	Cutting speed (B)	Feed (C)	MRR
1	0.2	3000	1500	7.300
2	0.2	3000	2500	7.805
3	0.2	3500	1500	7.328
4	0.15	3500	2000	5.517
5	0.15	3000	2500	5.878
6	0.2	3500	2000	7.649
7	0.2	4000	2000	7.500
8	0.1	4000	2500	3.983
9	0.1	3000	2500	3.934
10	0.1	3500	2500	3.967
11	0.15	3000	2000	5.496
12	0.15	4000	2500	5.926
13	0.1	4000	1500	3.609
14	0.1	3000	1500	3.582
15	0.1	3000	2000	3.692
16	0.15	3500	2500	5.414
17	0.2	4000	1500	7.245
18	0.1	3500	2000	3.664
19	0.2	3000	2000	7.500
20	0.15	3500	1500	5.361
21	0.1	4000	2000	3.840
22	0.2	4000	2500	7.895
23	0.15	4000	1500	5.373
24	0.1	3500	1500	3.561
25	0.15	4000	2000	5.760
26	0.2	3500	2500	7.934
27	0.15	3000	1500	5.333

Table 2. ANOVA for MRR

Source	DF	Seq SS	Adj SS	Adj MS	F	P
A	2	65.5170	65.5170	32.7585	3021.64	0.000
B	2	0.0345	0.0345	0.0172	1.59	0.262
C	2	0.9092	0.9092	0.4546	41.93	0.000
A*B	4	0.0945	0.0945	0.0236	2.18	0.162
C*C	4	0.0515	0.0515	0.0129	1.19	0.386
B*C	4	0.0355	0.0355	0.0089	0.82	0.548
Error	8	0.0867	0.0867	0.0108		
Total	26	66.7289	R-Sq = 99.87% R-Sq (adj) = 99.58%			

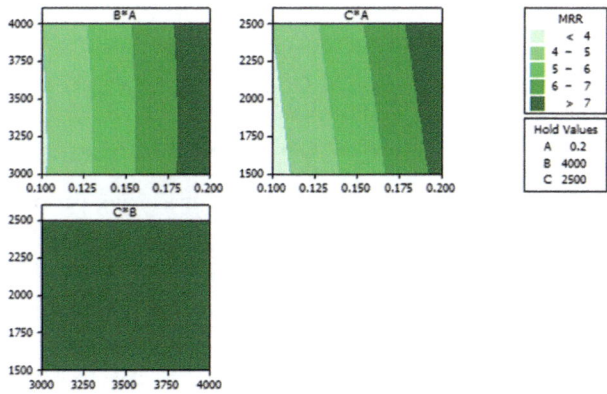

Fig. 1. Contour plot for MRR

5.3 Response Optimization by Desirability Function Approach (DFA)

In the present work, DFA from RSM in MINITAB 16.1 software applied on experimental data as shown in Table 1 to predict the optimum milling condition to maximize the MRR of D3 steel material. The processing and execution steps of DFA for calculating the desirability value d and calculating overall desirability function value and its optimization taken care by RSM approach. Finally, it provides the optimum input parametric setting and maximized MRR at optimum combination. The obtained optimization setting is: depth of cut (A) = 0.2 mm, cutting speed (B) = 4000 mm and feed (C) = 2500 rev/min and MRR = 7.90.

5.4 Conclusions

The followings are the conclusions drawn from the present investigation of optimization of surface roughness in VMC-5-axis machining of D3 steel material using integrated response surface methodology (RSM) and DFA:

1. From the ANOVA results, it is found that milling parameters: depth of cut and feed are most significant for MRR.
2. Mathematical model is generated by RSM to postulate the mathematical relationships between the input parameters and MRR.
3. Optimum VMC machining combination (i.e. depth of cut = 0.2 mm, cutting speed = 4000 rpm and feed = 2500 Rev/min) along with MRR = 7.90 value is obtained by desirability function approach.
4. From the present work, it is stated that integrated RSM and DFA is very useful for analyzing, modeling and predicting the MRR in VMC-5-axis machining of D3 steel.

References

1. Lamikiz, A., De Lacalle, L.N.L., Ocerin, O., Díez, D., Maidagan, E.: The Denavit and Hartenberg approach applied to evaluate the consequences in the tool tip position of geometrical errors in five-axis milling centres. Int. J. Adv. Manuf. Technol. **37**, 122–139 (2008)
2. Rishi, R.S., Singh, M.P., Sanjay, S.: Optimization of machining parameters for minimum surface roughness in end milling. Int. J. Innovative Comput. Sci. Eng. **3**(2), 28–34 (2016)
3. Zhang, C.Y., Li, W.D., Jiang, P.Y., Gu, P.H.: Experimental investigation and multi-objective optimization approach for low-carbon milling operation of aluminum. In: Sustainable Manufacturing and Remanufacturing Management (2019). https://doi.org/10.1007/978-3-319-73488-0_5
4. Rajeswari, B., Amirtha gadeswaran, K.S.: Experimental investigation of machinability characteristics and multi-response optimization of end milling in aluminium composites using RSM based grey relational analysis. Measurement **105**, 78–86 (2017)
5. Gopikrishnan, P., Aalim, A., Asokan, A., Bharath, B., Sumesh, C.S.: Numerical modelling and optimization of surface finish during peripheral milling of AISI 4340 steel using RSM. Mater. Today: Proc. **5**(11), 24612–24621 (2018)
6. Riadh, S., Brahim, B.F., Mabrouki, T., Salim, B., Mohamed, A.Y.: Modeling and optimization of the turning parameters of cobalt alloy (Stellite 6) based on RSM and desirability function. Int. J. Adv. Manufact. Technol. (2018). https://doi.org/10.1007/s00170-018-2816-x
7. Muhammad, M.L., Khaled, A.E.H.: Modeling and multi-response optimization of cutting parameters in SPDT of a rigid contact lens polymer using RSM and desirability function (2018). https://doi.org/10.1007/s00170-018-3169-1

Crest Factor Measurement by Experimental Vibration Analysis for Preventive Maintenance of Bearing

Ganesh Eknath Kondhalkar[(✉)] and Girikapati Diwakar

Department of Mechanical Engineering, Koneru Lakshmaiah Education
Foundation - Deemed to be University, Vijayawada, Andhra Pradesh, India
ganeshkondhalkar@gmail.com,
diwakar4236@kluniversity.in

Abstract. This paper elaborates about the vibration and its effect for the rotary machines and presents experimental vibration technique for fault detection in bearings [1]. Bearings are very important part in any rotating machine. The rotary machines are subjected to variable speed range according to the type of operations to be performed and other operative conditions [2]. One of the causes for the vibration is bearing wear which may occurs due to Pitting, scratching, misalignment of shafts etc. There are many techniques to find and measure the vibration in the system [4]. One of the methods i.e. measurements by calculating crest factor which is discussed in this paper. Crest factor is defined as the ratio of the peak value of a waveform to its RMS value; which is also called as "peak-to-RMS-ratio" [3]. This paper gives comparative analysis between damaged and healthy Roller and Ball bearings these may be taken the baseline for the preventive maintenance. To perform experimental measurements of vibration amplitude in axial and radial directions FFT analyzer and accelerometer is used. From the experimentation crest factors were found and compared with healthy bearings to conclude about the defective bearings. Bearings with damages showed higher crest values as compared with the healthy bearing [7].

Keywords: Crest factor · Rolling element bearing · Bearing elements defect · Vibration spectrum analysis

1 Introduction

Diagnosis of wearing parts of Rotating machines and engines is important as it improves the availability, safety and help to reduce material usage while repairing. The diagnosis of bearing defects can be used for predictive maintenance which prevents major damages in mechanical systems [3]. Each rotating machinery has a particular vibration signature identified with the development and the condition of the machine. Change in condition of the machine changes the vibration pattern [4]. The vibration signature obtained from the machine element can be useful for fault diagnosis as it is symptom of the condition of the machine. Condition monitoring will increase the maintenance efficiency, which will save the breakdown cost of the machines. Vibration

© Springer Nature Singapore Pte Ltd. 2020
V. K. Gunjan et al. (Eds.): *ICRRM 2019 – System Reliability, Quality Control,
Safety, Maintenance and Management*, pp. 133–138, 2020.
https://doi.org/10.1007/978-981-13-8507-0_21

analysis is a fundamental tool for the condition monitoring which is used for maximum utilization of existing recourses.

During working, the bearings are subjected to heavy and dynamic loadings generated by machines and transmitted through the components of rolling element like bearings [8]. The vibration produced by healthy bearing is low as it has good surface finish and no wear and tear takes place, it also create less noise. The vibration produced by the bearing changes as a fault begins to development. A pulse of vibration is created when rolling element encounters a discontinuity in its path & resulting pulses of vibration repeats periodically at a rate determined by the location of discontinuity and by the bearing geometry. These repetition rates of pulse are known as the bearing frequencies [6]. The ball passing frequency outer race (BPFO) for a fault on outer race: the ball pass frequency inner race (BPFI) for fault on inner race: the ball spin frequency (BSF) for a fault on ball; and the fundamental train frequency (FTF) for a fault on cage are used in vibration analysis [5].

Different methods can be used for the diagnosis of defects in the rolling element i.e. bearings like measurement of acoustic, vibrations, monitoring of temperature changes and wear debris analysis. From all these methods the simple and more effective method is to analyze is measurement of crest factor of vibration. This can be calculated by using vibration signature [3].

2   Methodology and Experimentation

2.1   Methodology for Diagnosis of the Bearing

Methodology for finding the Crest Factor for fault diagnosis is as follows. This is carried out by using the experimental set up, where the healthy bearing is tested creating different test conditions. The same set up consists of the arrangement for testing the defective bearings and the readings are taken for the same test conditions. The defects are artificially created for the testing purpose to check the effect of the faults in bearings. Defects may be present in the form of indentation, scratch, pits, abrasive particles embedded in lubricants etc (Fig. 1).

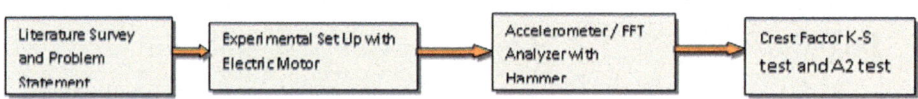

Fig. 1.   Methodology

2.2   Types of Bearing Defects [3, 5]

1. Distributed Defects:- Mostly distributed defects, which include not only surface roughness, waviness; misaligned races and off-size rolling elements but also errors due to manufacturing, abrasive wear, improper installation and usage. Similarly the variation in contact forces between rolling elements and raceways results in

vibration. To recognize damages the amplitude of bearing defect frequencies and their harmonics are important.

2. Localized Defects:- A localized defect includes pits, cracks, spalls which may develop over the rolling surfaces due to any reason.
3. Vibration:- Vibrations produced in the bearing assembly causes the defects in it. The causes of vibration may be misalignment etc.
4. Geometrical Imperfection:- Geometrical imperfections may present in the bearing components due to nature of the
5. Manufacturing Process:- When an axial load is acting on the bearing and rotating with moderate speed then shape from and surface finish of rolling elements are the largest sources of vibration.
6. Surface Roughness:- Surface roughness compared with lubricating film thickness between the rolling element raceway contacts will cause vibrations. The asperities break through the film and interact with the opposing surface resulting in vibration.

3   Experimental Vibration Analysis

3.1   Crest Factor

The Crest factor is defined as the ratio of the PEAK value of a waveform to its RMS value it is also called as "peak-to-rms ratio", a pure number, without units. The standard value of Crest factor for sine wave is 1.414; i.e. the peak value is 1.414 times the RMS value. A rotating machine element with only large imbalance will have a crest factor of about 1.5, when the bearings begin to wear, and some impacting start, the crest factor will increase drastically. The crest factor is very much sensitive with the vibration; peaks do not last very long in time, and therefore do not contain very much energy. The RMS value of waveform is proportional to the amount of energy in the vibration signal.

3.2   Relation Between Crest Factor and Amplitude

The following diagrammatical representation (Fig. 2) shows the relationship between crest factor and amplitude for normal bearing and the bearing with the fault. This shows that crest factor for the Normal Bearing is smaller than the bearing with the fault. This also represents the effect of vibration on the crest factor. The value of the crest factor gives idea about the impacting, which is associates with wear, cavitations etc. The higher value of crest factor indicates the less remaining life of the bearing, which is indication of replacement of the bearing.

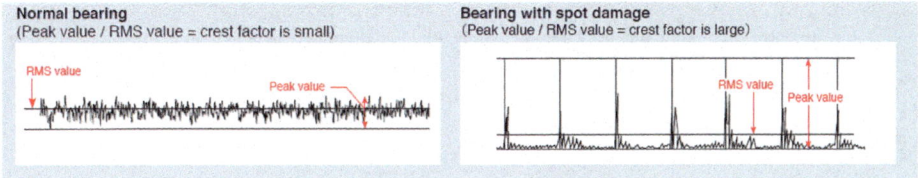

Fig. 2.   Crest factor and amplitude

3.3   Experimental Setup

Experimental set up is prepared for testing healthy and defected bearing. The setup consists of the electric motor with the power transmission pulleys and v belts. The loading arrangement is also attached to apply the load on the bearing, set of readings are taken at different loads and accordingly graphs are plotted.

Figure 3 shows the photograph of experimental set up for testing to get spectrum for healthy and defective bearings. This vibration signature obtained will indicate the change in the spectrum due to the defects in it. This will be useful for the health monitoring of the bearings, which will increase the life of the bearing.

Fig. 3. Photographic view of experimental setup

Using FFT analyzer readings were taken for different defective bearings and healthy bearing and the comparison is shown in the following graphs.

Test conditions:- The speed is kept constant and the healthy and defective bearings were tested for axial and redial directions. The defect is created on the outer race of the bearings.

3.4   FFT Analyzer

Digital signal processing techniques to analyze a waveform with Fourier transform to provide in depth analysis of signal waveform spectra is used in FFT or Fast Fourier Transform spectrum analyzer. FFT is an algorithm which samples the signals over a period of time and divides it into frequency component (Fig. 4).

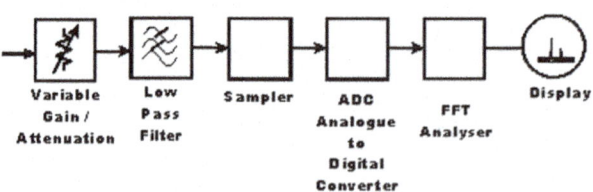

Fig. 4. FFT analyzer

4   Result and Discussion

Following are the output graphs obtained from the set of readings taken for Healthy and Defective ball and Roller bearings (Figs. 5, 6, 7, 8, 9, 10, 11 and 12).

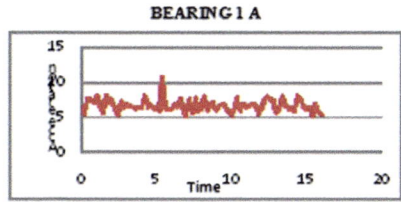

Fig. 5 Acceleration v/s time for healthy roller bearing in axial direction

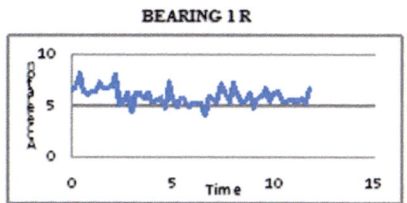

Fig. 6 Acceleration v/s time for healthy roller bearing in radial direction

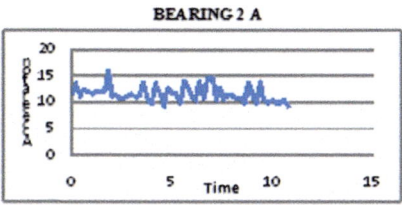

Fig. 7 Acceleration v/s time for defective roller bearing in axial direction

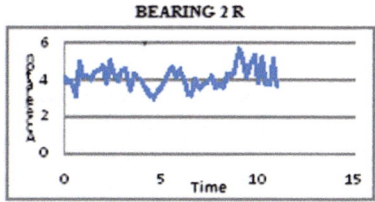

Fig. 8 Acceleration v/s time for defective roller bearing in radial direction

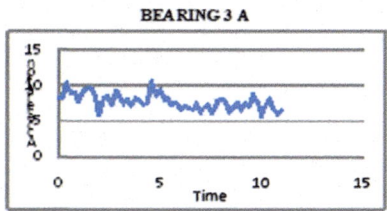

Fig. 9 Acceleration v/s time for healthy ball bearing in axial direction

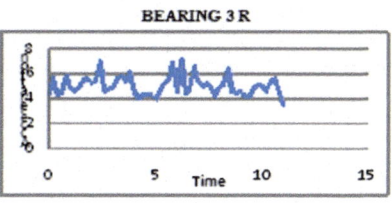

Fig. 10 Acceleration   v/s   time   for healthy ball bearing in radial direction

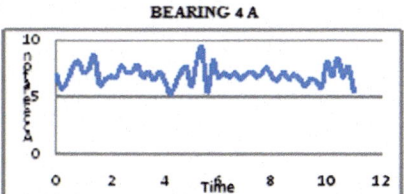

Fig. 11 Acceleration v/s time for defective ball bearing in axial direction

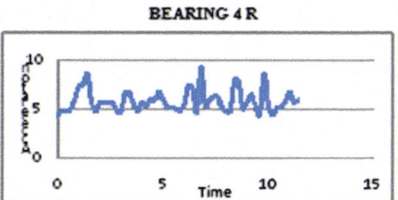

Fig. 12 Acceleration v/s time for defective ball bearing in radial direction

From the above spectrum RMS and Peak values are calculated and accordingly the values of the crest factors obtained are as follows (Table 1).

Table 1. Crest factors comparison

Bearing no.	Radial	Axial	Bearing no.	Radial	Axial
1	4.53	5.54	3	4.55	4.23
2	6.63	5.09	4	5.76	4.64

5   Conclusion

The value of the crest factor depends on the waveform obtained; the wave form will get disturbed due to irregularities. Value of Crest factor is more for defective bearings than Healthy bearings. As the defects in the Rollers and Balls bearing increases vibration acceleration values increases in amplitude. The increased in amplitude is indication for damage in Roller and Ball bearings. Hence FFT analyzers can be used to analyze health monitoring for rotational components.

References

1. Kulkarni, S., Wadkar, S.B.: Experimental investigation for distributed defects in ball bearing using vibration signature. Proc. Eng. **144**, 781–789 (2016)
2. Attoui, I., Fergani, N., et al.: A new time frequency method for identification and classification of ball bearing faults. J. Sound Vib. **397**, 241–265 (2017)
3. Gupta, P., Pradhan, M.K.: Fault detection analysis in rolling element bearing: a review. Mater. Today: Proc. **4**, 2085–2094 (2017)
4. Girikapati, D.: Diagnosis of misalignment in overhung rotor using K-S statistic and A2 test. J. Inst. Eng. India Series (C) **99**(1), 79–86 (2018)
5. Niu, L., et al.: A systematic study of ball passing frequencies based on dynamic modeling of rolling ball bearings with localized surface defects. J. Sound Vib. **357**, 207–232 (2015)
6. Patel, V.N., et al.: Vibration generated by rolling element bearings having multiple local defects on Races. ICIAMA **14**, 312–319 (2014)
7. Garad, A., et al.: Ananlysis of vibration signals of rolling element bearing with localized defects. Int. J. Curr. Eng. Technol. **7**(1), 37–42 (2017)
8. Durkhure, P., Loadwal, A.: Fault diagnosis of ball bearing using time domain analysis and fast furrier transformation. Int. J. Eng. Sci. Res. Technol. **3**(7), 711–715 (2014)
9. Djebili, O., et al.: Methodological approach of selecting a vibration indicator in monitoring bearings. Int. J. Phys. Sci. **8**(12), 451–458 (2013)
10. Lazovic, T., et al.: Mathematical model of load distribution in roller bearing. Faulty Mech. Eng. Belgrade, FME Trans. **36**, 189–196 (2008)

Reliable Approach for Management and Disposal of Electric Lamps for Pune Region of Maharashtra State

Sagar Mukundrao Gawande[1][✉], Ganesh E. Kondhalkar[2],
Sunil B. Thakare[3], and Shailesh S. Hajare[4]

[1] Civil–Environmental Engineering,
Anantrao Pawar College of Engineering and Research, Pune, India
gawande.sagar@gmail.com
[2] Mechanical Engineering,
Anantrao Pawar College of Engineering and Research, Pune, India
[3] Anantrao Pawar College of Engineering and Research, Pune, India
[4] E&TC, Anantrao Pawar College of Engineering and Research, Pune, India

Abstract. The most important environmental and health risk from electric lamp is due to unscientific disposal which relates to release of mercury. Any breakage or damage to these lamps will result in potential exposure to mercury vapor and powder. The heat and air movement will increase mobility of the mercury vapor as well as of the powder. The electric lamps are comes in various shapes and sizes. Standard straight lamps normally enclose with the 5 mg to 10 mg of mercury whereas high intensity discharge lamps contain around 30 mg.

The unscientific disposal of electric lamps will lead to reciprocate mercury level in surrounding environment. The toxicity from mercury reaches the human body due to consumption of water and food which contains mercury which leads to neurological disorder when it accumulates in brain and kidneys. On the other hand physiologically it hampers the normal working of human brain and mental instability. The main objective of this paper is to design and develop a low cost semi automated reliable disposal mechanism using Internet of Things for sustainable environment.

Corresponding Author and co authors are actively involved in collection and disposal of electric bulbs. They are trained and motivate the students and stakeholders to safe handling and disposal trashed bulbs. They often demonstrate the toxic effects of unscientific disposal of the electric waste in Pune region of Maharashtra.

Keywords: Electric lamp · Collection · Scientific disposal · Internet of Things · Recovery of metals

1 Introduction

The adoption of either new technologies or updation of existing technologies at industrial or domestic points creates the load of obsolesce products and burden on disposal facilities. These obsolesce products mostly found in term of electrical waste

© Springer Nature Singapore Pte Ltd. 2020
V. K. Gunjan et al. (Eds.): *ICRRM 2019 – System Reliability, Quality Control, Safety, Maintenance and Management*, pp. 139–144, 2020.
https://doi.org/10.1007/978-981-13-8507-0_22

which includes Cathode Ray Tubes (CRT), Electric Lamps and halogen, Electronic Screens, Smart phones etc. Most of these products contain the noxious elements. The mercury is most commonly used in all electrical appliances due to its feasibility and availability. The improper disposal to these obsolesce products into open environment offers hazard to human health and ecosystem [1]. In under develop or developing counties like India the scientific disposal of the electric waste is crucial issue. The unauthorized agencies openly dump their electrical debris which will further collect by the rag pickers. They burn this debris to collets the cash or value added products like copper, aluminum from it. The main objective of this paper is to design and develop a low cost semi automated reliable disposal and management through systematic mechanism by using Internet of Things for sustainable environment.

1.1 Objectives of Research

I. To design and develop semi automated mechanism to break electric lamp
II. To design the collection system using Internet of Things.
III. To demonstrate and evaluate the developed system

1.2 Electrical Lights

The lamp worked when an electric current was passed through a filament which heated and produced light. Electric bulb was improved in year 1920 when a carbon filament that was used until then is replaced with tungsten and a space inside a bulb was filled with argon gas, prolongs its life. Generally in both industrial and domestic sectors the neon and fluorescent lights are uses [3, 4]. On other side the car manufacturing industry and car owners have long wanted their headlights to be most compact while at the same time producing even more light. Conventional car Lamp technology was simply unable to meet these requirements. In the 21st century the zenrac the high pressure gas discharge lamp is likely to revolutionize in automobile headlights. This century will be witnessed ever increasing use of Fiber Optics in Automotive Lighting [6].

1.3 Bulb Industries

Electric bulb industry had annual growth of about 12% per annum in the last 4 years which is highest in the product sector of other electrical products and appliances. The consumption and contribution of Compact Fluorescent Lamps (CFL) has very high growth rate as high as 50% in the year 2006. The electrical bulb segment has registered total quantity of more than 100 million pieces during year 2006 and it was estimated that the figure crossed 140 million pieces in the year 2007 through existing and upcoming new plants [2].

Incandescent lamp (GLS) production was increased by more than 20% during the year 2006. Correspondingly, the fluorescent lamp market has shown a growth of 10% in year 2006. High Intensity Discharge (HID) lamp segment has equally shown good results of 24% growth in year 2006. It was estimated that the luminaries market has been up side trend at least 25% to 30% per annum for the last two years [2].

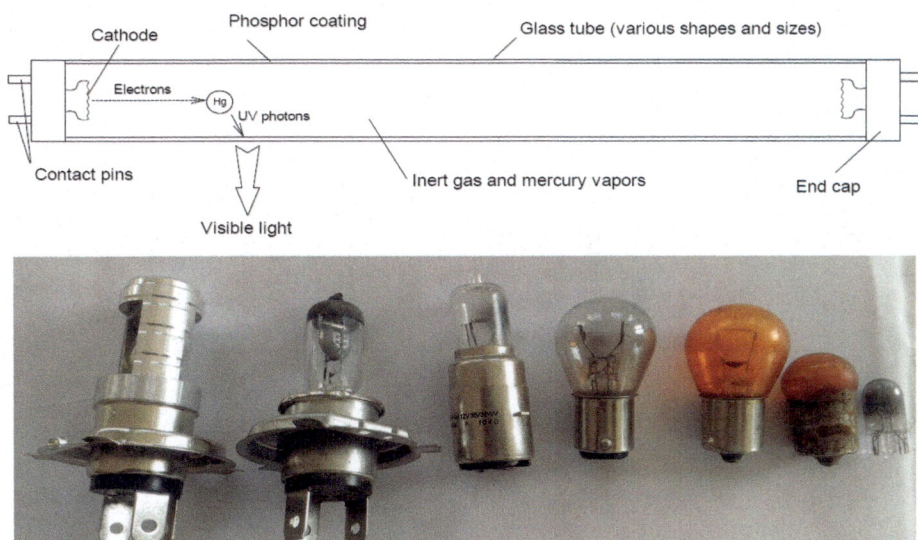

Fig. 1. Showing schematic diagram of typical fluorescent lamp [7] and collected vehicular light samples during research work

Table 1. Global manufacturing and Indian consumption of CFL [2]

Assessment year	World production (million pieces)	Growth	Indian consumption (million pieces)	Growth
2001	820	–	27	–
2002	880	7%	34	26%
2003	1144	30%	36	6%
2004	1500	31%	43	27%
2005	1930	29%	67	56%
2006	2650	37%	100	49%
2009 (Estimated)	6000	–	220	–

Types of lamps that includes mercury-

i. Fluorescent Tube Lamps (FTL)
ii. Compact Fluorescent Lamps (CFL)
iii. High Intensity Discharge (HID)
iv. Neon Lamps

As per World watch Institute of Washington estimates that 3.5 billion Compact Fluorescent Lamps were used globally in the year 2003. It says that in the United States between year 2000 and year 2004 CFLs sales went up from 21 million to 93 million pieces and in year 2007 sales reached 343 million pieces [5].

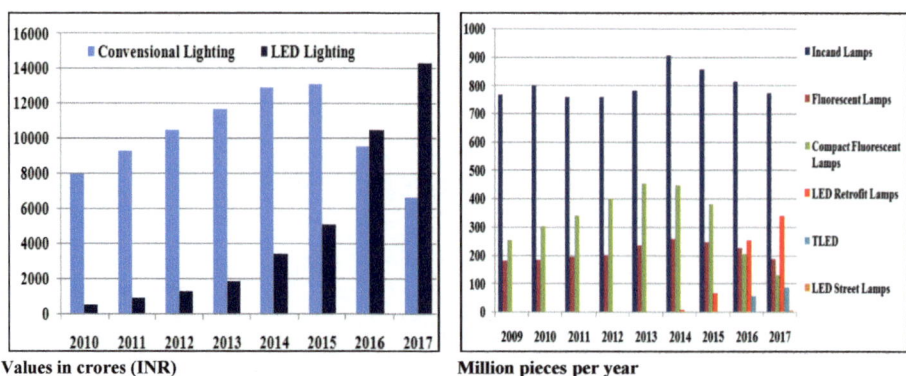

Fig. 2. Trends and categories in lighting industry in India from year 2010 to year 2017 [6]

2 Methods and Materials

(A) The samples of refused and discarded lights are collected in the bins available at collection points suggested for the Pune city nearby the Parvati area in which institute is located. The collection bins are equipped with weight sensors which send the signal to the microprocessor unit after collection bin collects the pre-defined quantity by weight of lights.

(B) The rotary drum with microprocessor and automatic control system allows crushing the electric lights inside the drum; the break glass and metals are collected and separated into chamber below it. as shown in Fig. 4(b) and as shown in Fig. 4(c)

(C) The execution of interface between the microprocessor and sensors attached to the collection bins by cloud computing as shown in Fig. 4(a). The system interface allows the user to get notification from the collection bins as and when required as shown in Fig. 4(a)

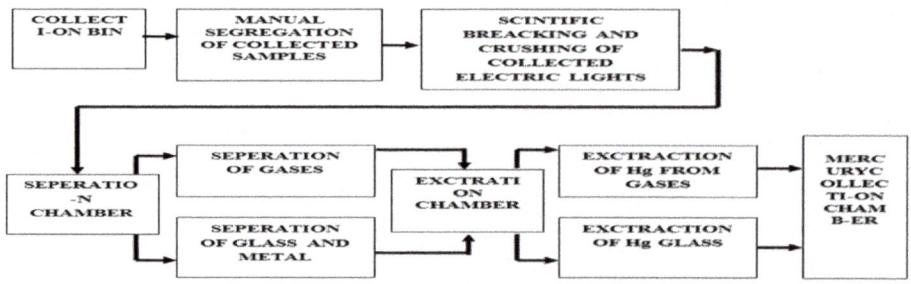

Fig. 3. Flow of the methodological activity during experimentation

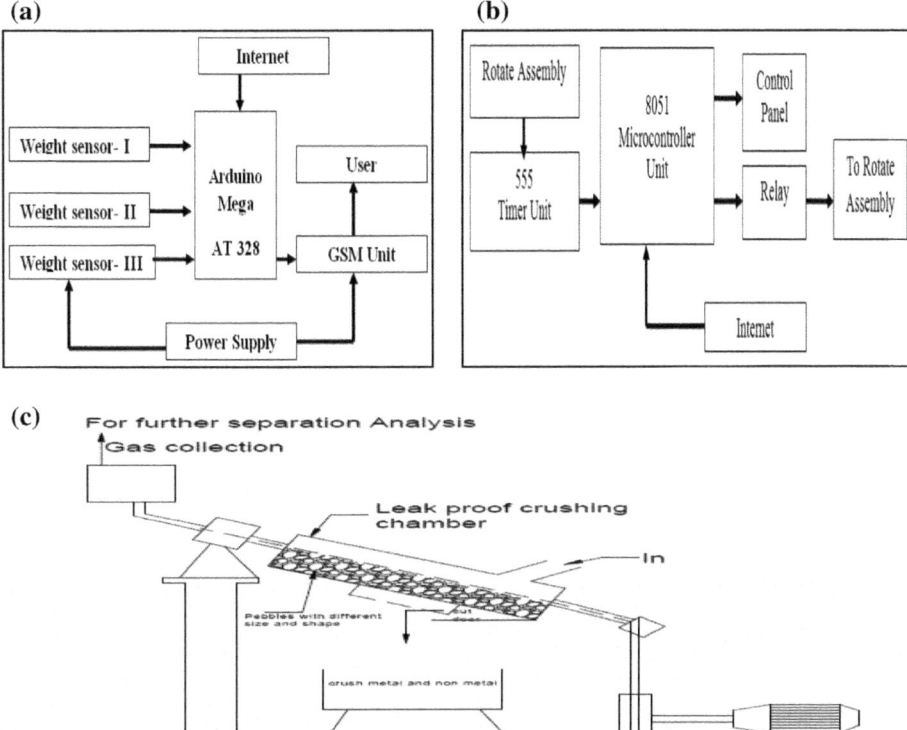

Fig. 4. (a) Bins & sensors with Arduino processor, (b) Controlling unit for drum crusher, (c) Crushing unit proposed lab scale model for crushing and disposal of lamps

(D) The collected samples stack at the central collection system at Anantrao Pawar College of Engineering and Research in Parvati area of Pune City. The light samples are cleaned manually with necessary safety precautions, then the cleaned sample send to the disposal system as shown in Fig. 4(c).

3 Conclusion

In the recent study India's Light Emitting Diode (LED) market is forecast to reach $1457.8 million by year the 2019, at a Compound annual growth rate (CAGR) of 35.9 per cent, during the period from year 2014 to year 2019. The government of India took initiatives to replace existing incandescent bulbs with LED bulbs to overcome on the increasing energy demand and supply gap, and to declining prices are the factors driving the growth of LED lighting in India [2]. The replacing the existing lights will creates heavy burden disposal on local governing authority.

To reduce this disposal load and its burden, this study attempts to develop efficient and economical solution for disposing the household, institutional, organizational electric lights. Also, the proper disposals of electric lights enable to protect the environment from contamination due to mercury. The automobile garages and electrical mechanics will also include in the coming stage of this research.

Acknowledgement. Corresponding author is thankful to Dr. Sunil Bhimrao Thakare the Principal of the engineering institute for his continuous motivation and guidance to take initiative to address the contamination of land and water sources by disposals of domestic and vehicular lights and his encouragement to develop systematic disposal system these lamps in Pune Region.

References

1. Udhayakumar, T.: Disposal methods of e-waste in India Survey conducted in Chennai. Int. J. Appl. Environ. Sci. **12**(3), 505–512 (2017). ISSN 0973-6077
2. Guidelines for environmentally sound mercury management in fluorescent lamp sector (CPCB), November 2008
3. https://www.electrical4u.com/electric-lamp-types-of-electric-lamp/
4. http://www.historyoflamps.com/lamp-history/history-of-electric-lamps/
5. Down to Earth- https://www.downtoearth.org.in/coverage/let-there-be-cfl-2908
6. Electric lamp and component manufacturers association of India (ELCOMA) (2017–2019)
7. Tunsu, C., et. al.: Sustainable processes development for recycling of fluorescent phosphorous powders – rare earths and mercury separation (2011)

A Reliable Solution for Treatment of River Water Using Hydrodynamic Cavitation in Combination with Chemical Additives

Sanket Tithe and Amravati Gode[(✉)]

School of Chemical Engineering, MIT Academy of Engineering, Alandi,
Pune 412105, India
sanketthite17@gmail.com, atamrakar@chem.maepune.ac.in

Abstract. Hydrodynamic cavitation treatment of Indrayani river water was carried out using hydrodynamic cavitation (HC) and in combination with soda lime. An in-house cavitation setup has been developed. As a cavitating device orifice plates of different geometries-single orifice plate diameter (4 mm), multiple orifice plates-6 holes diameter (1 mm) and 9 holes diameter (0.8 mm) were used in HC reactor. Tests were carried out with a sample size of (wastewater) 50 L, for 2 h duration. The effect of process parameters such as inlet pressure, cavitation number (CV), flow rates and soda lime loading on the extent of reduction of COD, TDS, permanent hardness, pH, color were studied. Observation shows that as cavitation no (CV) decreases from 4.25 to 0.55 percent removal for COD reaches up to 71.4%. At the same time percentage reductions for TDS and permanent hardness reaches up to 85% and 82.5% respectively. For 9 hole orifice plate (CV 0.55) there was the maximum percentage reduction of 71.4% COD, 85% TDS, 82.5% removal of permanent hardness at a pressure of 2 atm and flow rate 3 m^3/h. For 9 holes plate results shows that the COD, TDS, and permanent hardness levels reduced to the limit of drinkable water. With HC alone pH goes up to 6.7. Experiment with soda lime in addition to HC shows a further reduction in permanent hardness up to 87% but for COD results were the same as a 72% reduction. Hence hydrodynamic cavitation represents a reliable and cost-effective way for water treatment with the very less initial cost.

Keywords: Hydrodynamic cavitation · Cavitation number wastewater · Chemical additives · Hardness

1 Introduction

Due to increasing awareness about the environment and more stringent environmental regulations, treatment of industrial and domestic wastewater is a key aspect of current research. There is a lot of work done for the testing and developing new techniques for treating the wastewater as an individual method or as an added part to typical biotic and chemical methods. The composition of wastewater varies from place to place. Sometimes industrial wastes also mix with sewage. The type of process used in the treatment of wastewater thus depends upon its characteristics and the desired quality of water

© Springer Nature Singapore Pte Ltd. 2020
V. K. Gunjan et al. (Eds.): *ICRRM 2019 – System Reliability, Quality Control,
Safety, Maintenance and Management*, pp. 145–152, 2020.
https://doi.org/10.1007/978-981-13-8507-0_23

after treatment [1, 2]. Cavitation takes place once the pressure of liquid rapidly beads. In the furthermost elementary sense, cavitation happens once the pressure of fluid develops the identical with owned vapor pressure. The development of the vapor bubble is similar steaming but without accumulation of heat. Required vapor bubble is then rapidly compacted by the adjacent fluid pressure, as this process ensues right quickly which generates several confined energy [2]. As an when the required bubble is compacted adequately it reliefs the said energy inform the way of heat, sound, and light. The main issue with the cavitation is with pumping methods, but it can be avoided with appropriate pump sizing, pipe design, and attention of filters and strainers [2, 3]. Cavitation is generally distributed into two phases of comportment: transient and non-inertial cavitation. In an inertial cavitation development, a bubble in a liquid quickly breakdowns, with creating a shockwave; such type of cavitation takes place in nature in the incursions and in the vascular tissues of the plants [4–7]. Non-inertial cavitation is the method wherein a bubble in a liquid is enforced to fluctuate in size or shape because of the particular arrangement of energy involvement, for instance, an aural pitch. This type of cavitation is regularly engaged in ultrasonic cleaning bath-houses and can correspondingly be experiential in propellers, pumps, etc. The persistence of wastewater action is to lessen organic and inorganic constituents; supplements toxic ingredients exterminate pathogenic creatures etc. So as to the superiority of cleared water is enhanced to chance the allowable near of water to be settled in certain water form [8–13]. The current work is deals with the study of various diameters of plates in hydrodynamic cavitation with chemical additives for the treatment of river wastewater.

2 Methods

2.1 Hydrodynamic Cavitation Treatment

The procedure of gurgle group and the following development and failure of the cavitation bubbles consequences in very in height energy thicknesses and in actual all local temperatures and confined pressures at the superficial of the bubbles aimed at a short time. Measured cavitation can be used to improve chemical reactions or spread convinced unpredicted responses since free radicals are produced in the procedure due to disassociation of vapors surrounded in the cavitation bubbles [14, 15]. Regulatory the cavitation movements in liquids can be accomplished only by progressing the mathematical substance of the cavitation developments. These developments are demonstrated in dissimilar habits, the greatest public ones and capable of controller actuality bubble cavitation and wonderful cavitation [16]. Theory of throbs and constancy of lengthened axis symmetric cavities, etc. and in Dimensionality and resemblance approaches in the difficulties of the hydromechanics of containers [17].

2.2 Experimentation

Hydrodynamic cavitation was generated using an in-house constructed experimental model in which a feed vessel tank with a maximum capacity of 50 L and operates in a re-circulation mode. Effluent from the feed tank is pumped using a triplex plunger pump with a maximum discharge pressure of 4,500 psi and passes through an orifice plate which design in such a way that it creates maximum cavitations followed by a catalyst bed (optional) and finally back to the feed tank. An external heat exchanger (optional) unit is also provided to control the temperature in the feed vessel tank, which is necessary as cavitation results in the production of heat thereby increasing the temperature of the effluent stream. The diameter of pipes inlet to an outlet from the cavitation orifice plate is also designed to provide conditions to have maximum cavitations. The experiments were performed in the reactor of capacity 50 L in which effluent was lifted and circulate by the pump of capacity 0.5 hp for different intervals of time and chlorine was used, an oxidizing agent. The sample was kept for an inert condition for 1 h for the settlement of the precipitates. All experiments were carried out in batch mode and in normal atmospheric temperature at 28 °C. Several sets of experiments were carried out to check the optimum range of time. In the particular experiment, the local tap water is tested for the removal of temporary hardness and the local river (Indrayani) water is treated for the reduction of COD. No extra chemicals are added for pH adjustment, other treatment of wastewater in both cases. But in the case of Permanent hardness removal of tap water calculated amount lime and soda ash is used. The treated wastewater is then checked for the strength of wastewater.

In the above experimental setup (Fig. 1) there are three cavitation devices used with the dimensions of an orifice plate are summarized in Table 1.

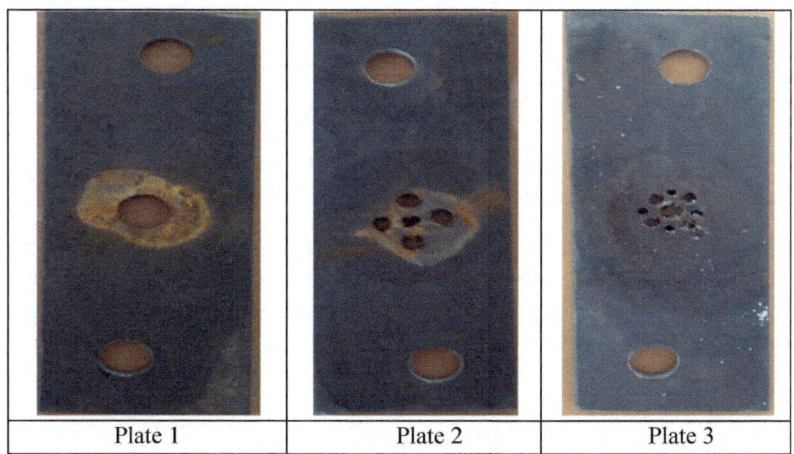

| Plate 1 | Plate 2 | Plate 3 |

Fig. 1. Various pore size plates in hydrodynamic cavitation device

Figure 1 shows a various pore size uses in the orifice plate which can be used in the hydrodynamic cavitation device for the treatment of wastewater. The pore size plays an important role in generating the bubbles with the desired intensity and removing the harmful components from the wastewater.

Table 1. Dimensions of an orifice plate

Plate number	No. of hole	Diameter of orifice (mm)	Total area of orifice (mm^2)	Total perimeter of orifice (mm)	Cavitation number
Plate 1	1	4	12.6	12.6	4.25
Plate 2	6	0.8	3.1	15.1	1.15
Plate 3	9	1	6.28	25.1	0.55

3 Result Discussion

3.1 Temporary Hardness Removal

Temporary hardness is caused by the chemical which unconcerned by the simple boiling process. So the 50 L of a river water sample is treated in hydrodynamic cavitation experimental setup and hardness of sample is measured in every 15 min. Also, the 50 ml of the initial sample is boiled for 15 min. to get the actual temporary hardness of the sample. Finally, the results of Time vs Hardness (ppm of CaCO$_3$ are summarized (Table 2).

Table 2. Comparison between the three plates

Plates	Percentage removal of the hardness of river water	Percentage removal of cod of river water
Plate 1	75.6	68.5
Plate 2	80.0	70.1
Plate 3	82.5	71.4

3.2 Permanent Hardness Removal

The hardness is characteristically produced by the occurrence in water of magnesium sulfates and calcium sulfate that do not experience rainfall at enlarged temperatures. The 50-liter river water sample is collected from the Indrayani River and treated in hydrodynamic cavitation set up with the addition of any chemical to it. The permanent hardness of testing water is checked after every 15 min. The hardness removal with respect to time in case of hydrodynamic cavitation experiment can be related as percentage hardness removal with respect to time.

3.3 COD Reduction of the River (Indrayani) Water

The chemical oxygen demand (COD) test is usually used to circuitously quantity the number of organic compounds in water. Most claims of COD regulate the number of organic pollutants found in surface water (e.g. lakes and rivers) or wastewater, manufacture COD a valuable amount of water quality. The COD of testing water is checked after every 15 min and the result. The COD removal with respect to time in case of hydrodynamic cavitation experiment can be related to percentage COD removal with respect to time.

3.4   Comparison Between the Plates

From the above study, we compare the result of three different plates. Compare the percentage removal of hardness and cod Indrayani river water of three plates. From the Table 2, plate 1 percentage removal of hardness is 75.6%, from plate 2% removal is 80% and from plate 3 is 82.5%. Also the results obtained as percentage removal of cod of river water are summarized in the Table 2.

   According to the above experimental data, it originated that the plate 3 was most suitable plate for water treatment by hydrodynamic cavitation method.

3.5   Study of the Different Flow Rate at Plate 3 (Flow Rate 3 m³/h)

Temporary Hardness Removal
By the definition of temporary hardness, it is caused by the chemical which can be unconcerned by the simple boiling process. So the 50 L of a river water sample is treated in hydrodynamic cavitation experimental setup and hardness of sample is measured in every 15 min. Also, the 50 ml of the initial sample is boiled for 15 min. to get the actual temporary hardness of the sample. Finally, the graph of time vs. hardness (ppm of $CaCO_3$) is plotted.

Permanent Hardness Removal
The 50-liter river water sample is collected from the Indrayani River and treated in hydrodynamic cavitation set up with the addition of any chemical to it. The permanent hardness of testing water is checked after every 15 min. The hardness removal with respect to time in case of hydrodynamic cavitation experiment can be related as percentage Hardness removal with respect to time.

COD Reduction of the River (Indrayani) Water
The chemical oxygen demand (COD) test is usually used to circuitously amount the number of organic compounds in water. Most requests of COD control the number of organic pollutants found in surface water (e.g. lakes and rivers) or wastewater, creation COD a beneficial amount of water quality. The COD of testing water is checked after every 15 min. The COD removal with respect to time in case of hydrodynamic cavitation experiment can be related to percentage COD removal with respect to time.

A Comparison Study of Different Flow Rate
From the above study, we compare the result of three different flow rates for optimum plate i.e. plate 3. Compare the percentage removal of hardness and COD of Indrayani River water of three different flow rate. The results are summarized in Table 3 for the comparison of percentage removal of hardness of river water. It was observed that from flow rate, 1% removal of hardness is 82.5%, from flow rate 2% removal is 50.00% and from flow rate 3 is 43.2%. Also the comparison of percentage removal of COD of river water is summarized. According to the above experimental data, it was found that the

flow rate 1 i.e. 3 m^3/h was optimum flow rate for plate 1 for water treatment by hydrodynamic cavitation method.

Table 3. Comparison between the three plates

Flow rates (m^3/h)	Percentage removal of the hardness of river water	Percentage removal of cod of river water
3	82.5	71.4
2	50.00	55.9
1.5	43.2	51.0

With Chemical Additive: - Soda Lime

In this development calcium and magnesium ions are triggered by addition of lime Ca (OH)$_2$ and soda ash (Na$_2$CO$_3$).

Lime addition

$$CO_2. + Ca(OH)_2 \rightarrow CaCO_3 + H_2O \tag{1}$$

$$Ca(HCO_3)_2 + Ca(OH)_2 \rightarrow 2CaCO_3 + 2H_2O \tag{2}$$

$$Mg(HCO_3)_2 + Ca(OH)_2 \rightarrow CaCO_3 + MgCO_3 + 2H_2OS \tag{3}$$

$$MgCO_3 + Ca(OH)_2 \rightarrow CaCO_3 + Mg(OH)_2 \tag{4}$$

Lime and soda ash addition

$$MgSO_4 + Ca(OH)_2 \rightarrow Mg(OH)_2 + CaSO_4 \tag{5}$$

$$CaSO_4 + Na_2CO_3 \rightarrow CaCO_3 + Na_2SO_4 \tag{6}$$

Percentage Removal of the Hardness of River Water

By the appropriate plate diameter in the hydrodynamic cavitation assemble the percentage of removal of the chemical oxygen demand (COD) (Fig. 2) as well as the hardness (Fig. 3) in the river wastewater get increases because of the acoustic cavitation phenomenon. Hydrodynamic cavitation is a capable relevance in wastewater treatment because of its easy reactor design and ability in the large-scale process. General studies with the basic device of water treatment, contaminant degradation and bubble dynamics models united with chemical additives reactions are assessed and it was found that by using chemical additive i.e. soda lime we can reduce the hardness up to 87% and COD of river water 72% respectively.

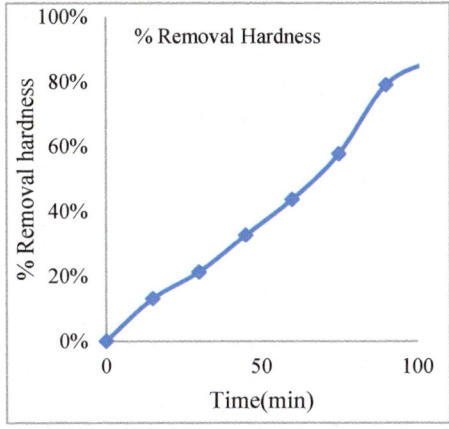

Fig. 2. Percentage removal hardness vs time

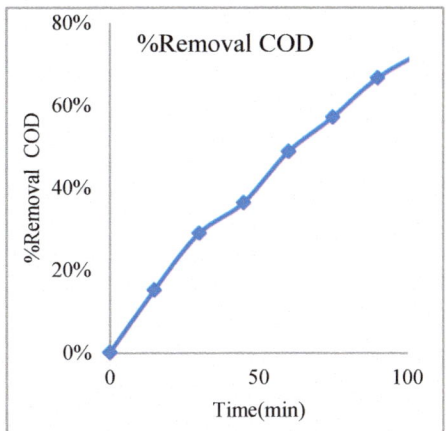

Fig. 3. Percentage removal hardness vs time

4 Conclusion

It has been observed that the hydrodynamic cavitation reduces hardness and COD of wastewater to a significant amount. The rate of reduction of hardness and COD is relative to the quantity of cavitation generated which depends upon the pressure of wastewater at the inlet of cavitation equipment and structure of cavitation equipment. It has been proved from the result that at the end of 120 min., the hydrodynamic cavitation equipment reduces the strength of wastewater to a significant optimum level. So we can conclude from the result as hydrodynamic cavitation can be rummage-sale for the treated wastewater with or without the addition of another chemical.

References

1. Bagal, M.V., Gogate, P.R.: Wastewater treatment using hybrid treatment schemes based on cavitation and Fenton chemistry: a review. Ultrason. Sonochem. **21**(1), 1–14 (2014)
2. Lopez, J.L.C., Reina, A.C., Gomez, E.O.: Bioreactor for pesticides degradation. Sep. Sci. Technol. **45**(11), 1571–1578 (2010)
3. Celalettin, O., Serkan, S., Mustafa, O.: Treatment of pesticide wastewater by physicochemical and fenton processes. Asian J. Chem. **20**(5), 3795–3804 (2008)
4. Chakinala, G.A., Gogate, P.R., Burgess, A.E., Bremner, D.H.: Industrial wastewater treatment using hydrodynamic cavitation and heterogeneous advanced Fenton processing. Chem. Eng. J. **152**(2-3), 498–502 (2009)
5. Chand, S.K.: Disintegration of sludge using ozone-hydrodynamic cavitation. Bangladesh University Proceedings (2008)
6. Botha, C.J.: A study of hydrodynamic cavitation as a method of water treatment. M.Sc. in Engineering, University of Natal Proceedings (2003)
7. Cairns, W.L.: Advances in UV disinfection technology for treatment of low-quality wastewater. In: For Australian Water and Wastewater Association (AWWA) 16th Federal Convention, Sydney, Australia, 2–6 April 2005 (2005)

8. Chang, J.C.H.: UV inactivation of pathogenic and indicator micro-organisms. Appl. Environ. Microbiol. **49**(6), 1361–1365 (2003)
9. Lehman, A.F., Young, J.O.: Experimental investigations of incipient and desinent cavitation. J. Basic Eng. **86**(2), 275–281 (2000)
10. Montgomery, J.M.: Water Treatment - Principles and Design, pp. 265–274. Wiley, New York (1985)
11. Neis, U.: Ultrasound in water, wastewater, and sludge treatment. Water **21**, 36–59 (2000)
12. Neppiras, E.A., Hughes, D.E.: Some experiments on the disintegration of yeast by high-intensity ultrasound. Biotechnol. Bioeng. **6**(3), 247–270 (2001)
13. Ransome, M.E.: Effect of disinfectants on the viability of cryptosporidium parvum oocysts. Water Supply **11**, 131–142 (2001)
14. Ransome, M.E.: Practical implications of disinfection. Water Supply **13**(2), 131–142 (2005)
15. Rice, E.W., Hoff, J.C.: Inactivation of giardia lamblia cysts by ultraviolet irradiation. Appl. Environ. Microbiol. **42**(3), 547–846 (2005)
16. Warne, S.: Energetic growth in UV sterilization. Ultra-Pure Water, World Water, pp. 38–40 (2002)
17. Winship, S.: Evaluation of different methods to produce free radicals for the oxidation of organic molecules in industrial effluents and potable water with reference to Cav-Ox. WRC Report No 388/1/99, July 1995

Stress and Deflection Analysis of Orthogrid and Isogrid Structure

Aniket S. Umap[1][(✉)], Vijay S. Pisal[1], Ghanshyam G. Rathod[1],
Ganesh R. Sonawane[1], R. R. Arakerimath[1], and Sajal Roy[2]

[1] Department of Mechanical Engineering, GHRCEM, Pune, India
aniketumap155@gmail.com
[2] R&DE (DRDO), Dighi, Pune, India

Abstract. A grid structures are the shell like structures such as orthogrid, isogrid which supports the skin of any structure. When made up with composite materials Grid structures find very good application in aerospace field. The properties of the skin can be uniformly distributed, thickness of the skin can be reduced which intern reduces the total weight of the structure. For the present skin stiffened structures orthogrid and isogrid, analysis is carried out with same skin and ribs dimensions, which gives the less stress and deflection in isogrid as compared to orthogrid but weight of isogrid is more. Inorder to get same weight of both structure skin thickness of isogrid is reduced and analysis is carried out which still gives less stress and deflection in isogrid structure. Therefore presently composite grid structure analysis is conducted to know the effectiveness of the isogrid and orthogrid structures in skin stiffening applications. After analysis it is found that isogrid is better than orthogrid for same weight.

Keywords: Grid structure · Isogrid · Orthogrid · Stress · Deflection

1 Introduction

1.1 Grid Structures

Grid structures are the shell like structures, which supports the skin of any structure.

1.1.1 Types of Grid Structures

There are several types of standard grid structures. Important among them are as follows:

- Grid structures with ribs running in four directions are referred to as quadri-directional grids.
- Grid structures with ribs that are in three directions are referred to as tri-directional grids. An Iso-grid is a special case of tri-directional grid structure in which the ribs form an array of equilateral triangles.
- Grid structures with ribs drawn in only two directions are referred to as angle grids and if the two directions are orthogonal then this structure is referred to as orthogrid.

© Springer Nature Singapore Pte Ltd. 2020
V. K. Gunjan et al. (Eds.): *ICRRM 2019 – System Reliability, Quality Control, Safety, Maintenance and Management*, pp. 153–158, 2020.
https://doi.org/10.1007/978-981-13-8507-0_24

1.1.2   Advantages of Grid Structures

- As all ribs are made of continuous and unidirectional fibers, they are intrinsically stiff, strong and tough.
- Composite grid structures are damage tolerant.
- Grid structures are open structures and thus easy to inspect or repair.

1.1.3   Application of Grid Structures

Grid structures are extensively used in aerospace, automobile and in civil structural applications. The grid structures consist of inherent resistance to impact damage, delimitation and crack propagation. Grid structure behavior study is inescapable, before implementation. Since the aerospace structures are subjected to combined loading situations, a proper study must be carried out for the grid structure model but not under single load case, but as multi-directional surface in failure space, which is termed as failure envelope.

1.2   Objectives of the Present Work

In the present work orthogrid and isogrid concepts will be used for skin stiffening applications. In the first step of our project a rectangular panel or plate is designed with Orthogrid and isogrid skin stiffener for some desired load conditions with same skin thickness. The designed plates are analyzed for deflection and stress using finite element analysis (FEA) software such as Ansys. In the second step the skin thickness of ortho-grids is changed to get the same weight of both the plates. Conclusions will be derived related to the deflection, stress distribution and stiffness to weight ratio of the designed structures.

2   FEA Analysis and Comparative Study of ISOGRID and ORTHOGRID

2.1   Material Properties for ISOGRID and ORTHOGRID

1. Material: Carbon composite
2. Transverse Tensile Strength: 600 MPa
3. Longitudinal Compressive Strength: 570 MPa
4. Density: $1.6 * 10^{(-6)}$ kg/mm^3.

2.2   Skin with Ortho-Grid

Present analysis carried out by using Ansys workbech consist of carbon composite sqaure plate of side 500 mm with skin thikness of 3 mm which support ortho-grid structure with rib thickness and rib depth are taken as 3 mm and 25 mm respectively (Weight: −16.48 N).

Load Case: Two ends fixed with transverse loading boundary conditions.

Two opposite ends that are of length 500 mm are fixed and a transverse load of 100 KN is applied on the skin. Stress and deflection find out are as below (Fig. 1).

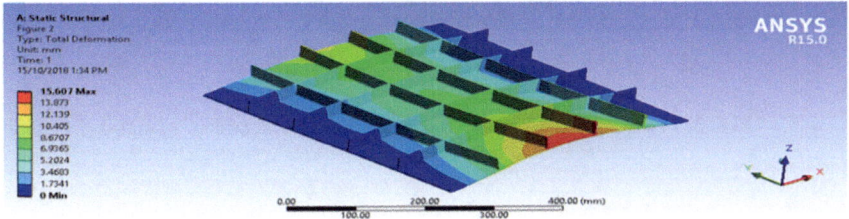

Fig. 1. Two ends fixed with transverse load (deflection) – Orthogrid

Result:
After the analysis it is found that the maximum deflection is 15.607 mm (Fig. 2).

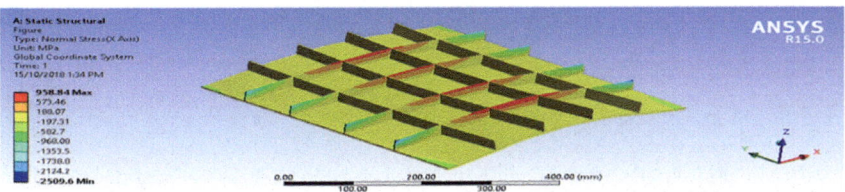

Fig. 2. Two ends fixed with transverse load (Stress) – Orthogrid

Result:
After the analysis a maximum stress of 958.84 N/mm² is observed at the center of the plate. Near the fixed ends the stress development is minimum.

2.3 Skin with Iso-Grid

Present analysis carried out using ansys workbench consist of carbon composite sqaure plate of side 500 mm with skin thikness of 3 mm which support iso-grid structure with rib thickness and rib depth are taken as 3 mm and 25 mm respectively (Weight: −18.67 N).

Load Case: Two ends fixed with transverse loading boundary conditions.

Two opposite ends that are of length 500 mm are fixed and a transverse load of 100 KN is applied on the skin. Stress and deflection find out are as below (Fig. 3).

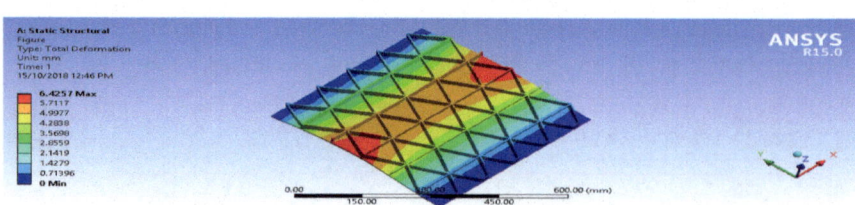

Fig. 3. Two ends fixed with transverse load (deflection) – Isogrid

Result:

After the analysis it is found that the maximum deflection is at the center of the plate and its magnitude is 6.4257 mm (Fig. 4).

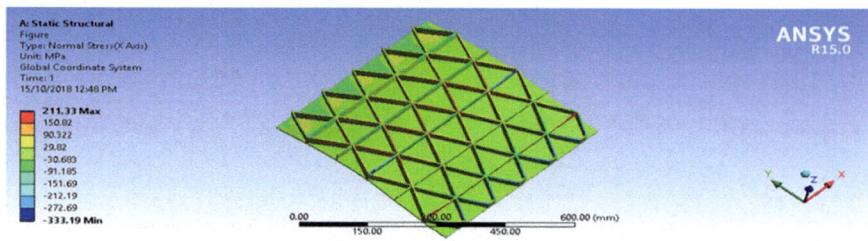

Fig. 4. Two ends fixed with transverse load (Stress) – Isogrid

Result:

After the analysis a maximum stress of 211.33 N/mm^2 is observed at the center of the plate. Near the fixed ends the stress development is minimum.

Reduced thickness of Iso-grid plate.

Now for the same amount of load the skin and rib dimensions for the Iso-grids structure are reduced to 2.64 mm to reduce the weight equal to orthogrid and the analysis is carried out.

Load Case: Two ends fixed with transverse loading boundary conditions (Fig. 5).

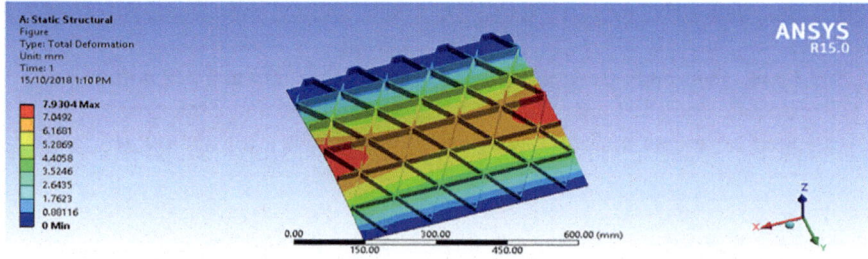

Fig. 5. Two ends fixed with transverse load (deflection) – modified Isogrid

Result:

After the analysis it is found that the maximum deflection is at the center of the plate and its magnitude is 7.9364 mm (Fig. 6).

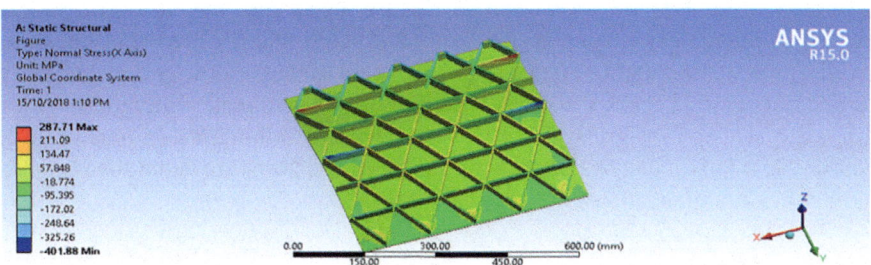

Fig. 6. Two ends fixed with transverse load (Stress) – modified Isogrid

Result:

After the analysis a maximum stress of 287.71 N/mm^2 is observed at the center of the plate. Near the fixed ends the stress development is minimum.

Results derived from the analysis are tabulated below (Tables 1 and 2).

Table 1. Analysis results for same skin thickness

Boundary condition	Ortho-grid structure	Iso-grid Structure	%Error
1. Weight	16.48 N	18.67 N	11.73
2. Two end fixed transverse load			
Deflection	15.607 mm	6.4257 mm	58.82
Stress	998.84 N/mm^2	211.33 N/mm^2	78.84
Stiffness	6407.38 N/mm	15562.5 N/mm	58.52

Table 2. Analysis results for reduced skin thickness of isogrid

Boundary condition	Ortho-grid structure	Iso-grid Structure	%Error
1. Weight	16.48 N	16.48 N	0
2. Two end fixed transverse load			
Deflection	15.607 mm	7.9346 mm	49.16
Stress	998.84 N/mm^2	287.71 N/mm^2	71.19
Stiffness	6407.38 N/mm	12600.17 N/mm	49.14

3 Conclusion

In the present work the effect of orth-grid and iso-grid on the composite skin structure is studied through finite element analysis. The conclusions made from the analysis are also discussed.

From the above results it can be observed that for the same skin thickness under transverse loading deflection in the iso-grid structure is 58.82% less than the orth-grid. Structural stiffness is enhanced by 58.52% in iso-grid. Also the maximum stress induced in the iso-grid plate is 78.84% lower than the ortho-grid plate. However the weight of iso-grid structure is found to be 11.73% higher than the ortho-grid structure. Therefore to reduce the weight of the iso-grid structure the ribs and skin thickness is reduced by. Now the weight of iso-grid plate is same as that of orth-grid plate. Analysis is carried out again to check the performance of the iso-grid plate with reduced weight. In this case also the iso-grids perform better than the ortho-grids in transverse load condition. So finally it can be conclude that iso-grid structures are better than ortho-grid in skin stiffening applications.

References

1. Hybrechts, S.M., Hahn, S.E., et al.: Analysis and behaviour of grid structure. Stanford Univercity, Stanford, CA (2007)
2. Khan, A.A., Hasham, H., et al.: A survey of recent development in optimization of ISOGRID. J. Space Technol. 5(1), 103–115 (2015)
3. Ghadi, N., Mathikalli, A., et al.: Design and FE analysis of composite grid structure for skin stiffening application I. RJET 4(8) (2017)
4. Antony, A., Resni, S.S., et al.: Effect of RIB ORIENTATION in isogrid structure: aerospace application. IJSTE 3(11) (2017)
5. Huybrechts, S., Meink, T.E., et al.: Advanced Grid Stiffned Structure for next Generation launch vehicle. Stanford University, Stanford, CA (2015)
6. Baker, D.J., Ambur, D.R., et al.: Optimal Design and Damage Tolerance Verification of an Isogrid Structure for Helicopter Application. American Institute of Aeronautic and Astronautic (2008)
7. Wegner, P.M., Higgins, J.E., et al.: Application of Advance Grid. Stiffened Structure Technology to the Minotapur payload Fairing. American Institute of Aeronautic and Astonautic
8. Marchetti, M., Sorrentino, L., et al.: Design and manufacturing of an isogrid structure in composite material: numerical and experimental result. Compos. Struct. 143, 189–201 (2016)
9. Kumar, S., Clint, J., et al.: Buckling analysis on aircraft fuselage structure skin. IJEDR 2(4) (2014)
10. Changsheng, R.F., Gan, C., et al.: Experimental Investigation of Energy Absorption in Grid Stiffened Composite Structure under Transeverse Loading. Wayne State University, Detroit (2005)

Modelling and Analysis of IC Engine Piston with Composite Material (AlSi17Cu5MgNi)

Shubham Jog, Kevin Anthony, Manasi Bhoinkar, Komal Kadam, and Mahesh M. Patil[✉]

Department of Mechanical, MGMCET, Navi Mumbai, India
maheshpatil.1646@gmail.com

Abstract. The Study of this research is to carry out how the Hypereutectic alloys can be used as a piston material rather than commonly used alloys of Aluminium, Cast Iron, AlSiC, Al2O3, etc. We carried out Structural and Steady Thermal Analysis on ANSYS to determine the properties of AlSi17Cu5MgNi (Hypereutectic alloy) which exhibits high performance durability, toughness and can be used in high performing engines, where the piston undergoes continuous dynamic loads and high stresses. Though this piston has not comparison in strong, but hypereutectic pistons are made as an ideal choice of selection for the engine producing power between 600HP to 700HP based on the application of use. In further technological developments, the new material of is made by the composition of aluminium i.e. the formation of meatal matrix composite on the basis of aluminium and production is done with the help of power metallurgy. Some other constituents are added to reduce the weight such as carbon and magnesium. The current research on piston material like cast and forged aluminium alloys provides better potential for optimization and plays a vital role in upcoming years.

Keywords: Brinell hardness · Manufacturing · Hypereutectic alloy · Temperature · Efficient

1 Introduction

In an automobile engines, the conversion of thermal energy to mechanical energy takes place with the help of piston. It is an important part in reciprocating engines and in other applications like pumps. Piston is connected to the crankshaft by interlinking with the connecting rod. When the air fuel mixture burn in combustion chamber the gas exerts pressure on the piston crown and piston transmit this force to the crankshaft converting reciprocating action into rotary motion. The modification of piston is necessary to increase the surface resistance of piston against thermal and mechanical stresses. At present, researchers are attracted in downsizing of the engine, which will help to reduce the emission of pollutants and consumption of less fuel. On the other way pressure boosters are used nowadays to increase the output of engine thus it causes the engine to produce high displacement and high stresses and the gas produced during the combustion causes increase in thermal stresses on the piston surfaces which leads to piston fatigue or failure of piston material.

© Springer Nature Singapore Pte Ltd. 2020
V. K. Gunjan et al. (Eds.): *ICRRM 2019 – System Reliability, Quality Control, Safety, Maintenance and Management*, pp. 159–169, 2020.
https://doi.org/10.1007/978-981-13-8507-0_25

Observing the continuous advancement for piston in combustion engines, the piston is developed by using certain parameters like reducing the weight of piston, rise in mechanical efficiency and capacity to sustain high thermal loads, low coefficient of friction and hence, rectified the scuffing resistance, etc. Also the other factors like reliability, durability, less oil consumption and reduced noise level which must be accounted. The objectives can be managed by using combination of such aluminium metal matrix composites, furnished piston designs and applying innovative idea.

2 Methodology

2.1 3D Modelling of Piston Design

The piston is designed according to proper design considerations. The three dimensional computer graphics deals with the generation of three dimensional model of a Piston using three dimensional software Autodesk Inventor. From the geometric modelling, the detail two dimensional drawings of the Piston can be created automatically (Fig. 1).

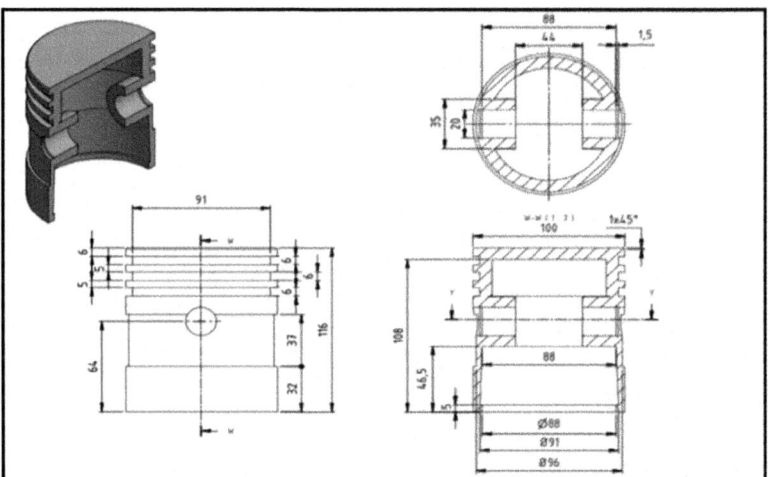

Fig. 1. Modelling of piston

2.2 Design Considerations

The piston that we are going to design should withstand a maximum temperature of 750–800 °C in the IC Engine and various other forces so we have certain design considerations of piston for the use in ICE. They are as follows:

1. The piston should have the ability to withstand gas Pressure, inertia and impulse.
2. Capable of dispersing the heat of combustion and avoid wear and tear due to distortion of heat.
3. It must be oil and gas sealant.

4. Sufficient bearing area to work for large number of reciprocating cycles.
5. Lower the weight so the piston travel at high speed.
6. Noiseless and smooth operation.
7. It should have sufficient support for piston pin.
8. The piston is constructed such that it can withstand mechanical and thermal distortion (Fig. 2).

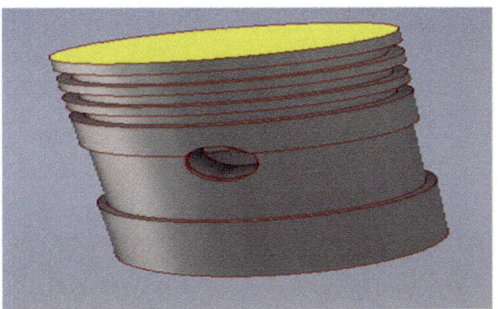

Fig. 2. Overview of piston model

3 Material Selection

The hypereutectic piston is also a cast piston but with an ideal amount of silicon to produce a much stronger version of the standard cast piston. The silicon material is added up to 12% to 19% to lower the value of coefficient of thermal expansion, increase hardness, strength, wear resistance properties and optimisation in weight. Although, the other general mechanical properties are basically improved but also the thermal fatigue failure is enhanced. The following are the materials commonly used for piston has been chosen to compare their result with our selected Hypereutectic alloy (AlSi17Cu5MgNi). Depending the various aspects of the materials like thermal conductivity, Coefficient of thermal expansion, Density Young's modulus, UTS and Brinell Hardness as shown in comparison below therefore we chose the material (**AlSi17Cu5MgNi**) for our piston (Fig. 3).

3.1 Comparison of Materials

Some other Materials are commonly used are as follows:

Material	Thermal conductivity (W/mK)	Coefficient of thermal expansion (μm/m − K)	Density (kg/m^3)	Young's modulus (Gpa)	Ultimate tensile strength (Mpa)	Brinell hardness (BHN)
Aluminium	205	21	2712	11	200–400	80
Cast Iron	79.5	10.8	6800–7800	14	160–450	60
AlSiC	170	20	3000	20	212	100 ·

(continued)

(*continued*)

Material	Thermal conductivity (W/mK)	Coefficient of thermal expansion (μm/m − K)	Density (kg/m³)	Young's modulus (Gpa)	Ultimate tensile strength (Mpa)	Brinell hardness (BHN)
Hyper-eutectic alloy	130	25–30	2700	75–80	210–300	110
Stainless steel	45	17	7400–8000	25–30	100–500	79

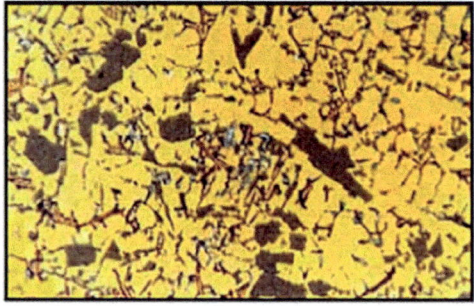

Fig. 3. Microstructure of hypereutectic alloy

3.2   Finite Element Analysis of Piston

The analysis was carried out by finite Element method and following the required steps f FEM.

1. Discretization.
2. Formation of Element and Global Matrices.
3. Formation of Global Load Vectors and Nodal. Displacement Vectors.
4. Forming an equilibrium equation.
5. Incorporation of specified boundary conditions.
6. Computation of element stress and strain.
7. Thermal Analysis.

4 FEM Analysis

Discretization

The main interest in the continuum mechanics is the deformation of a body by internal or external forces. The d(x, y, and z) deformation is expressed by the displacements. The displacement of a point in a three-dimensional elastic continuum is defined by three displacement components i.e. u(x, y, z), v(x, y, z) and w(x, y, z) respectively. It is the in direction of 3 Co-ordinates (x, y, z) such that.

$$\vec{d}(x,y,z) = \left\{ \begin{array}{c} u(x,y,z) \\ v(x,y,z) \\ w(x,y,z) \end{array} \right\} \tag{1}$$

The model is first given an input in ANSYS software. The model is extracted from Autodesk Inventor 2018 and imported to the ANSYS software for further calculations (Fig. 4).

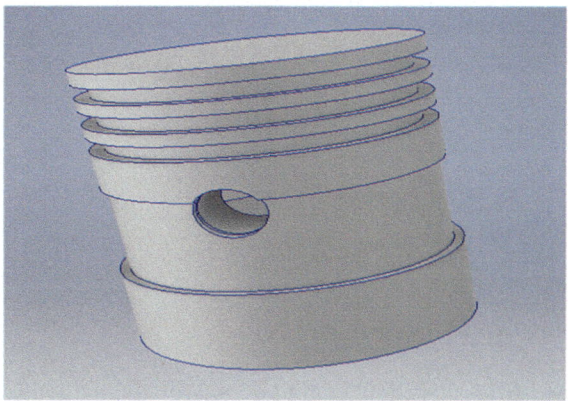

Fig. 4. Piston model

Element Type and Properties

In the Ansys program, a number of element types are given. These elements are used in this Ansys simulation. There are various characteristics of the element types and the Ansys uses these elements types and their grouping for further process to calculate the results. There are almost 100 different types or element formulations in this Ansys program. Thus the formal way an element is identified is by a name, It should have maximum 8 characters for example BEAM, consisting of a group Label (BEAM) and an unique identifying number is given to each of them. Thus, the elements are arranged in order of these numbers that can be identified easily. Thus, the element we choose or select from the library for analysis purpose is by giving input (Name of element type and its command). Thus in Ansys we have chosen the element from the pre-processor drop down menu.

Formation of Element and Global Matrices

The Global stiffness matrix has an order (n × n) where n is the degree of freedom. The main property of Global Stiffness matrix that it is a square symmetric matrix. Thus, in general the matrix is symmetric in this too. The element stiffness relation is given as [K (e)] [u (e)] = [F (e)], where K(e) is element stiffness matrix, u(e) displacement vector (nodal) and F(e) is force vector (nodal). There are many Finite element methods like Subdomain method which makes the weight function same for all the points, Galerkein method which uses coefficients of C_i (constant parameter) in y. These were direct approach methods, which is known for its simplicity to solve discrete problems. Another type is Variational Method, for example Rayleigh-Ritz Method that uses calculus of variation. Thus, ANSYS uses this differential equation to solve the analysis of piston. The meshing takes place after these methods are applied (Fig. 5).

Fig. 5. Meshing of piston

Formation of Global Load Vectors and Nodal

Displacement Vectors

In this, the direct stiffness method is used to identify the individual elements, which make up the whole structure. Each element is analysed individually after they have been disconnected from their nodes. Thus a matrix is formed. Once the individual element matrix if formed they are assembled together into the original structure. Thus construct the global stiffness matrix, Displacement vector, force vector, and once they have been constructed then, they are represented as a single matrix equation, which is solved further for solution (Fig. 6).

$$
\begin{bmatrix}
k^{(1)} & -k^{(1)} & 0 & 0 & 0 \\
-k^{(1)} & k^{(1)}+k^{(2)}+k^{(4)} & -k^{(2)} & -k^{(4)} & 0 \\
0 & -k^{(2)} & k^{(2)}+k^{(3)} & -k^{(3)} & 0 \\
0 & -k^{(4)} & -k^{(3)} & k^{(3)}+k^{(4)}+k^{(5)} & -k^{(5)} \\
0 & 0 & 0 & -k^{(5)} & k^{(5)}
\end{bmatrix}
\begin{Bmatrix} u_1 \\ u_2 \\ u_3 \\ u_4 \\ u_5 \end{Bmatrix}
=
\begin{Bmatrix} F_1 \\ F_2 \\ F_3 \\ F_4 \\ F_5 \end{Bmatrix}
\tag{2}
$$

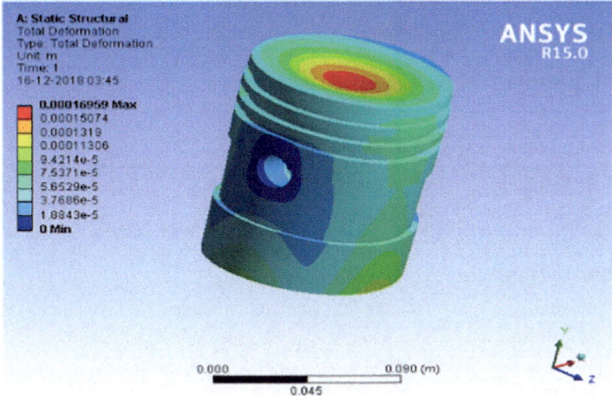

Fig. 6. Total deformation

Now, solving the matrix to find out unknowns from the system which are presented in displacements uij on global system of equations taking the following form:

$$[\mathbf{K}]\{\mathbf{u}\} = \{\mathbf{F}\} \tag{3}$$

Boundary Condition
By enforcing boundary conditions, such as temperature, pressure fixed supports etc., The equation becomes non-singular and the equation is solved for the reaction forces F and the unknown displacements $\{u1\}\{u2\}\{un\}$, for known (applied) Forces. The matrices are solved and various boundary conditions are applied to this equation and various solutions are obtained for the displacement and the forces are obtained from the boundary conditions that we have applied to it (Fig. 7).

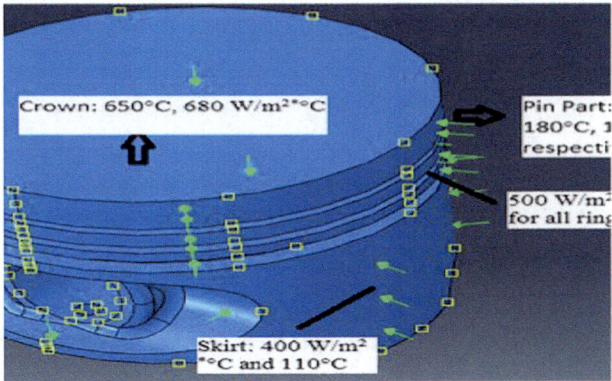

Fig. 7. Applied boundary conditions

The boundary conditions applied are shown above such that the forces and displacement are found out by these equations.

Computation of Element Stress and Strain

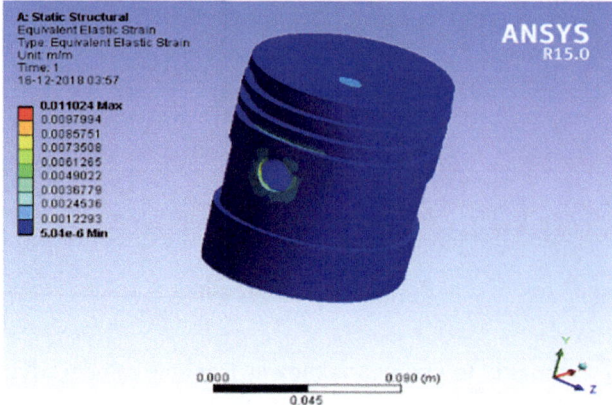

Fig. 8. Maximum strain

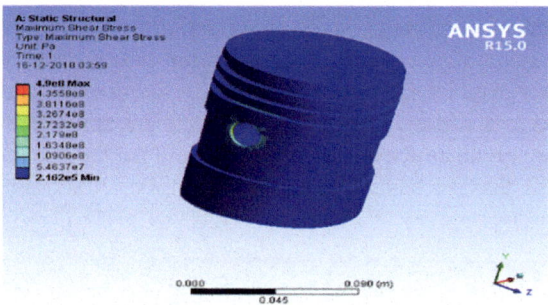

Fig. 9. Maximum stress

The maximum calculated stress and strain is shown in the above figures. The total calculated stress was to be 490 Mpa for the hypereutectic material and the piston made from the material Stainless steel and AlSiC had a maximum stress of 400 and 212 Mpa respectively. The boundary conditions that were applies not only increased the maximum stress but also the overall shear stress of the piston. The forces that were calculated resembles the piston going through a lot of pressure during the combustion process. Thus the max shear stress obtained for the hypereutectic material is much higher than any material used in piston manufacturing thus tremendously increasing its mechanical properties (Figs. 8 and 9).

Thermal Analysis

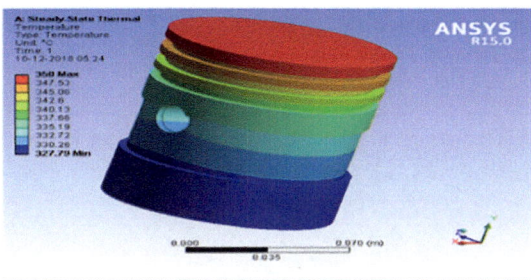

Fig. 10. Steady state temperature

A steady-state thermal analysis was calculated with the effects of steady thermal loads on a system or component. We have done a steady-state analysis before doing a transient thermal analysis, to help establish initial conditions. We have done steady-state analysis after the transient thermal analysis, after all transient effects have diminished. We have used steady-state thermal analysis to determine temperatures, thermal gradients, heat flow rates, and heat fluxes in an object that are caused by thermal loads that do not vary over time. Such loads include the following (Fig. 10):

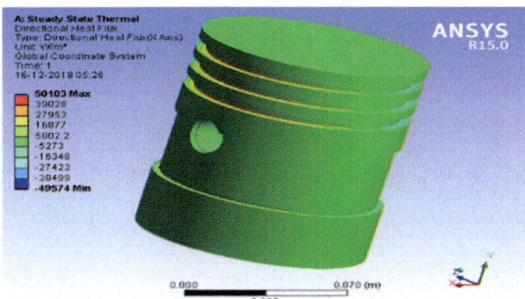

Fig. 11. Directional flux

- Convections Heat fluxes (heat flow per unit area).
- Heat generation rates (heat flow per unit volume).
- Constant temperature boundaries.

The figure shows the max steady state temperature obtained. Thus the max obtained temperature is 350 °C. With the advent of new engines the steady state temperature of normal piston elements like stainless steel and aluminium has much lower temperature. The reason for the hypereutectic material to have a greater temperature is that the use of additives like Cu, Mg, and Ni has increased its melting point as well as it has increased alpha and beta distribution in the solidus region in the TTT diagram. The use of our material at eutectic temperature tremendously has increased the total characteristic features and its properties (Structural, Mechanical and Thermal) are altered due to this reason (Figs. 11, 12, 13 and 14).

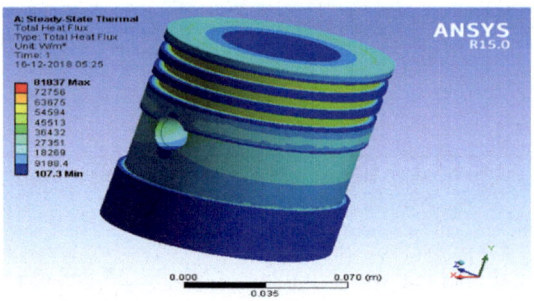

Fig. 12. Total flux

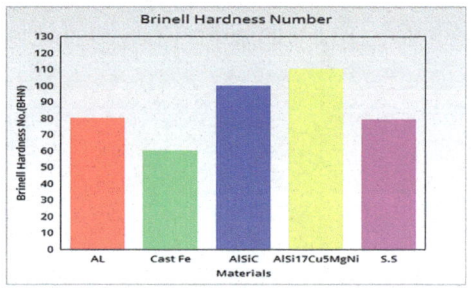

Fig. 13. Hardness of various material

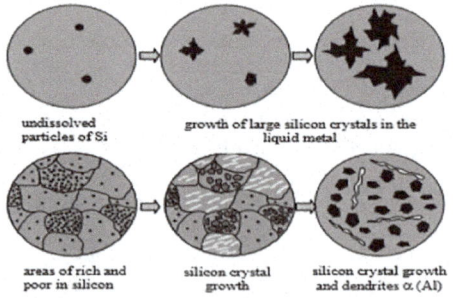

Fig. 14. Alpha and beta distribution

5   Conclusion

The design of the piston that we have made is in such a way that the flat design that we have chosen is the most efficient way for the combustion process in IC engine. With the changes in the normal element that is usually used in Automobile industry, we have selected the AL alloy which comprised of aluminium, silicon, magnesium, copper and Nickel with major percentage of aluminium and silicon present. The small additions of Mg, Cu, and Ni resulted in high melting point, high Brinell number and high ultimate tensile strength of the element. Here we performed stress analysis on aluminium piston and alloy of aluminium piston comprising of silicon, copper, magnesium, and nickel and found out that alloy of aluminium gave better results and high stresses as compared to aluminium. Similarly the results of strain analysis of the two materials showed that alloy of aluminium possessed better strain bearing capacity in comparison to aluminium. The alpha and beta constituents of Al, Si has led to a greater change in its mechanical, thermal properties which the essential backbone of the piston in IC engine.

The uses of such material resulted in not only greater melting point of the alloy but also increased thermal conductivity and increased strength to the material. Due to high temperature and high pressure in the combustion chamber normal elements used for piston results in fatigue failure, cracks on piston head etc. Thus the heating of the element resulted in high distribution alpha and beta constituents $(\alpha(Al) + \beta(Si))$ and thus resulting in increase in Brinell hardness number much higher than any other elements used for manufacturing piston. The problems of low fatigue strength and Low UTS have been identified and rectified by using this material.

Thus, this alloy has higher fatigue strength, Splitting of piston is reduced and the problem of incomplete combustion is addressed with this hypereutectic material. Thus, it increases the combustion, which in turn results to higher efficiency. The increase in efficiency is almost 5–8%, but even with only a small increase in efficiency due to much lesser price of the element, the manufacturers and consumers both have a win-win situation about price point as well as longevity of the piston.

References

1. Piątkowskia, J., Wieszałab, R.: Tribological properties of AlSi17Cu5Mg alloy modified with CuPMaster alloy with various speeds of friction, vol. 16 (2015). ISSN 1897-3310
2. Piątkowski, J., Kamiński, P.: Crystallization of AlSi17Cu5Mg alloy after time-thermal treatment, vol. 15 (2014). ISSN 1897-3310
3. Bhandari, V.B.: Design of Machine Elements Paperback, 3rd edn (2010)
4. Broutman, L.J., Chandrashekhara, K., Agarwal, B.D.: Analysis and Performance of Fiber Composites, 3rd edn. Paperback (2012)
5. Krishnan, S.B., Vallavi, M.S.A., Arunkumar, M., Haripraveen, A.: Design and analysis of an IC engine piston using composite material. Eur. J. Adv. Eng. Technol. 4(3), 209–215 (2017)
6. Shehanaz, M., Shankariah, G.: Design and Analysis of Piston Using Composite Material, vol. 6 (2017). ISSN 2319-8753
7. John, A., Mathew, J.T., Malhotra, V., Dixit, N.: Design and analysis of piston with SiC material. Int. J. Innovative Res. Sci. Technol. 1(12), 578–590 (2015). ISSN 2349-6010
8. Karl-Heinz, Z.G.: Microstructure and Wear of Materials (Tribology) (1987)

Seismic Response Control of Unsymmetrical RCC Framed Building Using Base Isolation Considering Soil Structure Interaction

Monika Jain[1(✉)] and S. S. Sanghai[2]

[1] Yeshwantrao Chavan College of Engineering, Nagpur, India
mnkjain26@gmail.com
[2] G. H. Raisoni College of Engineering, Nagpur, India
sanket.sanghai@raisoni.net

Abstract. In passive energy dissipation systems used for earthquake resistant structures, base isolation is one of the most powerful systems. For understanding the effect of base isolation system, G + 5 storey models are considered with planner asymmetry. Lead rubber isolator is used for controlling the response of building during earthquake effects and tremors. The paper deals with behavior of fixed base and isolated building in terms of natural time period, top floor acceleration, storey drift, base reaction and energy dissipation. The Non-linear Time history analysis has been performed by using FEM based software SAP2000. The paper also includes the effect of soil structure interaction on both fixed base and isolated base building. It was observed that with use of isolators, time period of structure becomes double as compared with the fixed base model. The soil parameters also plays vital role in performance of isolators.

Keywords: Lead rubber isolator · Seismic response · Soil-structure interaction

1 Introduction

Earthquake resistant building is basically to design a building which can withstand earthquake forces. As Seismic base isolation is a well-defined passive control system of earthquake for building. Earthquakes tremors are of greatest hazards; as due to these tremors there is serious damage to life as well as property also, the effect is most severe in manmade structures. To overcome these hazards caused due to earthquake waves there are so many modifications has to be done in current structural design by engineers and architects to nullify this effect. The main objective of seismic base isolation is to improve the response of building during earthquake forces i.e. during horizontal ground shaking of building. Seismic isolation provides horizontally flexible but vertically very stiff structure due to installation of isolation device between superstructure and substructure. Hence the nonlinear dynamic response of building is thus changed such that the natural fundamental time period of building is increased as compared with the fixed based building. As a result of this there is serious reduction in the response of building such as acceleration, forces and displacement in comparison with the non-isolated building.

© Springer Nature Singapore Pte Ltd. 2020
V. K. Gunjan et al. (Eds.): *ICRRM 2019 – System Reliability, Quality Control, Safety, Maintenance and Management*, pp. 170–178, 2020.
https://doi.org/10.1007/978-981-13-8507-0_26

Huge amount of literature is available on the isolated structures and their earthquake performance (Ribakov and Iskhakov [1], Monfared et al. [2], Chandak [3]). The most extensively studied base-isolation system is laminated rubber bearing (LRB) with and without a lead core (Kelly 1982, 1986; Kelly and Hodder 1982). More buildings are being built on laminated-rubber-bearing base-isolation systems all over the world [4]. Figure 1 shows c/s of lead rubber isolator and period shift of building after installation of isolator.

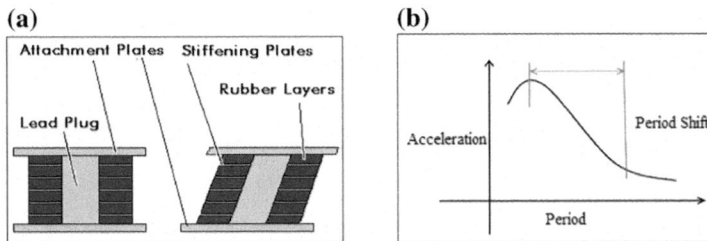

Fig. 1. C/S of (a) Lead rubber bearing [5] (b) Period shift induced by an isolator [6]

In recent years, many studies have been done on passive energy dissipation devices. The main objective is to understand the primary concept and behavior of the base isolated structures & to compare the responses of isolated building with fixed base building with the effect of SSI. The study is based on the comparison of base shear, time period acceleration, story drift member forces and energy dissipated by isolated building.

2 Mathamatical Model

A mathematical model of fixed base and isolated model is as shown in Fig. 2. The governing equation of motion for both fixed base and isolated base models are given below.

The equation of motion for the MDOF fixed base model can be written as,

$$[M]\{\ddot{u}\} + [C]\{\dot{u}\} + [K]\{u\} = -[M]\{r\}(\ddot{u}_g) \tag{1}$$

While for the MDOF isolated model can be written as

$$[M]\{\ddot{u}\} + [C]\{\dot{u}\} + [K]\{u\} + [D]\{F_b\} = -[M]\{r\}(\ddot{u}_g) \tag{2}$$

Where [M] is mass matrix, [C] is damping matrix, [K] is stiffness matrix, {u} is displacement matrix [D] is location matrix for isolator and {r} is vector of influence coefficient.

For this study, a G + 5 storey building frame is considered with Rectangular and C-shaped planner configuration. Also, it is well known that the soil beneath the structure also affect the performance of energy dissipater during earthquake [7]. Hence, in this study three types of soil conditions are also considered. The values of soil-spring stiffness are calculated by formulae given in ASCE 41-06 [8]. The Fig. 2(a) & (b) shows Rectangular and C-shaped building plan while Fig. 3 shows the elevation for both planner configurations.

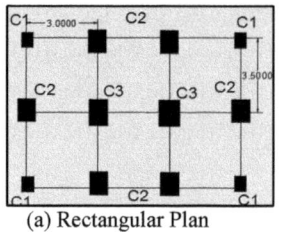 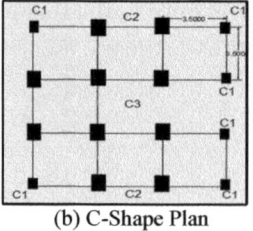 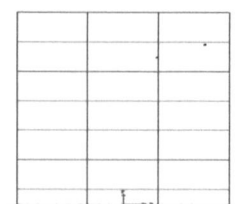

(a) Rectangular Plan   (b) C-Shape Plan

Fig. 2. Planner configuration of models   **Fig. 3.** Elevation of models

The properties of above Rectangular and C-shape building models are as given below in Table 1.

Table 1. Properties of building models

	Rectangular building	C-shape building
Mass of each storey (N.sec^2/m)	402300	435200
Stiffness of each storey (N/m)	16333.73 *10^3 N/m	16333.73 10^3 N/m
Inherent damping of structure (C)	5% for RCC structure	5% for RCC structure

For the building models considered for this study Lead Rubber Isolators are designed as per design approach discussed by Kelly and James (1993). For the modeling of this isolator in SAP2000, the link element 'Rubber Isolator' having same behavior as isolator is used. The properties of lead rubber isolator [9] are shown in Table 2.

Table 2. Properties of lead rubber isolator

Effective stiffness K_{eff} (kN/m)	830.89
Yield force F(kN)	16.24
Post elastic stiffness K_2 (kN/m)	765.67
Pre elastic stiffness K_1 (kN/m)	7656.7
Yield displacement Dy	0.002356

For plotting the step by step behavior of isolator, Non-linear Time History Analysis using SAP2000 is done. For seismic analysis three earthquake records are used as shown in Table 3.

Table 3. Earthquakes used for study

EQ	PGA	Year of occurrence
Loma Prieto	0.328 g	1989
Imperial valley	0.248 g	1940
Park40	0.141 g	1957

The effect of soil structure interaction is also been considered in this study. The stiffness of soil is calculated according to ASCE 41-06 formulas and the value of shear modules 'G' is calculated by penetration value 'N'. For the comparison of results, different parameters such as fundamental time period, Axial force, base shear, bending moment, interstory drift, acceleration & energy dissipation are studied for the fixed base and isolated building models with the effect of soil structure interaction for lead rubber isolator.

3   Results and Discussion

Using SAP2000 the analysis is carried out for rectangular and C-shape building model with the effect of soil structure interaction considering fixed base and isolated base. The natural time period for fixed based and isolated base for rectangular & C-shape building models are shown in Table 4.

Table 4. Natural time period of building models in sec

		Without SSI	With SSI		
			Hard soil	Medium soil	Soft soil
Rectangular shaped building	Fixed base	1	1.02	1.06	1.1
	Rubber isolator	2.44	2.44	2.45	2.46
C-Shaped building	Fixed base	1.83	1.84	1.89	1.96
	Rubber isolator	2.83	2.83	2.83	2.84

From the Table 4, it is clear that for rectangular as well as C-shape buildings model the time period increases compared with fixed base building after introducing isolator. The time period increases relatively higher after introducing the base isolator. When the effect of soil is considered, it is observed that, the time period of building further increases. As the soil is becoming softer, the time period increases accordingly. There is hardly any change in time period when underlying soil is of hard strata.

Table 5 shows the top floor acceleration of rectangular & C-shape models with the effect of isolator. For rectangular fixed base model the acceleration reduces from 7 m/s^2 to 2.25 m/s^2 due to installation of base isolator. After consideration of soil structure

Table 5. Top floor acceleration of building models in m/s²

			Loma Prieta		Park40		Imperial valley	
			Fixed base	Rubber isolator	Fixed base	Rubber isolator	Fixed base	Rubber isolator
Rectangular building	Without SSI		7.58	2.71	3.84	2.46	9.41	8.46
	With SSI	Hard soil	7.56	2.24	3.99	2.41	9.78	3.46
		Medium soil	7.38	2.25	4.11	2.41	10.16	2.25
		Soft soil	7.08	2.25	3.86	2.41	9.61	3.47
C-shape building	Without SSI		6.69	1.36	2.97	2.40	3.23	2.36
	With SSI	Hard soil	6.73	2.21	2.98	2.40	3.22	2.36
		Medium soil	6.78	2.21	3.04	2.40	3.17	2.36
		Soft soil	6.63	3.77	3.21	3.01	3.69	2.75

interaction the acceleration is increases from softer to harder soil. There is approximately 60% reduction in acceleration. Due to soil structure interaction the acceleration increases approximately up to 3–5% for softer soil. For C-shape building the acceleration reduces up to 35–40% after introduction of base isolator. Due to considering soil structure interaction the acceleration increases up to 80% for softer soil. This percent increases reduces from softer to harder soil.

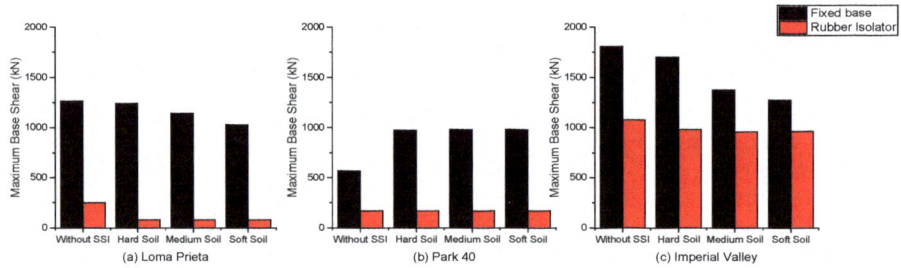

Fig. 4. Maximum base shear in X-direction for rectangular shaped building

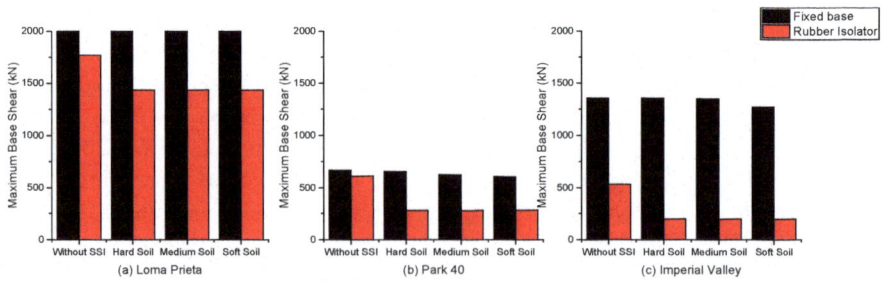

Fig. 5. Maximum base shear in X direction for C-shaped building

Figures 4 and 5 shows that the base shears in X-direction are maximum for fixed base building but after installation of base isolation the base shear reduces up to 80% for rubber isolator. From the graph, it is clear that the soil conditions are affecting the performance of isolator. Also, with considering SSI, the base shear reduces with the use of isolator. Due to use of SSI there is approximately 10–15% reduction in shear force for softer soil. As the soil gets harder the percentage of reduction reduces up to 1–2%. Figures 5 and 6 shows the base shear for building with and without SSI. After taking the effect of underlying soil the base shear reduces from harder to softer soil due to increase in acceleration for softer soil.

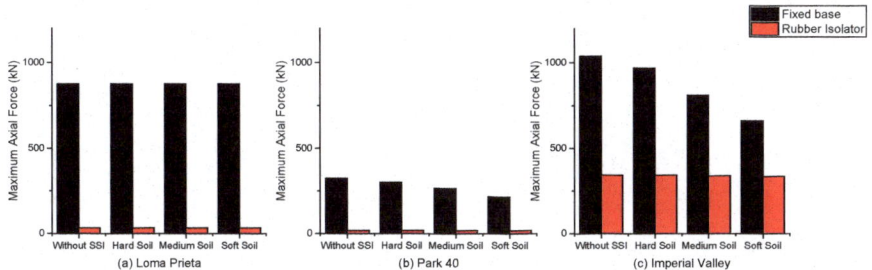

Fig. 6. Maximum axial force for rectangular shaped building

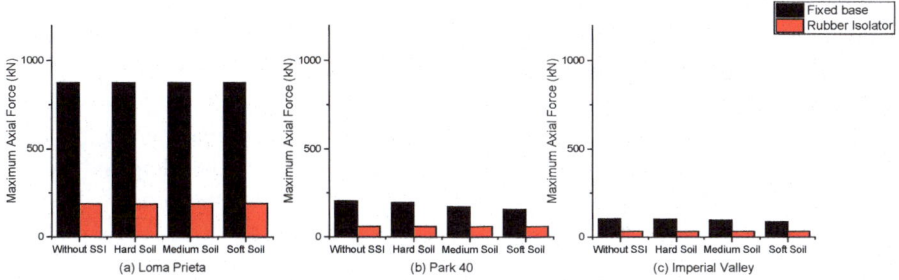

Fig. 7. Maximum axial force for C-shaped building

Figures 6 and 7 shows graph of maximum axial force. From graphs it is clear that the maximum amount of axial force is reduces due to isolator. The average % reduction in maximum axial force is up to 84% for rectangular building. Due to SSI the Base shear is increases from harder to softer soil. Figures 7 and 8 shows the maximum axial force for rectangular & C-shape building for fixed base and isolated base. The maximum axial force is in fixed base building is reduced up to 68% in case of isolated C-shape building. Soil structure interaction is showing more severe effect on softer soil compared with hard soil strata.

Figures 8 and 9 shows the maximum shear force for rectangular & C- shape buildings. The average reduction is shear force is 74% for isolated buildings. Also the

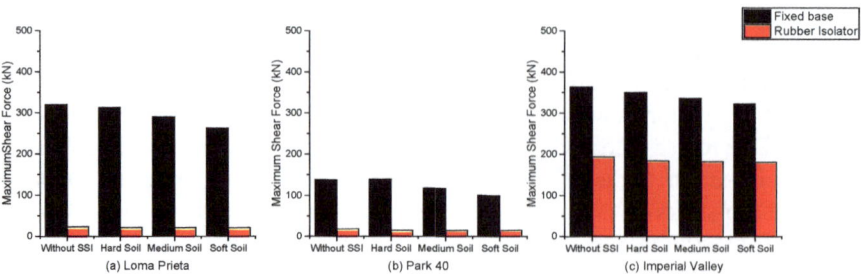

Fig. 8. Maximum shear force for rectangular shaped building

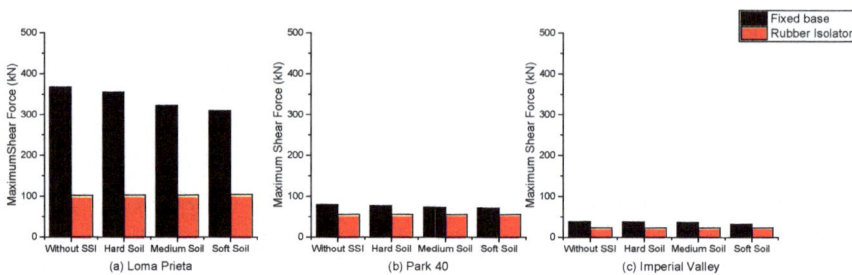

Fig. 9. Maximum shear force for C-shaped building

soil structure interaction shows significant variation in the result. Due to effect of SSI the shear force approximately reduces up to 5% from harder to softer soil. Figure 10 shows the graphs of maximum shear force for C-shape building. From graphs it is clear that the fixed base building is showing maximum shear force compared with isolated base building. The energy dissipation devices are performing for building and reducing up to 74% of shear force as the acceleration of building is also reducing.

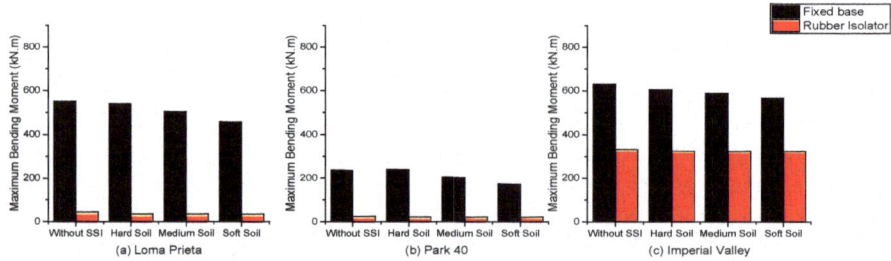

Fig. 10. Maximum bending moment for rectangular shaped building

Figures 10 and 11 shows the maximum bending moments of fixed base and isolated base building for both rectangular & C-shape model. The maximum bending moment

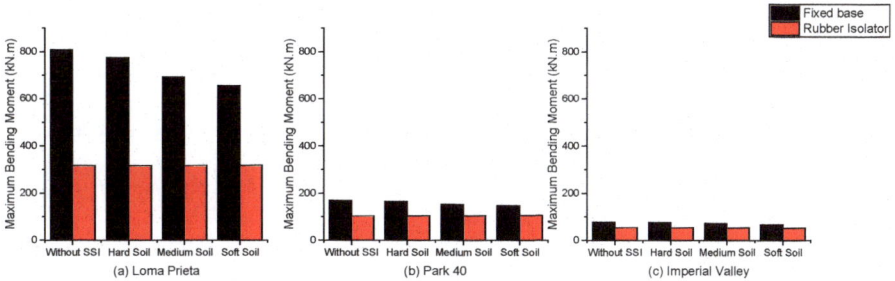

Fig. 11. Maximum bending moment for C-shaped building

is reducing up to 90% for rectangular isolated base building & up to 74% for C-shape building. Due to effect of soil structure interaction the bending moment reduces from softer to harder soil. Figure 11 shows the maximum bending moment for fixed base and isolated base building for C-shape planner configuration. Due to use of isolator the bending moment is reducing up to 44% as compared with fixed base building. The soil structure interaction changes the values of bending moment for all types of underlying soil. Due to softer soil the bending moment increases up to 5%, this percentage increase reduces as the nature of soil become harder.

4 Conclusions

From analytical results, it is observed that base isolation reduces the overall seismic response of building in comparison to fixed base building for both rectangular as well as C-shape building. The comparison of models show that the natural time period of vibration increases more than double for isolated building. Time period affects the earthquake response of the structure, as the time period increases the base shear and acceleration values found to be reducing. Introduction of seismic base isolation system at base level of building shows very low inter-story drift values for higher stories and makes the building to behave like a rigid body structure. The effect of soil structure interaction shows significant variation in results for isolated base building as for hard strata of soil there is not such change in natural fundamental time period but as the stiffness of soil reduces the time period increases from harder to softer soil. he result showed that there is hardly any change in response of structure if underlying soil is hard or stiff but there is considerable change if the soil beneath is soft or medium hard. Also the time period of structure get influenced due to soil type.

References

1. Ribakov, Y., Iskhakov, I.: Experimental methods for selecting base isolation parameters for public buildings. Open Constr. Build. Technol. J. **2**, 1–6 (2008)
2. Monfared, H., Shirvani, A., Nwaubani, S.: An investigation into the seismic base isolation from practical perspective. Int. J. Civil Struct. Eng. J. **3**, 451–463 (2013)

3. Chandak, N.R.: Effect of base isolation on the response of reinforced concrete building. J. Civil Eng. Res. **3**, 135–142 (2013)
4. Su, L., Ahmadi, G., Tadjbakhsh, I.G.: Performance of sliding resilient-friction base-isolation system. J. Struct. Eng. **117**, 165–181 (1991)
5. Shenton III, H.W., Lin, A.N.: Relative performance of fixed-base and base-isolated concrete frames. J. Struct. Eng. **119**, 2952–2968 (1993)
6. Anilduke, S., Khedikar, A.: Comparison of building for seismic response by using base isolation. Int. J. Res. Eng. Technol. (2015)
7. Yin, B.: The Effect of Soil Structure Interaction on the Behavior of Base Isolated Structures. Duke University (2014)
8. Raj, D.: Non-linear Static Analysis of Building-Foundation System considering Soil-Structure Interaction using SAP 2000 v 14. Research Scholar, Department of Earthquake Engineering, IIT Roorkee
9. Naeim, F., Kelly, J.M.: Design of Seismic Isolated Structures: From Theory to Practice. Wiley, Hoboken (1999)

Condition Based Maintenance of Gearbox Using Ferrographical Analysis

Ananda B. Gholap[1](✉) (ID) and M. D. Jaybhaye[2] (ID)

[1] Marathwada Mitra Mandal's College of Engineering, Pune 411052, India
anandagholap@gmail.com
[2] Department of Production Engineering, College of Engineering Pune,
Pune 411005, India
mdj.prod@coep.ac.in

Abstract. Ferrography technique is widely used for wear debris analysis present in the oil. A contamination and wear particle in lubricated oil gives trends of failure in future. Corrective remedies can be planned to avoid failure of the system. Alarms for normal, Marginal, Critical can be set using wear particle concentration (WPC). Percentage Large-scale particles (PLP) trend shows criticality of the system which can used to monitor the system. For said study a gearbox connected to three phase induction motor as input and a disk break as load. For varying load condition from no load to 5 Kgf/cm^2 is used. A speed of 710 to 2400 rpm is used. As trend of WPC and PLP are within the limit. Further samples will be monitored and failure can be predicted.

Keywords: Ferrography · Maintenance · Wear

1 Introduction

Ferrography is widely used in the industries to check the condition of machinery. Edge feature is used to identify types of wear, extracting parameters, wear mode and severity. However edge detection is crucial factor for wear debris analysis [1]. Detailed examination of direct reading ferrographic process is used to evaluate their applications as quantitative and qualitative failure prediction tool [11]. Particular emphasis is on the quantitative aspects allied with ferrography and the broad assumption analogous to them [2]. Amount of debris released in to the oil is taken into consideration against the background level wear debris [3]. For better understanding of wear process occurring, ferrography applications in laboratory and field engine is used. For creating ferrography quantitative and sensitive methods to wide range of debris concentration are used [4, 11]. Ferrography is effective technique in hydraulic fluid monitoring DR Ferrograph is used in monitoring industrial systems. DR values and ferrogram are used to detect nonmetallic material [5]. Texture based classification is more accurate than classification based on morphology in machine condition monitoring industry. Texture based classification is useful tool [6]. Histogram equalization is used for prediction using previous information [7]. Various wear mechanism like sliding, rolling, rubbing, abrasion can be investigated by wear debris analysis [8].

© Springer Nature Singapore Pte Ltd. 2020
V. K. Gunjan et al. (Eds.): *ICRRM 2019 – System Reliability, Quality Control,*
Safety, Maintenance and Management, pp. 179–184, 2020.
https://doi.org/10.1007/978-981-13-8507-0_27

2 Ferrography

Ferrography was first introduced in 1971 [9]. Researchers used ferrographic technique for monitoring condition of system which is under continuous operation [10, 11]. Till date various improvements and modifications are done. It is observed that, wear debris in lub oil shows constant density of particles in Normal behavior. Sudden increase in the density, shape or size indicates that Abnormality of system starts or at high severity. Data can be collected from direct reading ferrograph which will give quantitative results and prepared slide can be used as qualitative tool for analytical ferrography [11].

Ferrography analysis can be carried out using Density reading ferrograph, Dual Slide maker and Microscope (Fig. 1). Direct Reading Ferrograph is a trending instrument, providing readings for measurement of ferrous particles. Wear particle concentration is calculated as Density large (DL) and density small (DS). Ferrogram slide is prepared using ferrogram maker. It is used to analyze slide and represents the sample completely. This slide is used for microscopic examination using Microscope Olympus BX 52.

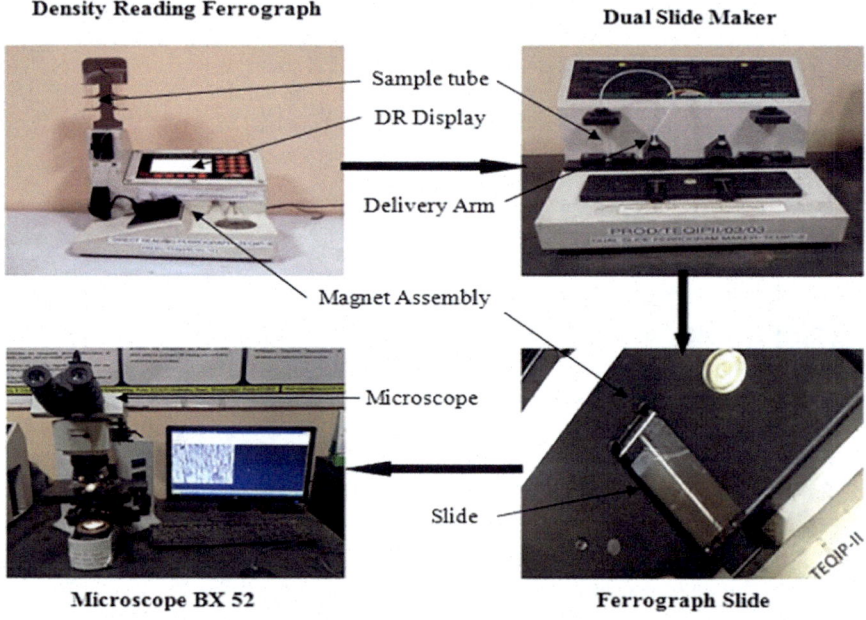

Fig. 1. Ferrography analysis flow diagram

3 Wear Debris Classification

Particles under analytical ferrography can be classified as ferrous, nonferrous, ferrous oxide and contaminants. Wear particles can be further sub classified. In normal Rubbing size of particle is from 1 μ to 15 μ. Incorrect lubrication, sever load condition and sever speed condition lead to sever sliding wear. Improper load is main cause of Gear wear. Bearing Wear particle are assosiated with bearing problem. Black oxide, Red oxide. Water contamination is a common cause of equipment failure.

4   Wear Particle Concentration

Wear particle concentration and Percentage large scale particles are two important measurement tools in quantitative ferrography analysis. WPC indicates sum of all magnetic particles present in the sample. Although the magnitude of the WPC is important, the trend of the values is the indicator of machine wear condition. PLP determine the percentage of large particles above 5 μ in the WPC. An increase in both WPC and PLP is an indication of an abnormal wear condition. WPC acceptance limit for various systems are shown in Table 1.

Table 1. Wear particle concentration limits for various systems [10].

Vacuum pumps	WPC = 1 to 5	Roller bearings	WPC = 5 to 100
Boiler feed pumps	WPC = 1 to 100	Engines transmissions	WPC = 10 to 150
Gas compressors, turbines, fans	WPC = 1 to 20	Extruder gearboxes	WPC = 100 to 600
Journal bearings	WPC = 1 to 20	Dragline gearboxes hoist gearboxes	WPC = 1000 to 50000

5   Alarms in Fegrrography

To detect significant changes in the Normal wear mode alarms in ferrography are set [12]. The Alarm A value is used to detect sudden high deviation in wear particle concentration. It is calculated by multiplying last highest value of WPC (With normal Range) by 1.8. The Alarm BH (High) value is based on the values of WPC which are rated Normal. It is calculated by multiplying 2.0 times standard deviation of WPC and adding to population mean of Normal wear readings. The Alarm BL (Low) value are suspect and result from faulty sampling techniques. It is calculated by multiplying 2.0 times standard deviation of WPC and subtracting from population mean of Normal wear readings.

6   Equipment Condition Report

Equipment condition is classified based on results obtained from analytical ferrography. Equipment condition report with normal rating shows expected wear debris particles. Marginal Ratings shows wear pattern is not within expected level so maintenance action is required. Critical rating shows serious wear condition for equipment.

7 Methodology

Figure 2 consists of a motor and compound two stage incremental gear box. The input shaft of gearbox is connected to 1 HP, 3 phase 60 Hz, induction electric motor through rubber coupling. All drive shafts are supported at its ends with antifriction bearings. Drive gear and Driven gear of 70 and 90 diameter are used in gearbox. At an interval of 200 running hours sample is collected. Speed ranges are, 420,710,1200,1600,2000 and 2400 rpm. Load from No load to 5 Kgf/cm^2 is applied at an interval of 40 h. Total 2400 h running of gearbox is considered for the study. Density readings for each sample is recorded using DR 5.

WPC and % LSP is calculated using following equations and summerised in Table 2.

Table 2. Wear particle concentration and % large particles in sample.

Sample	Running Hrs.	Date	Density large particles (DL)	Density small particles (DS)	Wear particle concentration	Large scale particles
1	200	12/03/2018	53	22	75	41.33
2	400	23/03/2018	16	7	23	39.13
3	600	05/04/2018	24	12	36	33.33
4	800	16/04/2018	40	26	66	21.21
5	1000	27/04/2018	15	7	22	36.36
6	1200	08/05/2018	27	11	38	42.11
7	1400	18/05/2018	33	15	48	37.50
8	1600	29/05/2018	10	6	16	25.00
9	1800	07/06/2018	38	12	50	52.00
10	2000	18/06/2018	27	8	35	54.29
11	2200	29/06/2018	22	6	28	57.14
12	2400	09/07/2018	14	6	20	40.00

Fig. 2. Gear box test rig setup

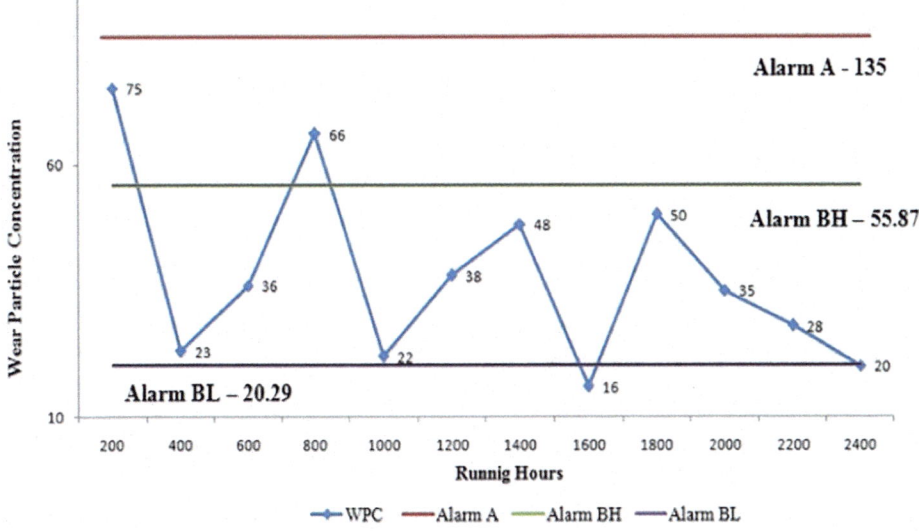

Fig. 3. Wear particle trend plot.

$$\text{Wear Particle concentration (WPC)} = \text{DL} + \text{DS} \qquad (1)$$

$$\% \text{ Large Scale Particle} = (\text{DL} - \text{DS/WPC}) \times 100 \qquad (2)$$

A graph of WPC verses time interval is ploted. For current study mean value of WPC is 38.08 and standard deviation is 8.89 so alarm values obtained are Alarm A: 135, Alarm BH: 55.87 and Alarm BL: 20.29 as shown in Fig. 3.

8 Results

As WPC trend and percentage large particles shows all readings falls under normal acceptance limit. First WPC analysis shows higher concentration as new system wear rate is high. Fourth and sixth observation is due to improper sampling. For further readings if WPC limit cross the BH value i.e. 55.87. A keen attention is needed for healthy condition monitoring. If any value of WPC obtained further is below 20.29 sampling procedure is faulty. Quantitative analysis indicated that the Alarms in Ferrography will be more helpful for condition monitoring and for further failure prediction.

References

1. Wang, J., Bi, J., Wang, L., Wang, X.: A non-reference evaluation method for edge detection of wear particles in ferrograph images. Mech. Syst. Sig. Process. **100**, 863–876 (2018)

2. Wakefield, G.R., Levinsohn, H.: An assessment of quantitative and qualitative Ferrography. Wear **126**, 31–55 (1988)
3. Yardley, E.D., Moreton, G.: An attempt to quantify the limits of failure detection by ferrography. Wear **90**, 273–279 (1983)
4. Johnson, J.H., Hubert, C.J.: An overview of recent advances in quantitative ferrography as applied to diesel engines. Wear **90**, 199–219 (1983)
5. Mccullagh, P.J., Campbell, W.E.: Application of ferrography to contamination control in fluid power systems. Wear **90**, 89–100 (1983)
6. Stachowiak, G.P., Stachowiak, G.W., Podsiadlo, P.: Automated classification of wear particles based on their surface texture and shape features. Tribol. Int. **41**, 34–43 (2008)
7. Jiang, G., Lin, S.C.F., Wong, C.Y., Rahman, M.A., Ren, T.R., Kwok, N., Shi, H., Yu, Y.-H., Wu, T.: Color image enhancement with brightness preservation using a histogram specification approach. Optik **126**, 5656–5664 (2015)
8. Kumar, S., Goyal, D., Dang, R.K., Dhami, S.S., Pabla, B.S.: Condition based maintenance of bearings and gears for fault detection – a review. Mater. Today Proc. **5**, 6128–6137 (2018)
9. Scott, D., Westcott, V.C.: Predictive maintenance by ferrography. Wear **44**, 173–182 (1977)
10. Dalley, R.J.: Ferrographic and oil analysis at vibe institute. Piedmont chapter, p. 29, May 2008
11. Biswas, R.K., Majumdar, M.C., Basu, S.K.: Vibration and oil analysis by ferrography for condition monitoring. J. Inst. Eng. India: Ser. C **94**(3), 267–274 (2013)
12. Predict guide to equipment and lubricant condition reports. Predict/DLI report, p. 5 (2014)

Green BIM for Sustainable Design of Buildings

Anju Ebrahim[✉] and A. S. Wayal

Civil and Environmental Engineering Department,
Veermata Jijabai Technological Institute (VJTI), Matunga, Mumbai, India
anjuebrahim@gmail.com

Abstract. Construction of Sustainable and Green certified buildings is gaining significance construction industry. The importance of Building Information Modeling (BIM) including the evaluation of Green Building strategies is being recognized by the construction sector. Effective decisions pertaining to sustainable design of a building facility can be only made during the early design and pre-construction stages. Traditional Building construction planning, however, do not support the possibility of such early decisions. Mostly, energy and performance analysis are carried out after the architectural design and further construction documents and records have been prepared. So, there is a lack of integration into the design procedure which leads to ineffective process of previously altering the design to attain an environment friendly sustainable building.

Green Building Information Modeling (BIM) is the application of BIM to provide data for energy performance evaluation and sustainability assessment. Green BIM also includes the application of Building Energy Modeling and deals with energy optimization to improve the energy efficiency of a building along with sustainable design and construction practices during its lifecycle.

This paper explores to give the steps required to assess building performance in the early design and preconstruction phase realistically such that a comprehensive set of information on buildings like buildings form, material, context, and technical systems can be simulated which then allows BIM for analyzing performance and suggest for various design alternatives and steps for making decisions for sustainable design.

Keywords: Building Information Modeling · Green Building Studio · Autodesk Revit

1 Introduction

Buildings consume a good amount of energy and also is a major contributor of Global Carbon dioxide emissions. Due to the rising energy prices and environmental concerns there is a demand for construction of sustainable building facilities with minimal environmental impact through the adoption of environmentally friendly architectural design and construction techniques.

Sustainability refers to fulfilling todays requirements without settling upon the future demands. Sustainability with respect to building industry can be described as how much greener the building is in terms of energy, materials, site planning, water

© Springer Nature Singapore Pte Ltd. 2020
V. K. Gunjan et al. (Eds.): *ICRRM 2019 – System Reliability, Quality Control,
Safety, Maintenance and Management*, pp. 185–189, 2020.
https://doi.org/10.1007/978-981-13-8507-0_28

usage etc. during its construction, after construction, upkeep and demolition. BIM can assist in the following areas:

i. Minimizing the energy consumption and thereby its costs by choosing a good orientation.
ii. Analyze building shape and form and improving the building envelope.
iii. Day-lighting analysis.
iv. Reducing water consumption in a structure.
v. Reducing energy requirements and assessing various renewable energy alternatives.
vi. Using of green, sustainable materials and using recycled materials.
vii. Planning and Site management to reduce wastage and carbon footprints.

Green BIM is the sustainability evaluation of buildings to provide data for assessment of energy performance. It consists of Building Energy Modeling which includes estimating project energy performance to determine various design alternatives to enhance the building energy efficiency during its period of operation.

This paper lays down the methodology for conducting BIM based performance analysis using Autodesk Revit and Autodesk Green Building Studio (GBS). Using these softwares one can optimize their energy consumption and work towards as a carbon neutral building.

Performance analysis with regard to this paper concentrates on economic savings and good environmental impact as a result of choices and policies adopted to minimize the energy requirement and CO_2 emissions.

2   Applications of BIM for Sustainable Design

As per previous studies, sustainable design can be carried out using Building Information Modeling by developing a framework which establishes the relationship between BIM based sustainability analyses and certification process like IGBC or LEED. These credit points can be directly or indirectly prepared using BIM based sustainability analysis software. BIM – based software can generate results faster when compared to traditional methods and thereby reduce substantial amount of resources and time.

Autodesk Green Building Studio software can be used to design a building plan in such a way as to achieve an energy-efficiency by considering multiple choices of alternative designs to design a sustainable building. The parameters like Energy Savings, Energy and Carbon emission values, Daylight and water efficiency parameters can be studied. GBS can be run for various scenarios to give an idea of how sensitive the parameters that affect energy usage are. The parameters like orientation, roof construction, wall construction, window glass, skylight glass, infiltration, equipment and lighting have enormous impact on energy performance of a building.

Building Information Modeling is an effective tool for integrating natural systems and technical features in an architectural design. The impacts of sustainable design methods in green buildings can be analyzed using Building Information Modeling including all aspects of energy efficiency, lighting, sustainability of materials and various other building performance parameters. Energy analysis in buildings including the analysis of air flow and buildings' sunshine ecosystems can be carried out through BIM.

BIM energy analysis tools like Ecotect, Green Building Studio etc. can be successfully integrated with BIM modeling softwares like Autodesk Revit and can be used for energy simulation and to find design alternatives. Potential Energy Savings, Energy and Carbon results, Daylighting and water efficiency can also be evaluated using BIM based energy analysis tools. Owners could also identify the potential IGBC or LEED points that can be earned based on the selected certification system during the design or conceptual stage itself.

BIM analysis tools provides many design alternatives to identify the sensitive parameters that affect the energy usage. By bringing variations in the parameter values, we can opt for the best option to create an energy efficient structure.

However, Building Information Modeling for energy or sustainability analysis also has its own disadvantages. The results of energy simulation are greatly affected by the accuracy of energy model, inputs given, size and complexity of the project (Table 1).

Table 1. Steps for carrying out green BIM using Revit and GBS

Step No.	Steps
Step 1	Review previous researches done in the field of BIM based performance analysis
Step 2	Identify the BIM based softwares and tools that can be utilized for the study. Here, Autodesk Revit and green building studio are selected
Step 3	Collect the drawings and specifications of the building selected for the study
Step 4	Create 3D model of the building in Autodesk Revit by the acquired drawings and information
Step 5	Export the 3D model from Revit to green building studio in gbXML format
Step 6	Run energy simulation for the model using green building studio
Step 7	Evaluate parameters like potential energy savings, daylighting, water efficiency, energy and carbon results
Step 8	Run alternative simulations in GBS by trying different settings on the base model or by adding various energy conservation measures to the base model. The changes can be made with respect to HVAC system type, glazing, adding insulation for roof, walls etc.
Step 9	Compare the simulation results to select the best energy efficient sustainable design

3   Methodology for Carrying Out Green BIM Using Revit and GBS

3.1   Input Needed for Energy Analysis in BIM

Energy Efficiency is a crucial parameter to decide whether a material is environment friendly or not. A certain amount of energy is vital to maintain the comfort and convenience of the users of a building. The major energy losses occurring in a building are the transmission and ventilation heat losses. Also, there are energy gains happening which are from appliances, users as well as solar energy gains through openings. These gains fully or partly compensate for the above said energy losses. Furthermore, power is essential to operate lighting, appliances and building service systems. The total energy demand of the building is then calculated by deducting gains from the overall losses.

Energy Analysis requires the knowledge of the following data:

- Dimensions, plan, areas and volumes of different spaces of the building.
- Arrangement of rooms in thermally similar zones.
- Building orientation.
- Thermal features of all construction units.
- Operational usage of the building (Residential/office/schools etc.)
- Appliances loads, Lighting and occupants loads.
- Heating, ventilating, and air conditioning system category and functioning properties.
- Power/Water rates and
- Climatic information.

4   Conclusion

The increase in the number of analysis tools is a proof of the increasing importance of sustainable design in architecture and the requirement to optimize building performance. The BIM-based design and documentation system can be used for providing information that can be used to improve design and building performance. The data required for supporting green design is obtained during the design process and the same is also taken from building information model as and when needed. Revit Architecture can be used for designing very complex process of sustainable design like orientation and solar study, and automates the activities like material takeoffs for Green Building Certification.

Revit further uses software such as Green Building Studio to further provide specialized functions like energy analysis, bounce light calculations, water efficiency, Carbon emissions and specification management.

Building location and orientation to maximize the energy efficiency can be carried out using BIM-integrated software.

References

1. Krygiel, E., Nies, B.: Green BIM: Successful Sustainable Design with Building Information Modeling. Wiley, Hoboken (2008)
2. Jalaei, F., Jrade, A.: Integrating BIM with green building certification system, energy analysis and cost estimating tools to conceptually design sustainable buildings. In: Construction Research Congress, pp. 140–149 (2014)

3. Azhar, S., Carlton, W.A., Olsen, D., Ahmad, I.: Building information modeling for sustainable design and LEED rating analysis. Autom. Constr. **20**, 217–224 (2011)
4. Maltese, S., Tagliabue, L.C., Cecconi, F.R., Pasini, D., Manfren, M., Ciribini, A.L.: Sustainability assessment through Green BIM for environmental, social and economic efficiency. Procedia Eng. **180**, 520–530 (2017)
5. Le, M.K.: Autodesk Green Building Studio for an energy efficient, Sustainable building. HAMK University of Applied Sciences, Finland (2014)
6. Wong, K., Fan, Q.: Building information modeling (BIM) for sustainable building. Facilities **31**(3), 138–157 (2013)
7. Bonenberg, W., Wei, X.: Green BIM in sustainable infrastructure. Procedia Manuf. **3**, 1654–1659 (2015)

Failure and Performance Elements of Catalytic Converter in Multi Cylinder Engine

Pavan B. Chaudhari$^{(\boxtimes)}$ and R. R. Arakerimath

GHRCEM Savitribai Phule Pune University, Pune 412507, MS, India
chaudharipb@gmail.com

Abstract. The automotive emission becomes harmful day by day. The engine performance and exhaust after treatments systems need to be optimized to reduce. The catalytic converters are essential component for this purpose. The optimized catalytic converter design is utmost important as it plays vital role in eliminating undesirable emissions from mixing into environment. As Euro VI is enforced and India is migrating from BS IV to BS VI by skipping BS V. Achieving tougher new standards is critical job. The catalytic converter geometry optimization, light off behavior and performance parameters is required to be modified to achieve lesser emissions. The current work is going on three way catalytic converters I which two stage oxidation and single stage reduction takes place. The present study focuses on multistage catalytic converters performance and their light off behavior and Idling behavior along with Washcoats reactions and Fault diagnosis model for sulfur poisoning and thermal damage.

Keywords: Catalytic converters performance · Geometry ·
Washcoats materials · Thermo fluidic performance · Fault diagnosis

1 Introduction

The Emission Norms are becoming more tougher. The catalytic converters are after treatment device that helps to reduce harmful gases [1], exhaust emissions of Internal combustion engine can be lowered by inserting catalytic converters in exhaust gases path, Delaying time for CC to reach maximum efficiency, Increasing catalyst conversion efficiency storing pollutants during cold start and releasing when CC activates. The three way catalytic converter which undertakes three reaction processes viz Oxidation of NOx in N2 and O2 and Oxidation of Hydrocarbon into Co2 and water vapours It works in closed loop with lambda or oxygen sensor to control air fuel ratio [7, 9] The Monoliths can be ceramic or metallic [1] Ceramics are more porus, has good coating adherence and temperature stability. And Metal monoliths have good pressure drop characteristics, mechanical stability and good heat transfer [1, 3]. The light off should be faster to achieve minimum emissions. The catalyst must be temperature resistance and durable. Electrically heated precatalyst is useful to heat up catalyst to kick of catalytic reaction early to reduce cold start emissions. The present paper focuses on various performance aspects of catalytic converters and the current work going on by various researchers and failure diagnosis techniques are discussed (Fig. 1).

© Springer Nature Singapore Pte Ltd. 2020
V. K. Gunjan et al. (Eds.): *ICRRM 2019 – System Reliability, Quality Control,*
Safety, Maintenance and Management, pp. 190–195, 2020.
https://doi.org/10.1007/978-981-13-8507-0_29

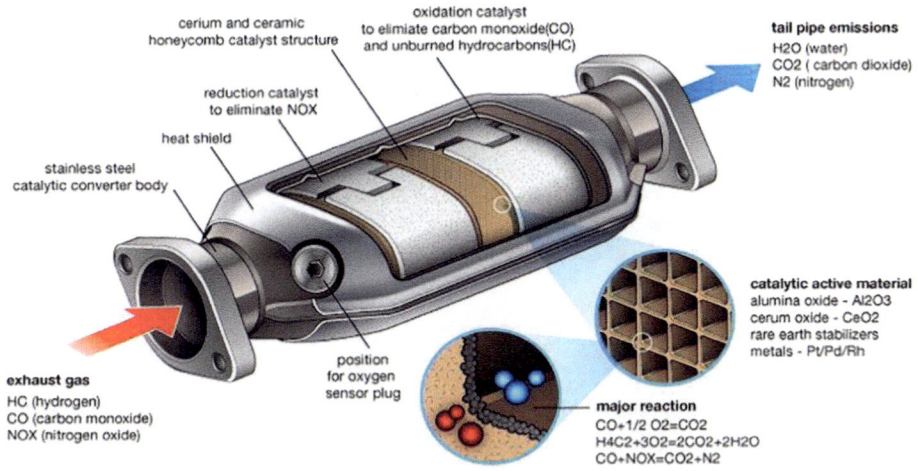

Fig. 1. A typical catalytic converter (www.crowndautomotive.com) [10]

2 Flow Through Catalytic Converter

It can be separated in three regions I and III are inlet and outlet sections with cones and II is monolith section. Single channel and multichannel models are simulated to evaluate the performance (Fig. 2).

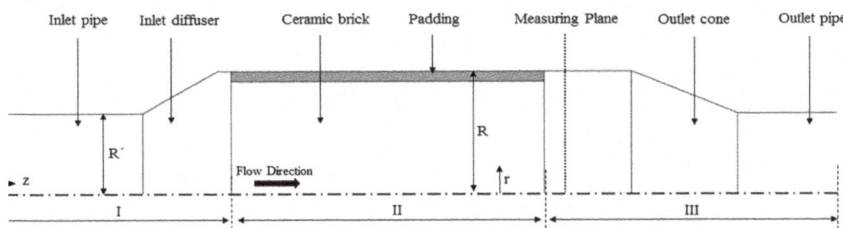

Fig. 2. Baseline substrate solution domain with all simulated components [11].

3 Geometry

Dinler et al. [8] studied numerical investigation of different channel geometries, inlet and wall temperature for performance of catalytic converter they found that increased wall temperature activates [8] the catalyst and improves performance during cold start period. The author has selected three different geometries square, circular and triangular with constant hydraulic diameters the set of boundary conditions were selected and the effects of inlet temperature are compared with wall temperature. Effects of channel geometries are investigated. It is found that triangular channel has more conversion efficiency. When wall temp is below 900 K. Increase in exhaust gas temperature

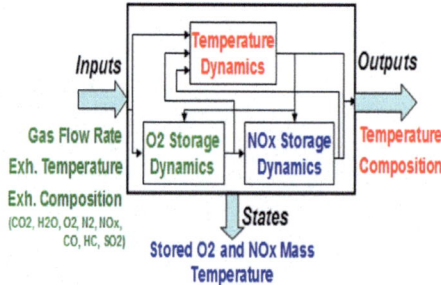

Fig. 3. Structure of LNT model implemented system model [10]

increases conversion efficiency. Emission Conversion efficiency is ratio of removal of mass amount of component to mass amount of component into catalytic converter. Use of catalytic converters causes pressure drop and high fuel consumption [4, 7]. Filtration efficiency is directly proportional to pressure drop. Substrate length cross section area and cell shape are important geometrical parameters [2, 3].

4   Catalytic Converter Performance Thermal Performance and Hydraulic Performance

Cold start and idling causes most of the emissions. During cold start the catalytic reactions doesn't starts and harmful gases remain unaltered causing them to pass without change. Thermal performance is dependent on monolith ability to heat up light off temperature very fast. It can be improved by optimizing converter design. The temperature, velocity and static pressure distribution are important as well. Light of temperature is of prime importance for thermal performance. If the pressure drop increases load on engine will increase which will cause increase in fuel consumption and emissions of harmful gases [4, 8]. Many researchers studied hydraulic performance. Cleanalytic is new converter developed by Vida fresh air corp [11]. Which divides substrate in two small chambers thermally insulated allowing fast light off to improve thermo fluidic performance and improves conversion efficiency [11]. Inlet flow conditions, substrate properties and converter geometry effects are studied the findings are pressure drop decrease during pulsating exhaust. Studied pressure drop across monolith brick with different cell density, wall thickness and coating and found that pressure drop is mainly due to viscous and inertial forces [5].

5   Flow Distribution and Hydraulic Performance

Flow in monolith depends on manifold design, monolith properties and inlet flow conditions. Inlet pipe length, inlet diffuser geometry, bending angle of inlet pipe, brick geometry and its properties are important design parameters for better hydraulic performance [11]. Increasing diffuser angle causes non uniform flow.

6 Flow Distribution and Thermo Fluidic Performance

can be increased to avoid early failure of catalytic converter can be increased by dividing substrate into two chambers for faster light off. Many researchers have considered steady state flow under reactive flow conditions some as non reactive and some took transient flow conditions for studying cold off behavior. Titanium based and Cerium based catalytic converters can be used to improve conversion where three catalysts can be used titanium dioxide, copper nitrate and zirconium dioxide coated onto wire meshes The results shows improved emissions [9].

7 Fault Diagnosis

Pisu et al. [10] has focused on lean NOx traps devices which helps to reduce NOx in diesel and lean burn engines. It stores NOx during Lean engine working. Traps can be regenerated by controlling exhaust air fuel ratio for creating rich gas mixture. Then NOx is released in rich condition and catalytically converted. And emissions are reduced to very large extent. LNT operates at high conversion efficiency efficient control of regeneration scheduling is required. LNT requires fault diagnosis to detect and isolate faults related to sulfur and thermal damages.

8 Model Based Fault Diagnosis Approach for LNT System

The diagnostic approach consists of generating residuals using system models and comparing actual and predicted variables like air fuel ratio, catalyst out temperature. and NOx concentration at output. The main intension is to detect and isolate controller faults and LNT parametric faults related to sulfur and thermal damage. In this the mathematical model of LNT system is created and fault diagnosis scheme is applied to it. This fault diagnosis is based on model of LNT catalyst which captures physical and chemical phenomenon. To characterize system dynamics. It is based on grey box approach. As shown in figure LNT system dynamics is generated from interaction of three subsystems viz oxygen storage dynamics, NOx storage dynamics and temperature dynamics Exhaust gas mass flow and temperature are given as input with feed gas consisting CO_2, H_2O, O_2, N_2, NO, CO, HC, SO_2 (Fig. 4).

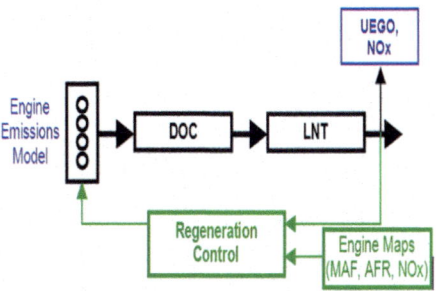

Fig. 4. Diesel NOx after treatment

The outputs of the models are output gas temperature, outlet mixture composition Proper air fuel ratios are selected Oxygen storage and release affects air fuel ratio. Various model equations proposed by different researchers are considered.

In LNT fault modeling approach characterizing failure modes of system is one feature. The sulfur content in fuel combines with oxygen during combustion causes SO2 which is harmful sulfur storage and release is dependent on temperature. And the reactions are predominant over O2 and NOx So trap capacity is calculated. Mass of sulfur is subtracted from that of N2 and O2 [10].

The model in Fig. 3 is created by combining LNT and quasy steady engine emissions model. Derived from steady state engine emission data and model of diesel oxidation catalyst also simple models of O2 NOx and temp sensors are developed.

9   FAULT DIAGNOSIS SCHEME for Thermal Damage and Sulfer Poisoning Fault Conditions by LNT

This scheme assumes operation is lean, NOx sensor calculates flow .regeneration control is saturated. Sensors are fault free, in out temp is known. The fault detection is done by comparing air fuel ratio, NOx and system temperatures. The proposed system is effective in isolating sulfur poisoning and thermal damage in certain time period before occurrence of fault (Fig. 5).

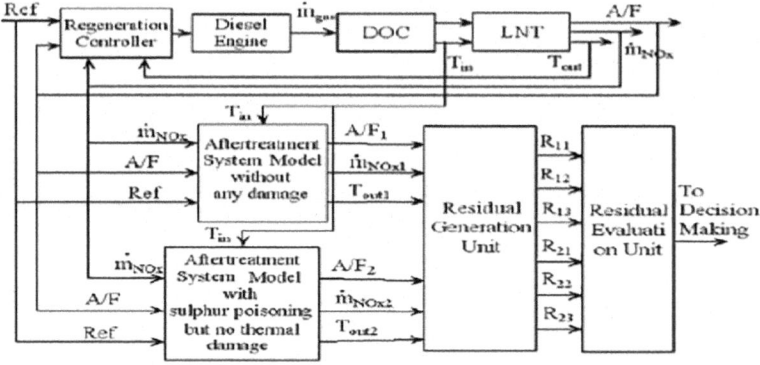

Fig. 5. Model based diagnostic scheme [10]

10   Summary and Conclusion

The stricter emissions standards the catalytic converters are becoming more and more important. The automotive industries are struggling to current achieve them. In this paper the attempt was made to discuss the current performance trends in catalytic converters The effect of Geometry change on performance, catalyst are contributing much. Fault diagnosis give satisfactory predictions of fault well before its occurrence. Triangular geometry shows better conversion efficiency than square geometry [8]. If wall temperature is increased as compared to inlet temperature the catalyst gets activated during cold start and performance improves Thermal performance is monolith ability to heat catalyst for fast light of temperature. Hydraulic performance depends on inlet flow conditions, inlet pipe cone angle and brick geometry and properties, diffuser angle. Thermo fluidic performance is increased to avoid early failure of catalytic converter by dividing substrate into two chambers. Lean NOx traps are useful for reducing NOx from lean diesel engine its fault diagnosis related to sulfur poising and thermal damages. Model based fault diagnosis for LNT consists three subsystems oxygen storage dynamics, NOx storage dynamics and temperature dynamics. The fault detection is done by comparing air fuel ratio, NOx and system temperatures in thermal damage and sulfur poisoning to avoid fault.

References

1. Govender, S., Friedrich, H.B.: Monoliths: a review of the basics, preparation methods and their relevance to oxidation. Catalysts, 1–69 (2017). https://doi.org/10.3390/catal7020062
2. Leman, A.M., et al.: Emission treatment towards cold start and back pressure in internal combustion engine against performance of catalytic converter: a reviewed sciences. In: MATEC Web Conferences ENCON 2016 (2016). https://doi.org/10.1051/matecconf/20178702021
3. Zainal, N.A., et al.: Effect of inlet cone pipe angle in catalytic converter. In: IOP Conference Series: Materials Science and Engineering, vol. 328, p. 012029 (2018). https://doi.org/10.1088/1757-899X/328/1/01202
4. Kapatkar, V.N., et al.: Design, analysis and testing of catalytic converter for emission reduction and backpressure optimization. IJESC (2017)
5. Ekstorm, F., et al.: Pressure drop of monolithic catalytic converters experiment and modeling. (SAE international) 2002-01-1010
6. Shi-Jin, S., et al.: Study on flow characteristics of automotive catalytic converters with various configurations (SAE international) 2000-01-0208
7. Selvam, T., Warmuth, F., Klumpp, M., Warnick, K.G., Lodes, M.A., Körner, C., Schwieger, W.: Fabrication and pressure drop behavior of novel monolithic structures with Zeolitic architectures. Elsevier Chem. Eng. J. **288**, 223–227 (2016)
8. Dinler, N., Aktas, F., Yucel, N.: Effects of channel design and temperature on the performance of the catalytic converter. Int. J. Green Energy **15**, 813–820 (2018)
9. Ranganathan, M., Remo, S.A.R., Kishore, U., Yuvaraj, S., Arun, S.: Development and performance analysis of new catalytic converter. ISSN 2319-5991. www.ijerst.com. Special Issue International Conference on "Advance Research and Innovation in Engineering, Science, Technology and Management" ICARSM 2015, vol. 1, no. 3, May 2015
10. Pisu, P., Canova, M., Soliman, A.: Model-based fault diagnosis of a NOx after treatment system. In: Proceedings of the 17th World Congress the International Federation of Automatic Control Seoul, Korea, 6–11 July (2008)
11. Ibrahim, H.A.: Experimental and Numerical Investigations of Fluid Flow through Catalytic Converters, Canada, September 2017

Design and Testing of Adsorption Column for PSA Process for CI Engine Performance and Emission

A. R. Patil[1][✉], A. D. Desai[2], Rohit Angre[3], and Harshal Shinde[3]

[1] Mechanical Department, G. H. Raisoni College of Engineering
and Management, SPPU, Pune, India
Patilamit20@hotmail.com
[2] Mechanical Department, SRES's Shree Ramchandra College of Engineering,
SPPU, Pune, India
[3] Mechanical Department, MES College of Engineering, SPPU, Pune, India

Abstract. The C I engines are commonly used applications made by engineering researchers. It has an effective efficiency of 30–35%, which suggests that just about seventieth percent of the energy within the fuel is lost within the exhaust gases, within the coolant, and in incomplete combustion of fuel and as radiation. Use of oxygen-enriched air in compression ignition engines has potential for low exhausts emission and particulate maters. There are various methods and experiments are developed in the field of reducing the emissions in the IC engines. There is no any other better technology for complete combustion and zero emissions from the engine. Oxygen enriched combustion is one of the attractive combustion technology to control pollution also it helps to improve the performance of the engine. Oxygen enrichment is achieved by the PSA system. PSA means Pressure Swing Adsorption which work on cyclic adsorption process for generation and purification of air. A PSA process has been alternative for different gas separations process. In the present work PSA System is developed using Zeolite and tested for CI Engine performance and emission at different load conditions. Its observed that PSA system improves BSFC, BTE and reduces HC and Smoke while resulting in increases in NOx formation.

Keywords: Adsorption · Combustion · Oxygen enrichment ·
Pollutant emission · PSA

1 Introduction

Diesel engine emission regulations demands for substantial reduction in emission and currently stringent emission regulations like EURO-6, TIRE-4 final, etc., The usual method of fueling the CI engine for combustion is to draw the air from the atmosphere to mix with the fuel, such as gasoline or other petrol products to combustion chamber so as to ignite the mixture therein. The major problem in this method is the emission of hydrocarbons, carbon monoxide, nitrous oxide and sulphur gases. It all caused due to the combustion of hydrogen and nitrogen which present in the air in large quantities along with the fuel. Incomplete combustion of air-fuel mixture because of presence of

© Springer Nature Singapore Pte Ltd. 2020
V. K. Gunjan et al. (Eds.): *ICRRM 2019 – System Reliability, Quality Control,*
Safety, Maintenance and Management, pp. 196–205, 2020.
https://doi.org/10.1007/978-981-13-8507-0_30

the unwanted gases in atmospheric air, since oxygen is the only necessary for burning, which results in reduced efficiency and also emission of poisonous gases which leads to environmental problems and health hazards. This problem can be reduced by supplying the pure oxygen instead of the atmospheric air intake. [1] PSA system is evaluated on the basis of purity of output and recovery in consideration with economy. Extra steps of bed coupling through recycling and cascading its product help in obtaining high purity. Mathematically represent as a set of partial differential equations [2] Adsorbents are chemical substance with high surface area/mass with porous solid structure. Since different gas molecules interact differently with adsorbent surface, hence make it possible to separate them. When it is in contact with fluid phase, an equilibrium state is achieved after certain time which put limit on its loading capacity for given fluid phase composition, temperature and pressure. Hence knowledge about equilibrium of different species is important in design and modelling of adsorption process.

2 Literature Review

Baskar, Senthilkumar studied the engine performance and emission parameters when affected by the use of oxygen enrichment. The HC emissions were drops to 10% at 23% oxygen to maximum of 40% at 27% oxygen enrichment level [1]. Jitin Yadav, Dr. Prof. Dhananjay Gupta, Dr. Prof. Manu Gupta has determined suitable value for an engine's compression ratio in which additional oxygen is used to lean the combustion. Based on the higher value of oxygen concentration and higher compression ratio are favorable in terms of performance for a diesel engine. Bhavin Mehta, Hardik Patel, et al. experimental results shows the fundamental consideration associated with the oxygen-enriched air – fuel ratio in combustion process. By using oxygen enriched air the brake thermal efficiency of the engine is considerably increased and it's obviously reduce the fuel consumption and CO, HC, PM emission. S. Jain, et al. has attempted to develop easy-to-use rules for design of PSA based on the analysis of inherent properties of adsorbate – adsorbent system [3]. Mohammed Fakhroleslam et al. proposed a hybrid observer which is employed for estimation of the active mode of the two beds, six-step PSA process. Dynamic behaviour of the PSA along Carlos A. Grande provides an overview of fundamental of PSA process while focusing specifically on different innovative engineering approaches that contributed to continuous improvement of PSA performance. Shivakumar, Singh P done several experiments on single cylinder diesel engine by providing oxygen enriched air inlet. Increasing the oxygen content with the air leads to faster burn rates and the ability to burn more fuel at the same stoichiometry. Added oxygen in the combustion air leads to shorter ignition delays and offers more potential for burning diesel. In their experiments studies the effect on Engine combustion by varying the oxygen at the inlet 21% to 27% with an interval of 2 to check their hypothesis that air with more oxygen and less nitrogen will result in reduced NOx emissions. [4] Dr. Hussein H. Hamed had designed the pressure swing adsorption system by using the molecular sieves zeolite 5A with the case of 2-column, and 4-step operation were used. Their results showed that an optimum concentration product of oxygen was 76.9% purity, at the adsorption pressure 4 bar, Temp 17.4 °C. Mahdi Asgari1, Hossein Anisi1, et al. have designed a commercial scale pressure swing adsorption system for hydrogen purification where a practical approach is proposed to estimate the breakthrough time of a commercial PSA process. They solve the mathematical model of PSA plant using Aspen adsorption software where the calculated breakthrough time for the hydrogen purification is compared with the industrial data. Zhang et .al have proposed in their work that oxygen enrichment in combustion process might help to improve the emissions, thermal efficiency and break power output of the diesel engine. The have performed study sing water diesel emulsion and oxygen enriched combustion. The result indicated that lower BSFC, higher cylinder pressure and shorter ignition delay period were observed when OEC was applied. [5] Zhang et al. had studied the effect of the EGR and oxygen enriched on smoke and NO emission where it is found that using a higher EGR rate can achieve low NO emission when oxygen enriched combustion is applied due to the proper combination of oxygen concentration and EGR rate can achieve low NO-smoke emission. [6] Shamal Indulkar, Sayali Dongare et al. performed study to compare different methods of enrichment of the oxygen like using air separation membrane, pressure adsorption theory (PSA) with the help of the zeolite, different additives such as karanja oil and proposed that air separation membrane is the most convenient method. Based on literature survey mentioned above along with other papers studied that there are many effort on emission control of CI Engine using different techniques of additives, biodiesel, engine design modification, exhaust treatment and supplementary devices like PSA which show different results based on situation. So in the present work we are going to test the applicability of PSA system in improving engine performance and specially it effect on engine emissions [7–12].

3   PSA (Pressure Swing Adsorption) Process

PSA system based on fact that due to high pressure, gas attracted to solid surface or adsorbed and this tendency increase with increase in pressure. When pressure is relived, this gas is released or desorbed. PSA system works on fact that different gas has different absorbing tendency with surface strongly or less strongly. In PSA with zeolite, air containing different gas, out of which nitrogen has attracted more strongly by zeolite thus trapped in it and remaining air is released thus raised the oxygen richness in air. When bed reaches end of its capacity, it can regenerated by reducing pressure thus releasing nitrogen. Hence become ready for next cycle. With the help of two adsorbent vessels makes the whole cycle continuous along with pressure equalization where leaving gas partially pressured incoming air [13] (Figs. 1, 2, 3, 4 and Table 1).

4   Design of Adsorption Column for PSA

Using mathematical calculation dimensions of Cylinders as per required pressure is decided. First important dimension is thickness of shell which we designed for required pressure and chosen material as shown in following. Assembly and individual component are drawn (Figs. 5, 6, 7 and 8).

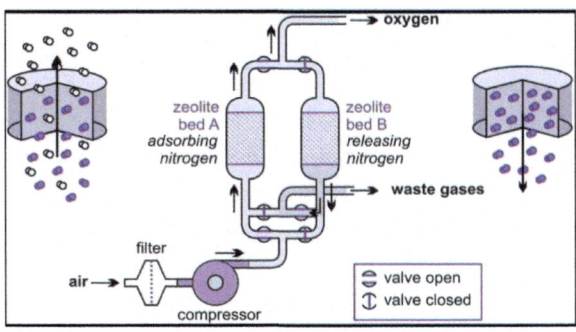

Fig. 1. PSA system [1]

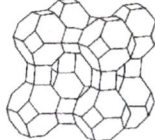

Fig. 2. Microstructure of zeolite

Fig. 3. Zeolite 5A molecular structure

Fig. 4. Zeolite 5A

Table 1. Properties of Zeolite 5A

Component name	Specification
Chemical name	Na12[(AlO2)12(SiO2)12]•27H2O
Pore diameter	5Å
Mesopore volume	0.062 cm^3/g
Micropore volume	0.176 cm^3/g
Surface area	571 m^2/g
Mass of crushed sample	2.046 g

1. Calculate the thickness of the cylinder hea

Design pressure, $P_i = 0.7$ N/mm^2

Length of column = 500 mm

Inner diameter of the shell, $d_i = 100$ mm

Yield strength of the material, $S_{yt} = 200N/$

Weld efficiency, $\eta_t = 85\% = 0.85$

Factor of safety, FOS = 2

Allowable stress, $\sigma_{all} = \dfrac{Syt}{FOS}$

$$= \dfrac{200}{2}$$

$$\sigma_{all} = 100 \text{ N/mm}^2$$

$$t_s = \dfrac{Pi \cdot di}{2 \cdot \sigma all \cdot \eta l - Pi} + 3$$

$$t_s = \dfrac{0.7 \cdot 100}{2 \cdot 100 \cdot 0.85 - 0.7} + 3$$

$$t_s = 4\text{mm}$$

Therefore the thickness of the shell is 4mm

2. Calculate the thickness of the head, t_h :

$$t_h = 0.7 * d_i \sqrt{\left(\dfrac{Pi}{\sigma all}\right)} + c$$

$$= 0.7 * 100 * \sqrt{\left(\dfrac{0.7}{100}\right)} + 3$$

$$t_h = 9 \text{ mm}$$

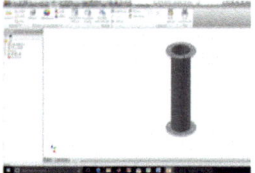

Fig. 5. AutoCAD model of cylinder

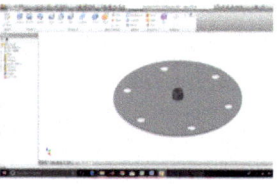

Fig. 6. AutoCAD model of flat plate head

Fig. 7. AutoCAD model of PSA system

Fig. 8. Actual photograph of PSA system

4.1 Components of PSA

See Fig. 9.

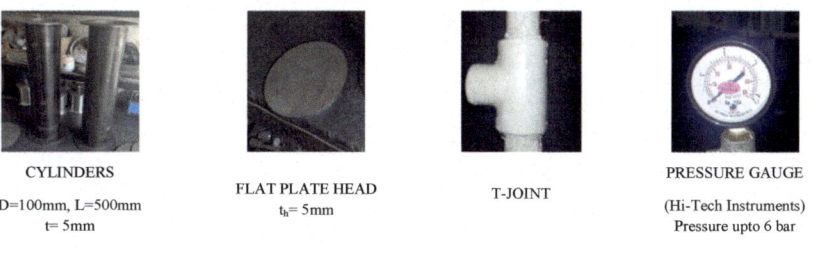

CYLINDERS	FLAT PLATE HEAD	T-JOINT	PRESSURE GAUGE

CYLINDERS
D=100mm, L=500mm
t= 5mm

FLAT PLATE HEAD
t_h= 5mm

T-JOINT

PRESSURE GAUGE
(Hi-Tech Instruments)
Pressure upto 6 bar

Fig. 9. Component of PSA system

5   Experimental Procedure

Air compressor started until the tank pressure reached the desired pressure (5 bar for O_2). Operate the control panel. Then the air enters to dryer to drying from moisture and cleans it from impurities. Regulate the air by the air regulator to the require pressure. The air enters to the PSA bed (1) from the bottom, the zeolite adsorbs nitrogen and the other gases and oxygen rich gas flow from the top. The air from the compressor is passed through the valve (V1) through the adsorption column where the nitrogen is trapped and the oxygen is released by opening the valve (V5). The oxygen through the valve (V5) is passed into the air box of the engine. While when the bed 1 is saturated with nitrogen the air is passed through the bed. Close the valve (V1) and open the valve (V2) the air passes through the bed 2 and oxygen is passed by opening valve (V6). By opening the valve (V3) the nitrogen in the bed 1 is released to atmosphere. By opening the valve (V34 the nitrogen in the bed 2 is released to atmosphere. Now during the oxygen enrichment the diesel engine is operated at varying load and the observation are noted (Fig. 10).

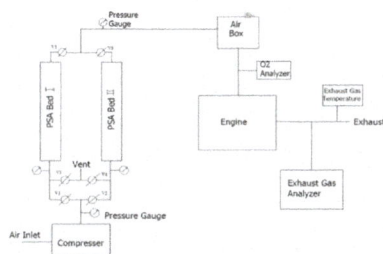

Fig. 10. Block diagram of experimental setup

6   Observations

As evident from Fig. 11 drowned using Tables 2 and 3 thermal efficiency is increased by oxygen enrichment and increases with load. Without PSA system the thermal efficiency is low as compared to that with PSA system. PSA system shows substantial improvement of around 20% at high load conditions. The reason for this may be

Table 2. Emission trial on engine without PSA system

Sr. no.	Load (Kg)	Speed (Rpm)	Time req. for 25 cc (sec)	CO	HC	CO$_2$	O$_2$	NO
1	0	1500	156	0.04	26	0.9	19.46	11
2	4	1500	85	0.05	30	1.9	17.87	52
3	8	1500	55	0.08	40	2.7	16.64	63
4	12	1500	42	0.4	74	4.01	14.59	102
5	15	1500	29	1.55	120	4.7	12.5	121

Table 3. Emission trial on engine with PSA system

Sr. no.	Load Kg	Speed Rpm	Time req. for 25 cc (sec)	CO	HC	CO$_2$	O$_2$	NO
1	0	1500	159	0.01	22	0.6	24.46	12
2	4	1500	105	0.02	24	1.6	23.1	65
3	8	1500	80	0.04	28	2.7	21.54	87
4	12	1500	57	0.15	64	3.5	19.16	129
5	15	1500	47	1.2	112	4.1	17.33	170

Table 4. Performance trial on engine without PSA system

Sr. no.	Load (kg)	SFC (kg/KW-s)	BTE (%)	CO (% vol)	HC (% vol)	CO$_2$ (% vol)	O$_2$ (% vol)	NO (ppm)
1	0	0	0	0.04	26	0.9	19.46	11
2	4	2.19	10.85	0.05	30	1.9	17.87	52
3	8	1.69	14.06	0.08	40	2.7	16.64	63
4	12	1.47	16.10	0.4	74	4.01	14.59	102
5	15	1.718	13.91	1.55	120	4.7	12.5	121

Table 5. Performance trial on engine with PSA system

Sr. no.	Load (kg)	SFC (kg/KW-s)	BTE (%)	CO (% vol)	HC (% vol)	CO$_2$ (% vol)	O$_2$ (% vol)	NO (ppm)
1	0	0	0	0.01	22	0.6	24.46	12
2	4	1.77	13.41	0.02	24	1.6	23.1	65
3	8	1.16	20.43	0.04	28	2.7	21.54	87
4	12	1.08	21.84	0.15	64	3.5	19.16	129
5	15	1.057	22.51	1.2	112	4.1	17.33	170

oxygen enrichment of air mixing with rich air fuel mixture hence abundance of oxygen at high load condition resulting in complete combustion resulting in high energy release and also its complimented by drop in SFC with PSA system compared with that without PSA system as evident from Tables 4, 5 and Fig. 12.

From Fig. 13 of Load Vs CO, CO formation increases with load due to increase in excess carbon which is due incomplete combustion of air fuel mixture while CO formation is less in engine with PSA system as sufficient oxygen is available for each Carbon molecules. It is evident from Fig. 13 that there is around 10% drop in CO at high load conditions.

When we compare the effect of PSA for HC formation using Graph of Fig. 14, it's observed that there is considerable drop in HC formation during idle and low load conditions which drop little as load increases.

The given below is the graph of carbon dioxide with load. As the load increase the CO_2 also increases. The CO_2 level in oxygen enriched system is less than the CO_2 level in the system without PSA (Fig. 15).

When we study the effect of PSA on NO_x formation using Fig. 16 and Tables 3, 4, it's observed that PSA has adverse effect shown by increase in NOx formation the reason for this may be due to leanness of air fuel mixture and high reactivity of nitrogen at high load condition. It was expected that NO_x should drop due to absence of nitrogen from air we are supplying which present for further studies and improvement in PSA system (Table 5).

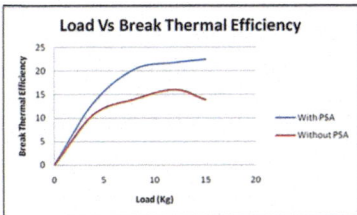

Fig. 11. Load vs Break thermal efficiency graph

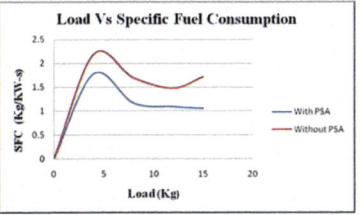

Fig. 12. Load vs Specific fuel consumption graph

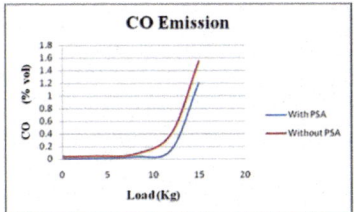

Fig. 13. Load vs CO emission graph

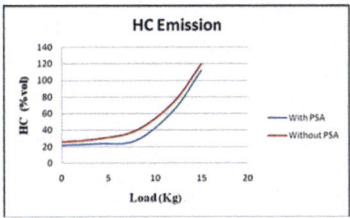

Fig. 14. Load vs HC emission graph

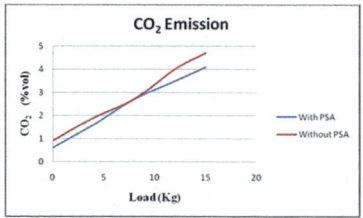

Fig. 15. Load vs CO_2 emission graph

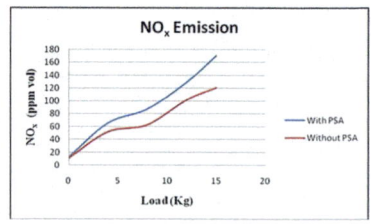

Fig. 16. Load vs CO_2 emission graph

7 Conclusion

The effect of oxygen enrichment on engine performance and emission characteristics are studied using single cylinder direct injection diesel engine and the results can be summarized as follows:

- There is increase in break thermal efficiency and reduced break thermal fuel consumption due to higher combustion rate, which in turn due to the high flame temperature. The air-fuel mixture becomes leaner due to increase in oxygen percentage which is result in lower break specific fuel consumption.
- The formation of the carbon monoxide and unburned hydrocarbon is reduced by the oxygen enrichment which in turn results in the complete combustion of the fuel. The oxygen enrichment in the diesel engine can help to reduce the smoke opacity effectively and suppress the growth of the soot particle diameter while reducing the number of the large soot particles.
- A high value of the NO_x emission with increase oxygen concentration is the main drawback of the technology which represents scope for future work.

References

1. Senthilkumar, A., Baskar, P.: Effects of oxygen enriched combustion on pollution and performance characteristics of a diesel engine. Eng. Sci. Technol. Int. J. **19**, 438–443 (2016)
2. Na, B.-K., Song, H.K., Chung, Y.: Short-cut evaluation of pressure swing adsorption systems. Comput. Chem. Eng. **22**(1), S637–S640 (2000)
3. Moharir, A.S., Li, P., Wozny, G., Jain, S.: Heuristic design of pressure swing adsorption: a preliminary study. Sep. Purif. Technol. **33**(1), 25–43 (2003)
4. Shivakumar, S.P.: Engine enhancement using enriched oxygen inlet. Int. Lett. Chem. Phys. Astron. **48**, 37–49 (2015)
5. Shu, G., Wei, H., Zhang, W., Liang, Y., et al.: Effect of oxygen enriched combustion and water-diesel emulsion on the performance and emissions of turbocharged diesel engine. Energy Convers. Manag. **73**, 69–77 (2013)
6. Chen, Z., Li, W., et al.: Influence of EGR and oxygen-enriched air on diesel engine NO–Smoke emission and combustion characteristic. Appl. Energy **107**, 304–314 (2013)
7. Climent, H., Miguel, L., Luján, J.M., et al.: Volumetric efficiency modelling of internal combustion engines based on a novel adaptive learning algorithm of artificial neural networks. Appl. Therm. Eng. **123**, 625–634 (2017)

8. Bupesh Raja, V.K., JayaPrabakar, J.: Performance and emission characteristics of cashew nut shell oil on the CI engine. Int. J. Ambient Energy (2018). https://doi.org/10.1080/01430750.2017.1421584
9. Desai, A.D., Madavi, A.D., Kamble, S.A., Navale, S.B., Dhutmal, V.U., Patil, A.R.: Comparative study on effect of biodiesel on CI engine performance and emission characteristics. In: Materials Today: Proceedings, vol. 5, pp. 3556–3562 (2018)
10. Portugal, A.F., Magalhães, F.D., Santos, J.C.: Simulation and optimization of small oxygen pressure swing adsorption units. Ind. Eng. Chem. Res. 43, 8328–8338 (2004)
11. Boozarjomehry, R.B., Fatemi, S., Fakhroleslam, M.: Design of a dynamical hybrid observer for pressure swing adsorption processes. In. J. Hydrogen Energy 42, 1–13 (2017)
12. Shen, Y., Zhang, D., Sun, W.: A systematic simulation and proposed optimization of the pressure swing adsorption process for N/CH separation under external disturbances. Ind. Eng. Chem. Res. 54(30), 7489–7501 (2015)
13. Gupta, D., Gupta, M., Yadav, J.: To study the performance of oxygen enriched diesel engine by varying compression ratios. IRJET 4(6), 5654–5657 (2017)

Prediction of California Bearing Ratio by Reliability Analysis: A Review

Harshita Bairagi[✉], Shreyas Mutkule, Pranali Malunjkar,
and Madhur Jain

Department of Civil Engineering, GHRIET, Pune 412207, India
harshita.bairagi1@gmail.com,
shreyasmutkule@gmail.com,
pranalimalunjkar23@gmail.com, 21mjl996@gmail.com

Abstract. Reliability analysis is used as a vital tool to predict the uncertainty and take right decisions. It is very important for engineers to predict the behaviour of soil and construction-materials (after construction) used in infrastructures. California Bearing Ratio (CBR) test is performed to measure the strength of soil, is often used as a design parameter of sub-grade for the design of flexible pavement. CBR test is a laborious test; therefore, it is vital to develop the models for quick assessment of CBR value. This study presents a review to determine the CBR value from reliability analysis which is quicker to estimate from their standard method of testing. In this study authors review predictive models using different sets of soil samples containing different index and engineering properties.

Key points: Reliability analysis · California Bearing Ratio ·
Index and engineering properties

1 Introduction

To design flexible pavement and quality control of unbound granular material, California Bearing Ratio test is the most suitable test which is used worldwide. It is an empirical test done by CBR test apparatus on sub-grade soils and unbound granular materials in the laboratory or field. CBR is a measure of the relative resistance of the unbound granular base/sub-base materials and sub-grade soils to uniaxial penetration. The strength of the material is completely dependent on the CBR value, so it requires high precision for sampling, test conditions (soaked or un-soaked sample) and for Penetration test. If the number of samples is more, it will be difficult to maintain the same precision for each sample testing. Paper here is to review the correlation between the basic properties of soil to CBR test by using Probabilistic Approach. Also the paper reviews the prediction of CBR value of soil after addition of admixtures with different length and amount. It's been concluded that better results for pavement design can be achieve by using Probabilistic approach.

© Springer Nature Singapore Pte Ltd. 2020
V. K. Gunjan et al. (Eds.): *ICRRM 2019 – System Reliability, Quality Control,
Safety, Maintenance and Management*, pp. 206–212, 2020.
https://doi.org/10.1007/978-981-13-8507-0_31

2 Need of Reliability-Based Design for Pavements

The uncertainty to predict the variation of California Bearing Ratio results in the field varies the performance of pavement. This uncertainty is the main cause of failure of pavement before the life span after designing with all parameters. Normally deterministic design optimisation was used for most of the pavement design. In this method the pavement were face the limits of design constraints. To reduce the design constraints, factor of safety has been used which is margin between the applied stress on the pavement and the sub-grade strength at failure. In this method the optimal pavement thickness is associated the high chance of failure because of uncertainties associated with the CBR. Deterministic design optimisation without including uncertainties is not reliable and may lead to failure. To overcome the limitation of deterministic approach reliability based design has been used. In this method the chances of failure will be less as compared to the deterministic approach. So it is important that the engineers adopt the reliability approach in place of deterministic design to reduce the chances of failure in the pavements. It reduces the cost of construction as the margin between response and capacity has been reduced and ensures safety. This method presents a model that shows the relationship between independent and dependent variable in statistical form.

Reliability analysis can be done manually using statics. As the equations are complicated, takes time to develop the relation between the dependent and independent variable. Now a day's number of software's has been developed to predict the regression equation. Some are given as:

- SPSS
- Stata
- JMP
- RATS
- JASP
- Weka
- Eviews

Several methods have been proposed to evaluate Probabilistic approach.

3 Methods Used in Reliability Analysis

Statistics is the base for Reliability analysis. Analysis does have two parts non-linear regression and linear regression. The data are fitted by a method of best fit method. It includes many techniques for modelling and analyzing multiple variables when the focus is on the relationship between a dependent variable and one or more independent variables. In the present review paper CBR is a dependent variable which the authors need to predict with the help of index and engineering properties (independent variable) of soils. In some paper admixture amount, type has been considered as an independent variable for the prediction of CBR (dependent variable).

3.1 Linear Regression

It is a linear approach to modelling the relationship between two variables, in which one is explanatory variable and other is dependent variable.

A linear regression is in the form of:

$$Y = a + bX_1 + cX_2 + \dots\dots$$

Where, X_1 and X_2 are explanatory (independent) variables and Y is dependent variable and $a, b, c, \dots\dots$ are the coefficients.

3.2 Non-linear Regression

It is a form of regression analysis in which observational data are designed by a function which is non-linear combination of model parameters and depends on one or more independent variables. In non-linear regression statistical model is in the form of

$$y \sim f(x, \beta)$$

Where x is one or more independent variable and y is dependent variable, f is non-linear function in the component of vector of parameters ß.

4 Literature Based on Reliability Analysis for Pavements

The reliability analysis for design of pavement was introduced first in 1990s. Agrawal and Ghanekar (1970) first recognised the importance of reliability based design. After that Chua, et al. (1992) and Kenis and Wang (1998) prepared the numerical simulation of pavement designs which is based on reliability principles. AASHTO in 1993 had introduced reliability based procedure for the design of flexible pavements. In addition Livneh et al. (1996) used reliability approach to determine the CBR for design of pavement. To evaluate pavement performance reliability, Kim, et al. (1998) preferred first-order second moment, Kim and Buch (2003) used point estimate methods and Retherford and McDonald (2010) performed First Order Reliability Method (FORM). Sani, et al. (2014) notified the method of reliability analysis for CBR using FORM on black cotton soil treated with cement kiln dust.

Semen (2006) reported the prediction models for CBR value to the different soil properties collected from specific locations worldwide. He has collected soil sample from 46 different locations to cover the all types of soils and determine the gradation, moisture content, density, specific gravity, plasticity and CBR and compared the CBR to the index properties of soil while using Machine learning methods. It includes nonlinear relationship mapping, nonparametric distribution treatment and superior generalization and implicit modelling were applied. He has prepared different sets of plastic and non-plastic soils and the data is compared with CBR by using conventional regression and existing CBR prediction methods. Karen S. Henry, Joshua Clapp, William Davids, Dana Humphrey and Lynette Barna in October 2009 reported the studies of soil types, characteristics of CBR values and correlate the CBR with the grain size distribution and plasticity of soil.

Depend on grain size distribution of soils, plasticity characteristics of soils, compaction characteristics of soils regression equation has been developed by number of researchers to predict the CBR value of soil (Begum et al. 2014; Shirur et al. 2014; Carter 1991 and Rehman et al. 2017). Some of researcher tried to establish the relation between CBR and field strength parameters (Kaur et al. 2012; Smith et al. 1983; Harison 1989; Webster et al. 1992; Webster et al. 1994; Chua 1992; Livneh et al. 1996; Ese et al. 1994; Coonse 1999, Sahoo and Reddy 2009 and Rakareddy et al. 2015).

Mechanistic Empirical Pavement Design Guide (NCHRP 2001) reported a model to predict the relation between CBR value and soil passing through 75 μ sieve and plastic limit of soil. Ramasubbarao and Siva Sankar (2013) have studied papers related to correlations/models to predict the soaked CBR and gave a simple correlation equation for predicting of soaked CBR of compacted soils. The equation is shows that Soaked CBR is dependent on percentage fine (F), sand (S), gravel (G), liquid limit (LL), plastic limit (PL) and compaction characteristics (OMC and MDD).

$$CBRs = 0.064F + 0.082S + 0.033G - 0.069LL + 0.157PL1.81MDD - 0.061OMC$$

The author concluded that the above equation to determine the soaked CBR is applicable only for particular range of soil parameters. Deepak et al. (2014) used multiple regression analysis method to develop model between soaked CBR and index properties of soil. To develop the correlation, soil samples have been collected from different road projects on Madhya Pradesh (India) wearing different plasticity characteristics. The soil samples cover low, medium and high compressible soils.

Alawi and Rajab (2013) have tested the sub-base layer of roads in Makkah area, Saudi Arabia. The prediction of CBR value is based on simple test such as sieve analysis, Los angeles abrasion test and compaction characteristics. With the reference of sieve analysis and compaction characteristics for the estimation of CBR multiple linear regression models are investigated.

Nguyen et al. 2015 conducted CBR test in the laboratory with fine grained subgrade soil collected from different locations from Victoria. It has been concluded that the satisfactory empirical correlation can be achieved for CBR by using physical properties of soil. Later the paper has been reviewed some researches related to prediction of CBR.

Srinivasa et al. 2016 presented the regression equation to predict the CBR using index properties and field strength parameters for black cotton soil collected from different locations of Karnataka (India). In this study 26 different locations has been selected for collection of black cotton soil. Standard Penetration test, Dynamic Cone Penetration test and CBR were performed with the soil samples and regression equation was developed using the statistical analysis by SPSS Software.

Moghal et al. 2017 has been evaluated the performance of two synthetic fibres (fibre cast and fibre mesh) used as a reinforcement for pavement sub-grade and CBR has been performed as a indicator of pavement performance. Number of CBR tests were carried out with varying percentage and length of the fibres (fibre cast and fibre mesh) with lime treated and without lime treated soil samples. In lime treated fibre reinforced soil sample also has been evaluated for 14 days of curing. All the tests has been analysed by using both approaches deterministic as well as probabilistic. In the probabilistic approach CBR has been used as a dependent variable while Fibre length, amount used as an independent variable and the design charts were developed to determine the optimum percentage of fibre length and amount using Targeted Reliability Index.

5   Conclusions

The study for the Reliability analysis has given some of the following conclusions:

1. Preferring reliability analysis instead of going for the tradition deterministic approach is much better to predict the CBR value for the pavement in all the way, still there is no sufficient Standards are available for the design of pavement using probabilistic approach.
2. It has been concluded that when the number of independent variable used for prediction of CBR value with reliability analysis increases, it gives better fit and the results obtained by reliability gives almost correct values (R^2 value is nearly equal to 1).
3. Sometimes it is observed that simple linear regression equation is not fit for the particular soil type, in that case multiple-linear equations will be more reliable.
4. Reliability analysis is a trial and error method to get the best fit equation for a particular parameter.
5. The accuracy of the analysis is determined on the value of R^2, which means if it is less than 1, other parameters should also be included for Reliability analysis and if R^2 value is 1, the analysis is considered as most accurate.
6. The model developed for a correlation of CBR is applicable only for particular soil for which it has been developed. It is not applicable globally or sometimes in soil from different regions of the country.
7. Moghal et al. 2017 determined the optimum percentage of fibre dosages and length by using Targeted Reliability Index which reduces the imbalance of fibre length and percentage amount in soil.

References

Moghal, A.A.B., Chittoori, B.C., Basha, B.M.: Effect of fibre reinforcement on CBR behaviour of Lime blended expansive soils: reliability approach. Artic. Road Mater. Pavement Des. **19**, 690–709 (2017)

Semen, P.M.: A generalized approach Technical Report ERDC/CRREL TR-06-15. Engineer Research and Development Centre, Cold Regions Research and Engineering Laboratory, Hanover, NH (2006)

Agrawal, K.B., Ghanekar, K.D.: Prediction of CBR from plasticity characteristics of soil. In: Proceedings of 2nd South-East Asian Conference on Soil Engineering, Singapore, June 11–15, pp. 571–576 (1970)

Carter, M., Bentley, S.P.: Correlation of Soil Properties. Pentech Press, London (1991)

Livneh, M., Ishai, I., Livneh, N.A.: Effects of vertical confinement on dynamic cone penetrometer strength values in pavement and sub-grade evaluations. Transport Research Record 1473, Washington, DC (1996)

Chua, K.H., Der Kiureghian, A., Monismith, C.L.: Stochastic model for pavement design. J. Transp. Eng. **118**(6), 769–786 (1992)

Kenis, W., Wang, W.: Pavement variability and reliability. In: International symposium on Heavy Vehicle Weights and Dimensions, Maroochydore, Queensland, Australia, Part 3, pp. 213–231 (1998)

Kim, H.B., Buch, N.: Reliability-based pavement design model accounting for inherent variability of design parameters. In: 82nd Annual Meeting on Transportation Research Board, Washington DC (2003)

Kim, H.B., Harichandran, R.S., Buch, N.: Development of load and resistance factor design format for flexible pavements. Can. J. Civ. Eng. **25**, 880–885 (1998)

Sani, J.E., Bello, A.O., Nwadiogbu, C.P.: Reliability estimate of strength characteristics of black cotton soil pavement sub-base stabilized with bagasse ash and cement kiln dust. Civ. Environ. Res. **6**(11), 115–135 (2014)

Retherford, J.Q., McDonald, M.: Reliability methods applicable to mechanistic-empirical pavement design method. Transp. Res. Rec. J. Transp. Res. Board **2154**, 130–137 (2010)

Webster, S.L., Grau, R.H., Williams, T.P.: Description and application of dual mass dynamic cone penetrometer. Instruction Report. GL-92-3, U.S. Army Engineer Research and Development Centre, Waterways Experiment Station, Vicksburg (1992)

Webster, S.L., Brown, R.W., Porter, J.R.: Force: projection site evaluation using the electronic cone penetrometer (ECP) and the dynamic cone penetrometer (DCP). Technical report GL-94-17, U.S. Army Engineer Research and Development Centre, Waterways Experiment Station, Vicksburg, MS (1994)

Sahoo, P.K., Reddy, K.S.: Evaluation of sub-grade soils using dynamic cone penetrometer. IIT Kharagpur, India (2009)

Rakaraddi, P.G., Gomarsi, V.: Establishing relationship between CBR with different soil properties. Int. J. Res. Eng. Technol. **04**(02), 182–188 (2015)

Kaur, P., Gill, K.S., Walia, B.S.: Correlation between soaked CBR value and CBR value obtained with dynamic cone Penetrometer, February (2012)

Shirur, N.B., Hiremath, S.G.: Establishing relationship between CBR value and physical properties of soil. IOSR J. Mech. Civ. Eng. Circ. Rep. North Carolina Department Of Transportation (NCDOT)IOSR-JMCE), October (2014)

Smith, R.B., Pratt, D.N.: A field study of in-situ California bearing ratio and dynamic cone penetrometer. Test. Road Subgrade Investig. **13**(4), 285–294 (1983)

Coonse, J.: Estimating California bearing ratio of cohesive piedmont residual soil using the scala dynamic cone penetrometer. Master's thesis (MSCE), North Carolina State University, Raleigh, N.C. (1999)

Ese, D., Myre, J., Noss, P., Vaernes, E.: The use of dynamic cone penetrometer (DCP) for road strengthening design in Norway. In: The Fourth International Conference on the Bearing Capacity of Roads and Airfields, Proceedings (1994)

Harison, J.A.: In situ CBR determination by DCP testing using a laboratory- based correlation. Aust. Road Res. **19**(4) (1989)

Srinivasa, R.H., Naagesh, S., Gangadhara S.: Development of regression equationsto predict C. B.R of black cotton soils of Karnataka, India. Int. J. Civ.Eng. Res. **7**(2), 135–152 (2016). ISSN: 2278–3652

Deepak, Y., Jain, P.K., Kumar, R.: Prediction of soaked CBR of fine grained soilsfrom classification and compaction parameters. Int. J. Adv.Eng. Res. Stud. (2014)

Divinsky, M., Ishai, I., Livneh, M.: Simplified generalized California bearing ratiopavement design equation. Transp. Res. Rec. **1539**, 44–50 (1996). National Research Council, Washington, D.C.

Begum, N., Sharma, B.: Determination of CBR value from compactioncharacteristics and index properties of fine grained soil. In: Proceeding of IndianGeotechnical Conference (2014)

Ramasubbarao, G.V., Siva Sankar, G.: Predicting soaked CBR value of fine grainedsoils using index and compaction characteristics. Jordan J. Civ.Eng. **7**(3), 354–360 (2013)

Rehman, Z.U., Khalid, U., Farooq, K., Mujtaba, H.: Prediction of CBR value from indexproperties of different soils. Tech. J. **22**(2) (2017). University of Engineering andTechnology (UET) Taxila, Pakistan

Alawi, M.H., Rajab, M.I.: Prediction of California bearing ratio of subbaselayer using multiple linear regression models. **14**(1), 211–219 (2013). Taylor & Francis

Thach Nguyen, B., Mohajerani, A.: Determination of CBR for fine-grained soilsusing a dynamic lightweight cone penetrometer. Int. J. PavementEng. **16**(2) (2015)

A New Approach to Control Assembly Variation in Selective Assembly Using Hierarchical Clustering

S. V. Chaitanya$^{(\boxtimes)}$ and A. K. Jeevanantham

Department of Manufacturing, School of Mechanical Engineering,
Vellore Institute of Technology, Vellore 632014, Tamil Nadu, India
svcaissms@gmail.com, akjeevanantham@vit.ac.in

Abstract. Complex assembly constitutes more than two parts. Tolerances assigned to individual components decide precision of assembly. Clearance and variation resulted in assembly decides precision of assembly and affect performance during working of assembly. During high precision mechanical assemblies, many parts become surplus due to more variation on component tolerances. Then, selective assembly is only solution to control the clearance variation. In this paper, a new methodology of hierarchical clustering approach is developed to predict the precision in assembly variation so that an assembly can confirm the desired clearance specifications. A valve train assembly of an IC engine that consists of cam-tappet-stem, is considered for the case analysis. The proposed methodology can be implemented in any number of components in real situations.

Keywords: Selective assembly · Clustering · Clearance variation ·
High precision assembly

1 Introduction

Selective assembly is a method of classifying the components by the selective number of groups (bins) according to their dimensional deviations, and assembling them with respect to a purposeful strategy rather than being at random, so that equal and small clearance can be obtained at the assembly level with lowest manufacturing cost. The method of partitioning is usually adopted in selective assembly. However, if partitioning is clubbed with consideration of the clustering then desired combination of parts becomes easy. The technique of selective assembly have been used in the industries for many years, notably for the assembly of ball bearing, nozzle unit of fuel injection pump, piston and cylinder fit, cutting drilling bit, electric drives, optical devices, etc.

© Springer Nature Singapore Pte Ltd. 2020
V. K. Gunjan et al. (Eds.): *ICRRM 2019 – System Reliability, Quality Control,
Safety, Maintenance and Management*, pp. 213–222, 2020.
https://doi.org/10.1007/978-981-13-8507-0_32

2   Literature Review

Many efforts have been made in improving the quality of selective assembly process. Mansoor [1] addressed the mismatching of number of components in the selective groups. Arai and Takeuchi [2] simulated a selective assembly process by assembly accuracy and the variation in storage based on a geometrical design model using a transformation matrix and a manufacturing error model of probability distribution function. Fang and Zang [3] developed an algorithm to minimize the surplus components in selective assembly using two methods *viz.* balanced probability and unequal tolerance zone. Wang [4] established a new dimensioning and tolerancing scheme for selective assembly by using semantic tolerance modelling. Kannan et al. [5] developed a new method to identify the best combination of selective group components in order to produce uniform and minimum assembly clearance to replace traditional methods. In another study, Kannan et al. [6] implemented the concept of Taguchi's quality loss function in selective assembly and modelled to derive the assembly loss due to the deviation in assembly clearance mean. Matsuura and Shinozaki [7] presented the study for optimal binning strategies under squared error loss when measurement error is present. Fischer [8] suggested methodology to obtain the Tolerance stake up in mechanical assemblies. Desrochers and Riviere [9] presented a matrix approach coupled to the notion of constraints for the representation of tolerance zones and clearances within CAD/CAM systems. Singh et al. [10] attempted for optimal tolerance design in mechanical assemblies involving interrelated dimension chains using genetic algorithm and demonstrated with the help of suitable examples. Marziale and Polini [11] compared two tolerance analysis models, the vector loop and the matrix. Khodaygan et al. [12] expressed geometrical and dimensional tolerances by small degrees of freedom of various geometric entities. Cao et al. [13] demonstrated a tolerance modelling method of complex features by dividing them into intrinsic and situation deviations. Weihua and Zhenqiang [14] modelled the cylindricity error using L-F functions and evaluated by particle search optimization (PSO) algorithm. Bo et al. [15] elaborated three main models for tolerance analysis, the Jacobian, the vector loop, and the torsor and compared. Yang et al. [16] used variation propagation control to reduce the assembly variations to prove that the average radial variations and angular errors decrease with the number of available orientations for the same stage. Chen et al. [17] elaborated the study of three dimensional tolerance analysis methodology for mechanical component using computer aided design. Calvo et al. [18] proposed a novel accurate method of minimum zone tolerance in which the non-linear minimax formulation of the original flatness or straightness problem was transformed into a set of linear problems and the optimal solution of the envelop planes or lines was derived through vectorial calculus of point coordinates. Laosiritaworn et al. [19] applied clustering for die storage, a case of paper packaging business, using k means clustering. Söderberg et al. [20] studied to increased quality in welded components using information and simulation framework.

Hence, existing research contribution mainly focuses on minimization of assembly clearance variation, assembly loss and surplus parts. Many researchers have been dedicated to the tolerance analysis of assemblies that incorporate different types of tolerances.

3 The Problem

The tolerance offers inevitable variation in the assembled part. If the part variation is not within control, then effects on clearances within parts in assembly. The individual components have significant spread of dimensions due to variation in process capability, and in assembly process such parts offers more clearance variation. The uniform partitioning of tolerance spread lead to results in surplus parts, and increases cost of manufacturing. For complex assembly it is necessary to obtain the close control on clearance variation between mating parts. The proposed method suggests controlling the variation in the clearance between parts, and helps to obtain the required precision in assembly without resulting surplus parts.

4 Proposed Methodology

In the proposed method, component individual tolerances are divided in to "n" partitions (say 6). Geometric tolerances partitioned, segregates parts as per tolerance specification. After the components are partitioned, then within partitioned group parts are clubbed together using cluster technique. For this initially population of all components is considered to 100. However one can perform the clubbing to any real number in manufacturing. The clubbing of component is determined between the close distances between them, within the respective partitioned group. This group is termed as cluster. The close clusters further clubbed together to form a group of components with desired clearance. The distance between the groups is referring as the clearance specification of the component. And finally all clusters together form the assembly having minimum distance between the components. Thus we will get the desired assembly specification. Clustering can be visualized as a Dendrogram- a tree like diagram that records the sequences of the merges or splits (divisions). The traditional method of clustering is represented in Fig. 1.

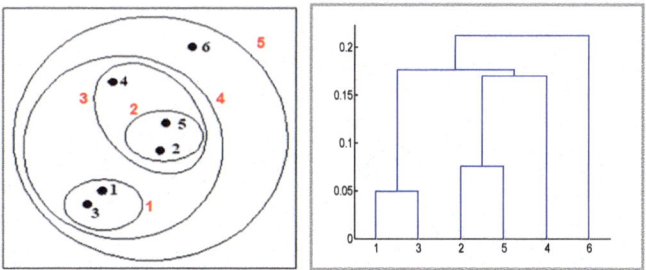

Fig. 1. Traditional hierarchical clustering and its dendrogram

The hierarchical algorithm is as follows.

1. Compute the proximity matrix (for all selective groups)
2. Let each data point (selective group) be a cluster
3. **Repeat**
4. Merge the two closest clusters
5. Update the proximity matrix
6. **Until** only a single cluster remains-.

Here the key operation is to obtain proximity of two clusters. For this it is important to obtain distance between the two clusters. Thus to measure distance between two clusters many methods are available. One of them is single link method.

$$sim(c_i, c_j) = \max_{x \in c_i, y \in c_j} sim(x, y).$$

This will result in "straggly" long thin clusters due to chaining effect. After merging C_i and C_j, the similarity of the resulting cluster to another cluster C_k is:

$$sim\left((c_i \cup c_j), c_k\right) = \max(sim(c_i, c_k), sim(c_j, c_k))$$

The significant use of this methodology is to obtain desired selective assembly with the consumption of all components. The present practice of selective assembly results in the surplus parts which can be conveniently avoided with desired precision using proposed methodology.

5   The Case

A complex linear valve train assembly considered here consists of three mating parts, namely, camshaft, valve and tappet and is shown in Fig. 3. The camshaft is shown with the 'nose up' position, but as the camshaft rotates the nose of the camshaft repeatedly makes contact with the tappet and pushes it down for a period of time causing the valve to open. The functional clearance is the variation from the bottom of the camshaft to the tappet. Deviation from the target clearance means that for each cycle, the valve will remain open for a longer or shorter period of time than specified by the design, causing engine performance to suffer. Selective assembly method can be used to reduce this variation to accomplish lesser assembly clearance variation without sacrificing the benefit of manufacturing components with wider tolerance. The quality characteristics of Cam-valve- tappet assembly and the tolerances are as follows.

1. Base radius of cam $(P) = 25.000 \pm \frac{0.012}{0.000}$ and Circularity tolerance $= 0.006$ mm.
2. Tappet thickness $(Q) = 8.000 \pm \frac{0.018}{0.000}$ and parallelism of 0.010 mm.
3. Valve stem length $(R) = 12.000 \pm \frac{0.012}{0.000}$ and cylindricity tolerance 0.003 mm.

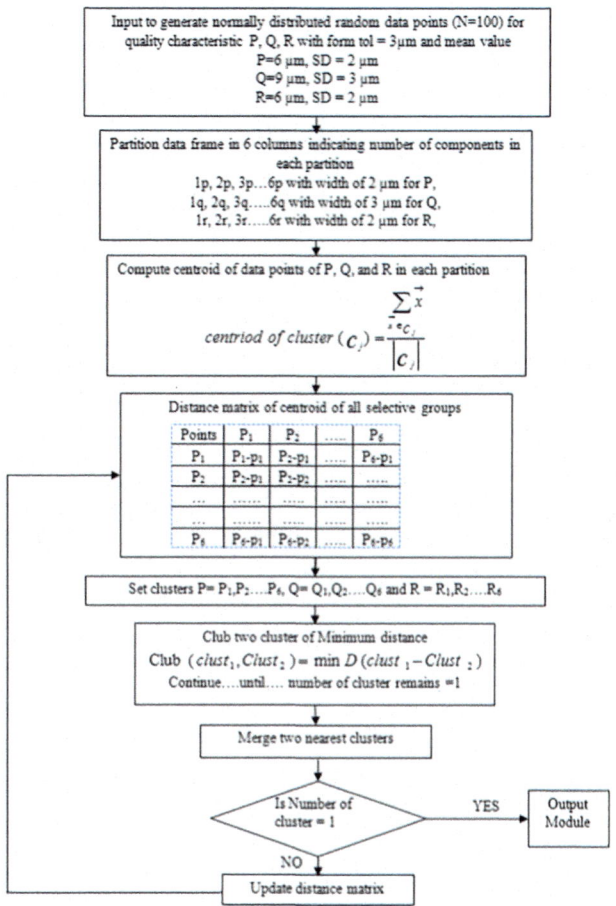

Fig. 2. Flow chart of proposed methodology

5.1 Surface Relation in Valve-Train Assembly

The cam incorporate faces F_1 and F_2. F_1 represents part of the cylinder and F_2 specifies surface of contour of the cam which operates tappet. The variation in the surfaces F_1 and F_2 affects the duration of valve operation resulting in a longer or shorter period. Between the faces F_1 and F_2 flatness and circularity of shape is considered for the analysis. For tappet F_3, F_4 and F faces are applied to form tolerance consideration. Between F_3 and F_4 straightness of surface and perpendicularity decided amount of deviation of surface F_3 and the variation of the distance between F_2 and F_3 results in displacement of stem which finally reflects on opening of valve time. Between tappet and valve stem perpendicularity and cylindricity of the stem is considered for analysis. Deviation in giving limit of these tolerance values reflect on functional performance of the cam-valve system. The surfaces and their importance are represented in the Table 1.

Table 1. Surfaces and their importance in valve train assembly

Description		Form				Profile		Orientation			Location	
Surfaces		—	□	○	/⊘/	⌒	⌓	∠	//	⊥	⊕	⊕
Cam	F1—F2	NA	Y	Y	NA	Y	NA	NA	NA	NA	NA	NA
Cam-Tappet	F2-F3	Y	Y	Y	Y		NA	NA	NA	Y	Y	NA
Tappet	F3-F4	Y	NA	NA	NA	NA	NA	NA	NA	Y	NA	NA
Tappet	F3-F5	Y	NA	NA	NA	NA	NA	NA	y	NA	NA	NA
Tappet-stem	F5-F6	Y	Y	Y	Y	NA	NA	NA	NA	Y	Y	NA
Valve stem	F6-F7	Y	NA	NA	Y	NA	NA	NA	NA	y	NA	NA

The tolerances for these three quality characteristics are the natural tolerances of the manufacturing process. The dimensions of the P, Q, R are initially generated as random numbers up to $n = 100$ and these 100 dimensions divided into '6' number of selective groups as per the specification of selective group. *i.e.* dimensions of P are clustered in the selective group with 2 μm. Dimensions of Q characteristics are clustered as a selective group with group specified as 3 μm. Similarly dimensions of R characteristics are clustered as a selective group with group specified as 2 μm. The steps followed are given in flowchart (Fig. 2) and the quality characteristics of valve train components are represented in Fig. 4.

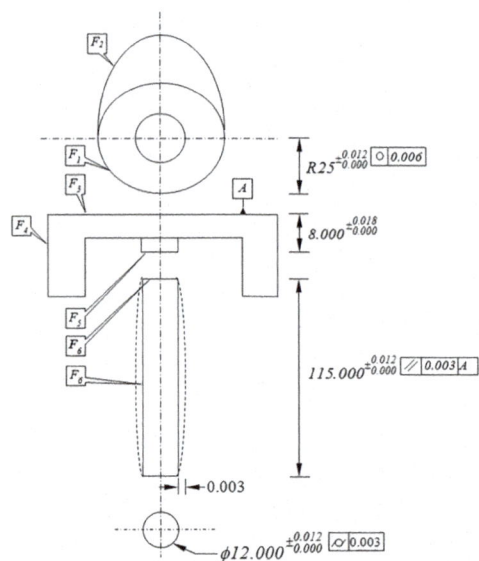

Fig. 3. Surfaces in valve train assembly

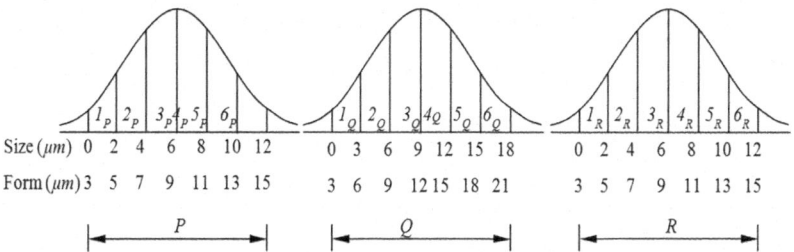

Fig. 4. Dimensional distributions of quality characteristic in cam, tappet and stem

Table 2. Centroid for quality characteristics P Q and R

Component	Point	Selective group	Number of components	Centroid
Cam	P_1	1_P	1	25.00456
	P_2	2_P	18	25.00611
	P_3	3_P	36	25.00813
	P_4	4_P	32	25.01000
	P_5	5_P	11	25.01186
	P_6	6_P	2	25.01369
Tappet	Q_1	1_Q	3	8.00490
	Q_2	2_Q	14	8.007757
	Q_3	3_Q	31	8.010708
	Q_4	4_Q	36	8.013316
	Q_5	5_Q	12	8.016187
	Q_6	6_Q	4	8.018774
Stem	R_1	1_R	1	12.00447
	R_2	2_R	18	12.00617
	R_3	3_R	36	12.00806
	R_4	4_R	32	12.01013
	R_5	5_R	11	12.01187
	R_6	6_R	2	12.01367

6   Result and Discussion

An algorithm is developed using R code to form the clusters from database of 'n' components. (n = 100) in case analysis).In this process the selective groups are clubbed together using hierarchical clustering on the basis of the minimum distance between the two groups. This is achieved using centroid method. The centroid is a centre point of the dimensions of the respective group. For example centroid of 2_P is obtained as follows. All other centroid are listed in Table 2.

$$\text{Centroid} = \frac{\text{sum of all points in respective group}}{\text{Total number of points in respective group}}$$

Form the table of distribution for 100 components for P the centroid for 2_P is obtained as

$$\text{Centroid} = \frac{25.00526 + 25.00582 + 25.0063 + 25.0064 + 25.00655}{5} = 25.00611$$

7   Assembly of Cam Tappet and Stem

The assembly of cam tappet and stem components is obtained according to cluster formation. The 100 number of each component is segregated as per cluster Dendrogram which will help to obtain the assembly with maximum and minimum clearance variation. The centroid of each selective groups indicate the average dimension of that group. Based on the centroid dimensions the assembly of component is clustered. For assembly of six groups of P, Q, R they are termed as Assembly-1 (A_1), Assembly-2 (A_2) and so on.

The Assembly dendrogram is obtained as per centroid method. This means the components having close distance are clubbed together. The assembly dendrogram for component resulted is indicated in Fig. 5 (Table 3).

Table 3. Component centroid for clustering

Cluster Point (Assembly)	Cam(P)	Tappet(Q)	Stem(R)
A_1	25.00456	8.004900	12.00447
A_2	25.00611	8.007757	12.00617
A_3	25.00813	8.010708	12.00806
A_4	25.01000	8.013316	12.01013
A_5	25.01186	8.016187	12.01187
A_6	25.01369	8.018774	12.01367

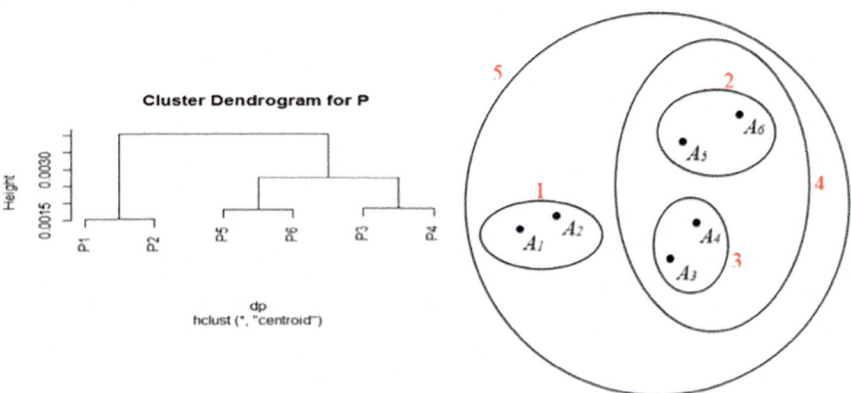

Fig. 5. Assembly dendrogram of hierarchical clustering

After making the assembly of three components, the total assembly clearance resulted will be obtained. Table 4 Represents the total assembly clearance.

Table 4. Assembly clearance variation

Cluster Combination		A_1	A_{12}	A_{15}	A_{16}	A_{13}	A_{14}
P		1_P	2_P	5_P	6_P	3_P	4_P
Q		1_Q	2_Q	5_Q	6_Q	3_Q	4_Q
R		1_R	2_R	5_R	6_R	3_R	4_R
Assembly clearance variation (μm)	C_l max	14	22	29	36	43	48
	C_l min	7	17	23	30	37	46
Clearance range (μm)	C_{range}	7	5	6	6	6	2
Clearance Variation (μm)		$7-2 = 5$					

8 Conclusion

From the case study it is clear that, the total clearance variation obtained in the assembly resulted 5 μm. The individual component tolerance can be set to 1.51 μm, 2 μm and 5 μm. The methodology used in the case is hierarchical clustering which will give benefit to manufacturer to set desired assembly clearance depends on customer requirement and can be obtained using selective assembly. For the dimensional distribution of components the distribution will be divided into selective groups. By obtaining the best combination of selective groups the assembly variation will be controlled to required extent.

References

1. Mansoor, E.M.: Selective assembly—Its analysis and applications. Int. J. Prod. Res. **1**, 13–24 (1961). https://doi.org/10.1080/00207546108943070
2. Arai, T., Takeuchi, K.: A simulation system on assembly accuracy. CIRP Ann. **41**(1), 37 (1992). STC A
3. Fang, X.D., Zhang, Y.: A new algorithm for minimizing the surplus parts in selective assembly. Comput. Ind. Eng. **28**(2), 341–350 (1995)
4. Wang, Y.: Semantic tolerancing with generalised intervals. Comput.-Aided Des. Appl. **4**(1–4), 257–266 (2007)
5. Kannan, S., Jayabalan, V., Jeevanantham, K.: Genetic algorithm for minimizing assembly variation in selective assembly. Int. J. Prod. Res. **41**, 3301–3313 (2003). https://doi.org/10.1080/0020754031000109143
6. Kannan, S.M., Jeevanantham, A.K., Jayabalan, V.: Modelling and analysis of selective assembly using Taguchi's loss function. Int. J. Prod. Res. **46**, 4309–4330 (2008). https://doi.org/10.1080/00207540701241891
7. Matsuura, S., Shinozaki, N.: Optimal binning strategies under squared error loss in selective assembly with measurement error. Commun. Stat.-Theory Methods **36**, 2863–2876 (2007). https://doi.org/10.1080/03610920701386984

8. Fischer, B.R.: Mechanical Tolerance Stackup and Analysis, 2nd edn. CRC Press, Boca Raton (2011)
9. Desrochers, A., Riviere, A.: A matrix approach to the representation of tolerance zones and clearances. Int. J. Adv. Manuf. Technol. **13**, 630–636 (1997)
10. Singh, P.K., Jain, S.C., Jain, P.K.: Advanced optimal tolerance design of mechanical assemblies with interrelated dimension chains and process precision limits. Comput. Ind. **56**, 179–194 (2005)
11. Marziale, M., Polini, W.: A review of two models for tolerance analysis of an assembly: vector loop and matrix. Int. J. Adv. Manuf. Technol. **43**, 1106–1123 (2009)
12. Khodaygan, S., Movahhedy, M.R., Fomani, M.S.: Tolerance analysis of mechanical assemblies based on modal interval and small degrees of freedom (MI-SOF) concept. Int. J. Adv. Manuf. Technol. **50**, 1041–1061 (2010)
13. Cao, Y., Zhang, H., Mao, J., Xusong, X., Yang, J.: Study on tolerance modeling of complex surfaces. Int. J. Adv. Manuf. Technol. **53**, 1183–1188 (2011)
14. Weihua, N., Zhenqiang, Y.: Cylindricity modeling and tolerance analysis for cylindrical components. Int. J. Adv. Manuf. Technol. **64**, 867–874 (2013)
15. Bo, C., Yang, Z., Wang, L., Chen, H.: A comparison of tolerance analysis models for assembly. Int. J. Adv. Manuf. Technol. **68**, 739–754 (2013)
16. Yang, Z., Popov, A.A., McWilliams, S.: Variation propagation control in mechanical assembly of cylindrical components. J. Manuf. Syst. **31**, 162–176 (2012)
17. Chen, H., Jin, S., Li, Z., Lai, X.: A comprehensive study of three dimensional tolerance analysis methods. Comput.-Aided Des. **53**, 1–13 (2014)
18. Calvo, R., Gómez, E., Domingo, R.: Vectorial method of minimum zone tolerance for flatness, straightness, and their uncertainty estimation. Int. J. Precis. Eng. Manuf. **15**(1), 31–44 (2014)
19. Laosiritaworn, W., Kitjongtawornkul, P., Pasui, M., Wansom, W.: 'Die storage improvement with k-means clustering algorithm', a case of paper packaging business. In: 4th International Symposium on Computational and Business Intelligence, pp. 212–215 (2016). 7743286
20. Söderberg, R., et al.: An information and simulation framework for increased quality in welded components. CIRP Ann. – Manuf. Technol. (2018). https://doi.org/10.1016/j.cirp.2018.04.118

Application of Desirability to Find Out Ideal Input Parameter Setting in WEDM Operation

Himadri Majumder[✉], Santosh Hiremath, Subhash Kumar, and Pragat Kulat

G.H.Raisoni College of Engineering and Management, Pune 412207, India
himu.nita@gmail.com

Abstract. Among the different non-traditional machining practice wire electrical discharge machining (WEDM) is the most flexible, useful and high précised machining process which is used to machine conductive materials. Presence of several contradictory responses in WEDM makes it difficult to choose the optimum machining parameter setting. This research demonstrates a multiple-criteria decision analysis (MCDA), desirability function analysis to optimize key input parameters for some contrary outputs during WEDM of inconel 718. Several significant machining variables, like pulse on time (T_{ON}), pulse off time (T_{OFF}), pulsed current (WF) and servo voltage (SV) were considered as machining inputs for this investigation. The selected WEDM outputs are kerf width (KW), material removal rate (MRR) and root mean square roughness (Rq). The ideal input setting for multi-performance features has been found as $T_{ON} = 120$ µs., $T_{OFF} = 53$ µs., $I = 210$ A. and $SV = 25$ V. The current research concentrated on the application of MCDA desirability function analysis as a precarious selection approach to deal with multi criteria optimization atmosphere.

Keywords: Inconel 718 · MCDA · DFA · Optimization · WEDM

1 Introduction

The problems accompanying with a conventional machining process like chatter, vibration and stress can be considerably eradicated using non-conventional machining process like wire electrical discharge machining (WEDM). WEDM utilises electro-thermal manner which conveniently can cut electrically conductive "difficult-to-machine" materials. The theory of WEDM process was established by Soviet scientist Lazarenko in the 1940s [1]. In WEDM, the electric discharge jumps from the wire electrode to the workpiece and erodes material from the workpiece. Di-electric, flowing in between electrode and workpiece, flushes away the eroded debris. Electric discharge produces by the pulsed current causes the melting and vaporization of materials from the workpiece [2]. WEDM, a distinct form of non-conventional machining process, is one of the most promptly developed, accurate and essential manufacturing practices utilized to machine nickel-titanium alloys [3–5], shape memory alloys [6, 7], hard material [8], ceramics [9], metal matrix composites, miniature and micro-parts in metals etc.

© Springer Nature Singapore Pte Ltd. 2020
V. K. Gunjan et al. (Eds.): *ICRRM 2019 – System Reliability, Quality Control, Safety, Maintenance and Management*, pp. 223–228, 2020.
https://doi.org/10.1007/978-981-13-8507-0_33

Due to complicated process mechanism, plenty of contradictory input variables make it appallingly difficult to decide optimum input parameter grouping to get most favored outputs. Here comes the requisite of multiple-criteria decision analysis (MCDA). Quite a few unique MCDA are now available which can offer this selection process [10–13]. Desirability function analysis (DFA) does not comprise complex mathematical computation and is easy to implement among other MCDA methodology [14].

In this paper, MCDA methodology DFA has been adept to pick superlative input parameter combination for WEDM outputs known as kerf width (KW), material removal rate (MRR) and root mean square roughness (Rq) for inconel 718.

2   Materials and Method

2.1   Materials, Experimental Setup and Data Collection

Nickel chromium alloy inconel 718 choose as workpiece due to their widespread utilization in elevated temperature applications like gas turbine, nuclear reactors, electrical power generation equipment, chemical vessels and aerospace industries etc. In this study, an amount of tests were accomplished on four axes CNC WEDM following L_9 orthogonal array. Diverse noteworthy variables like pulse-off time (T_{OFF}), pulse-on time (T_{ON}), pulsed current (I) and servo voltage (SV) were taken as input parameters (see Table 1) to find out vital machinability features known as kerf width (KW), material removal rate (MRR) and root mean square roughness (Rq). Brass wire having 0.25 mm diameter and de-ionized water was utilized as wire electrode and dielectric medium respectively. From 5 mm thick Inconel 718 plate, 5 mm length was cut by respective input setting. For respective setting, the kerf width was measured utilizing Toolmaker microscope. Taylor Hobson 3D profilometer was used to find out root mean square roughness. The material removal rate has been calculated as per the Eq. 1.

$$MRR = V * KW * t \qquad (1)$$

here, V = Cutting speed.
KW = Kerf width.
t = Thickness of the workpiece.

Table 1. Input parameters with their levels

Input parameter	Indication	Unit	I	II	III
Pulse ON time (T_{ON})	A	μs	110	115	120
Pulse OFF time (T_{OFF})	B	μs	46	53	60
Pulsed current (I)	C	A	210	220	230
Servo voltage (SV)	D	V	15	20	25

2.2 Adopted Methodology

In this research, to optimize dissimilar correlated responses of WEDM of inconel 718, a MCDA desirability function analysis has been applied.

2.3 Desirability Function Analysis (DFA)

Following stages are adopted in desirability function analysis:

1st *Stage:* Identification of the objectives and its features. Decision matrix prepared to exemplify the performance characteristics with respect to different outputs.
2nd *Stage:* For individual response, the desirability index (d_i) is calculated. This value depends on the character of the particular output [14].

- When attributes characteristics is "Larger-is-better"

$$d_i = \begin{cases} 0, & x \leq x_{min} \\ \left(\frac{x-x_{min}}{x_{max}-x_{min}}\right), & x_{min} \leq x \leq x_{max}, r \geq 0 \\ 1, & x \geq x_{min} \end{cases} \quad (2)$$

where, d_i = Single desirability index;
x_{min} & x_{max} are the lowermost and uppermost values of x;
r = Weight.

- When attributes characteristics is "smaller-is-better"

$$d_i = \begin{cases} 1, & x \leq x_{min} \\ \left(\frac{x-x_{max}}{x_{min}-x_{max}}\right), & x_{min} \leq x \leq x_{max}, r \geq 0, \\ 0, & x \geq x_{max} \end{cases} \quad (3)$$

3rd *Stage:* The overall desirability grade (d_G) is assessed using the following equation:

$$d_G = \left(d_1^{w_1} * d_2^{w_2} * d_3^{w_3} * \ldots * d_i^{w_i}\right)^{\frac{1}{w}} \quad (4)$$

where, d_i = Single desirability index;
w_i = Weight allotted to distinct output individually;
w = Summation of all distinct weights.

Highest overall desirability grade value is replicated as optimum setting of corresponding inputs.

3 Results and Discussion

Total nine experiments were performed, varying different input variables, ensuring Taguchi's L_9 orthogonal array. Input variables grouping with their respective responses are presented in Table 2.

Table 2. Input variables grouping and their responses.

Exp. no.	Input variables				Responses		
	T_{ON}	T_{OFF}	I	SV	KW	MRR	Rq
1	110	46	210	15	326.84	784.42	3.41
2	110	53	220	20	339.47	1544.59	3.87
3	110	60	230	25	322.51	1209.41	3.74
4	115	46	220	25	332.14	863.56	3.42
5	115	53	230	15	353.69	1167.18	3.56
6	115	60	210	20	330.66	1421.84	3.98
7	120	46	230	20	332.54	1679.33	4.19
8	120	53	210	25	329.93	1567.17	3.97
9	120	60	220	15	358.25	1271.79	3.68

The optimum grouping of the input variables was resolute using desirability function analysis. From the dissimilar correlated responses, "smaller-is-better" principles were implemented for avg. kerf width (KW) and avg. root mean square roughness (Rq) whereas "larger-is-better" principles was implemented for material removal rate (MR). Using Eq. 2 the individual desirability index (di) were calculated for avg. kerf width (KW) and avg. root mean square roughness (Rq) on the contrary for material removal rate (MRR) was calculated using Eq. 1. The weightage were taken as 30%, 40% and 30% for KW, MRR and Rq respectively. Using Eq. 3, the overall desirability grade (dG) was estimated. Table 3 shows the individual desirability index and the overall desirability grade for all the responses. In desirability function analysis, maximum value of overall desirability grade implies the optimal combination. It is evidently explicable from Table 3 that experiment no. 8 has the maximum dG value with 0.57515. Desirability function analysis gives the optimal input variable combination as T_{ON} = 120 µs., T_{OFF} = 53 µs., I = 210 A. and SV = 25 V.

Table 3. Optimization using desirability function analysis.

Exp. no.	Individual desirability index (di)			dG	Rank
	KW	MRR	Rq		
1	0.87885	0	1	0	7
2	0.52546	0.84944	0.41026	0.53655	3
3	1	0.47490	0.57692	0.54237	2
4	0.73055	0.08844	0.98718	0.20501	6
5	0.12759	0.42771	0.80769	0.24727	5
6	0.77196	0.71227	0.26923	0.49621	4
7	0.71936	1	0	0	7
8	0.79239	0.87467	0.28205	0.57515	1
9	0	0.54460	0.65385	0	7

4 Conclusions

Using WEDM operation, inconel 718 plate was machined and the experimental outcomes were optimized utilizing desirability function analysis. Following major conclusions might be drawn:

- MCDM method desirability function analysis was found helpful approach because it involves fewer mathematical designs. So it will be very useful for them who does not have very strong mathematical contextual.
- For different preferred outcomes, the optimum input setting was found using desirability function analysis as T_{ON} = 120 μs., T_{OFF} = 53 μs., I = 210 A. and SV = 25 V.
- Desirability function analysis has the ability to irradiate different process variation which makes it added beneficial where abundant outputs are present.

The correctness of the projected optimization method is subjected to the studied parameter range and environment. As a future work, desirability function analysis can also be employed in different other conventional as well as non-conventional machining operations.

References

1. Lazarenko, B.: To invert the effect of wear on electric power contacts. Dissertation of the All-Union Institute for Electro Technique in Moscow/CCCP (1943)
2. Majumder, H., Maity, K.: Predictive analysis on responses in WEDM of titanium grade 6 using general regression neural network (GRNN) and multiple regression analysis (MRA), Silicon, pp. 1–14 (2018)
3. Kumar, A., et al.: NSGA-II approach for multi-objective optimization of wire electrical discharge machining process parameter on inconel 718. Mater. Today: Proc. **4**(2), 2194–2202 (2017)
4. Majumder, H., et al.: Use of PCA-grey analysis and RSM to model cutting time and surface finish of Inconel 800 during wire electro discharge cutting. Measurement **107**, 19–30 (2017)
5. Majumder, H., Maity, K.: Optimization of machining condition in WEDM for titanium grade 6 using MOORA coupled with PCA—a multivariate hybrid approach. J. Adv. Manufact. Syst. **16**(02), 81–99 (2017)
6. Majumder, H., Maity, K.: Application of GRNN and multivariate hybrid approach to predict and optimize WEDM responses for Ni-Ti shape memory alloy. Appl. Soft Comput. **70**, 665–679 (2018)
7. Manjaiah, M., Narendranath, S., Basavarajappa, S.: Wire electro discharge machining performance of TiNiCu shape memory alloy. Silicon **8**(3), 467–475 (2016)
8. Saha, A., Mondal, S.C.: Statistical analysis and optimization of process parameters in wire cut machining of welded nanostructured hardfacing material, Silicon, pp. 1–14 (2018)
9. Kumar, C.S., Patel, S.K.: Effect of WEDM surface texturing on Al2O3/TiCN composite ceramic tools in dry cutting of hardened steel. Ceram. Int. **44**(2), 2510–2523 (2018)
10. Khan, A., Maity, K.: A novel MCDM approach for simultaneous optimization of some correlated machining parameters in turning of CP-titanium grade 2. Int. J. Eng. Res. Afr. **22**, 94–111 (2016)

11. Khan, A., Maity, K.: Application of MCDM-based TOPSIS method for the optimization of multi quality characteristics of modern manufacturing processes. Int. J. Eng. Res. Afr. **23**, 33–51 (2016)
12. Naik, D.K., et al.: Experimental investigation of the PMEDM of nickel free austenitic stainless steel: a promising coronary stent material, Silicon, pp. 1–9 (2018)
13. Majumder, H., Maity, K.: Prediction and optimization of surface roughness and micro-hardness using grnn and MOORA-fuzzy-a MCDM approach for nitinol in WEDM. Measurement **118**, 1–13 (2018)
14. Majumder, H., Maity, K.: Multi-response optimization of WEDM process parameters using taguchi based desirability function analysis. In: IOP Conference Series: Materials Science and Engineering. IOP Publishing (2018)

Parametric Study of Plate Girder Using Different Stiffener Arrangement Stiffener Length Per Meter Span Using Cold Formed Material

Bhushan D. Jadhav[1][(✉)], Subodh S. Patil[1], Santosh S. Mohite[1], and Sharif H. Shaikh[2]

[1] Department of Civil Engineering, ADCET, Ashta 416301, India
jadhavbhushan0028@gmail.com
[2] Department of Civil Engineering, G.H.R.C.E.M, Wagholi, Pune, India

Abstract. Thin walled cold formed steel members have broad close attention in building structures. If girders are subjected to the combination of bending, shear, uniformly distributed load, concentrated loads and axial loads. The increasing stability behavior in the design should be taken into consideration.

This paper presents research analysis of the optimal design of cold formed stiffener arrangement under two point loading conditions. Firstly keeping all other parameters constants analysis is carried out for different stiffener arrangements in finite element program ANSYS, secondly analysis result are compared with experimental results. Finally a parametric study was undertaken to determine the different stiffener arrangements on structural behavior of I beam girder.

Keywords: Cold formed steel · Stiffener · ANSYS

1 Introduction

The plate girders are generally I-beams manufactured from separate steel plates, which are bolted or riveted or welded to each other to form web and flanges of I beam plate girder.

As is widely known, for economical section of plate girder to make a web as thin as possible. The plate girder are widely used for railway or highway bridge members beyond 20 m span. Because, the plate girder steel section are used to resist bending action and shear action as well as torsional effect on it. The longitudinal or transverse or diagonally stiffeners are used to resist the bending and shear action. Transverse stiffener used for bearing shear web buckling and longitudinal stiffener used for bearing compression or bending buckling of web. The diagonal stiffener improve local buckling of web under combination of shear and bending.

The light gauge structures are use after World War 2 and it has more importance in construction material. Light gauge can be used in industries as wall coverings and floor decking and purlins to roof sheeting, metal building construction. When we compare light

© Springer Nature Singapore Pte Ltd. 2020
V. K. Gunjan et al. (Eds.): *ICRRM 2019 – System Reliability, Quality Control, Safety, Maintenance and Management*, pp. 229–237, 2020.
https://doi.org/10.1007/978-981-13-8507-0_34

gauge steel section and hot rolled section the light gauge steel structure having less thickness. Due to less thickness it is light in weight and it has good yield strength (Fig. 1).

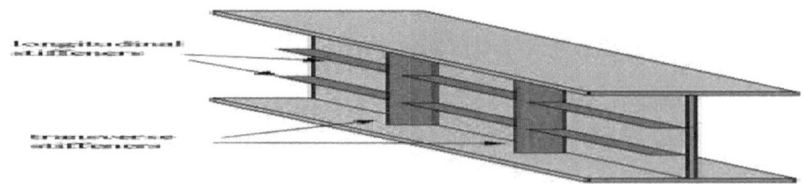

Fig. 1. Cross section of plate girder with stiffeners

1.1   Brief History

Degtyareva and Degtyareva [1]
In this paper they have developed an equation for prediction of coefficient of shear buckling of longitudinal web stiffened solid channel. Also equation for prediction of elastic shear buckling load and ultimate shear strength of channel with longitudinal stiffened slotted web. 92 and 1876 channel models with solid webs and longitudinal stiffened slotted analyzed in four stages. The author also did study on ultimate shear strength and elastic buckling load of the channels. The slotted channels failed in shear buckling with has analyzed for the equations without tension field action.

Xie et al. [2]
In this paper author studied cold formed Steel beam behavior subjected to bending and torsion. The analysis is done with the finite element program ABAQUS. Two methods are used to track an equilibrium path are modified risk method and arc length method. The 30 beam element was modelled and performed in ABAQUS. The beam element was restrained at both ends. The analysis results was compared with experimental results.

Gotluru et al. [3]
In this paper they have done the thin walled members can be used as a separate structural framing. The frame members are considered as beam element and shell element. The beam and shell elements are modelled using finite element program ABAQUS. The model was analyzed with 30 numbers of beam element along the length. Analysis is done with the help of modified risk method and for tracing equilibrium path used as arc length method. The material for model analysis is used elastic perfectly plastic. The comparison is done with the help of result obtained with experimental result. The author has suggested that the beam is loaded it displace horizontally and rotates gradually. Due to load lateral buckling of beam does not takes place. The failure is occur due to yielding of material.

2 Literature Gap

From above literature it is concluded that

- Cold formed sections are used as a lipped channel beams. They have analyzed for shear buckling, concentrated load, uniformly distributed load, axial load, and simple support condition, torsional and bending.
- Cold formed material never used to make plate girder.
- This project is going to introduce to a unique and innovative concept to make a plate girder by cold formed material.
- With full analysis of behavior of cold formed plate girder with different stiffener arrangements for stress variations and deformations for optimum design.

3 Problem Statement

The material used in girder and stiffener is hot rolled because it has a good strength and it is available in any thickness but cold rolled materials has about 15% to 30% more strength than hot rolled material. So there is a scope of improvement in a stiffener girder design. This work aims to replace a hot rolled material used to design a stiffener by cold rolled material and also analyze new material for different stiffener arrangement for minimum deformation at maximum load.

4 Methodology

- Collection of review of journals and articles to get idea of research work conducted earlier on this subject.
- Study and understand design analysis procedure using different codes and practices.
- Design plate girder and stiffeners by IS 800-2007, Keeping stiffener length constant.
- Design plate girder in cold formed section by using IS 801:1975.
- Import stiffener design to FEM software.
- Prepare a model in FEM software using cold form material properties.
- Keeping load on model constant perform analysis for different stiffener arrangements.
- Based on result, stiffener arrangement which gives minimum deformation at maximum load will be recommended.
- Conclusion will be drawn from results of analysis.

The cold formed section $140 \times 60 \times 15 \times 2$ mm is selected from IS 811:1987 for 140 mm depth.

- Total stiffener length for 25 m plate girder is 1,58,933.33 MM
- Converted total stiffener length for 1.5 m plate girder is 9,536 MM (Fig. 2)

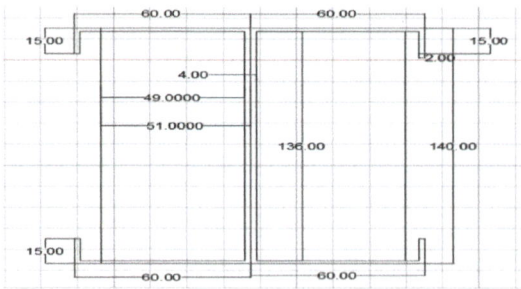

Fig. 2. Cross section of cold formed plate girder

5 Modeling in CATIA

CATIA software generally used for the 3d modeling (Figs. 3, 4, 5, 6, 7, 8, 9, 10, 11, 12, 13, 14, 15, 16 and 17).

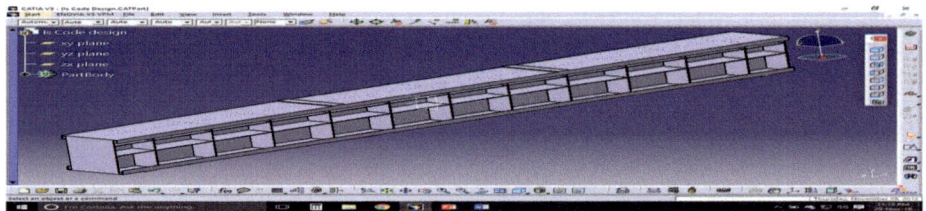

Fig. 3. IS code design standard stiffener arrangement

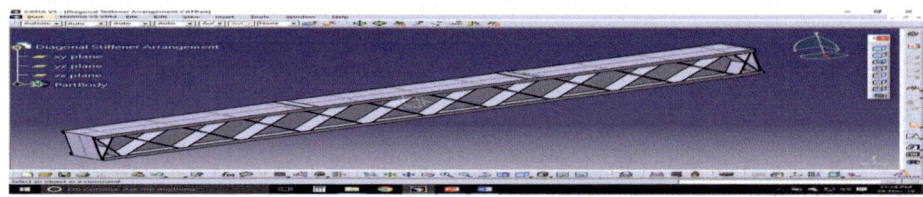

Fig. 4. Pattern 1. Diagonal stiffener arrangement

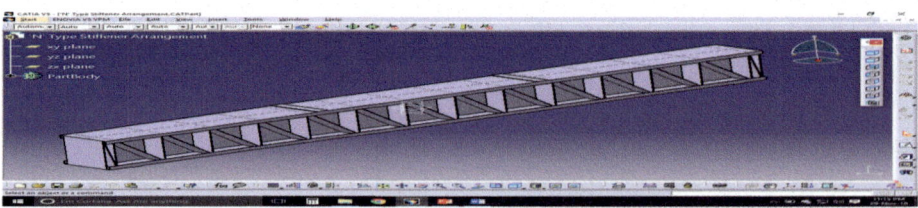

Fig. 5. Pattern 2. "N" type stiffener arrangement

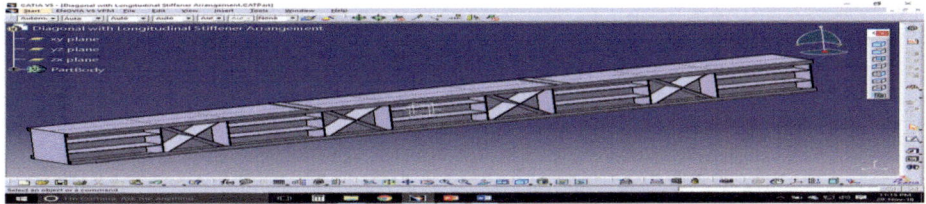

Fig. 6. Pattern 3. Diagonal with longitudinal stiffener arrangement

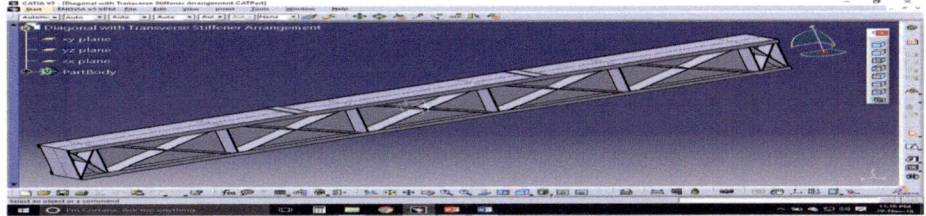

Fig. 7. Pattern 4. Diagonal with transverse stiffener arrangement

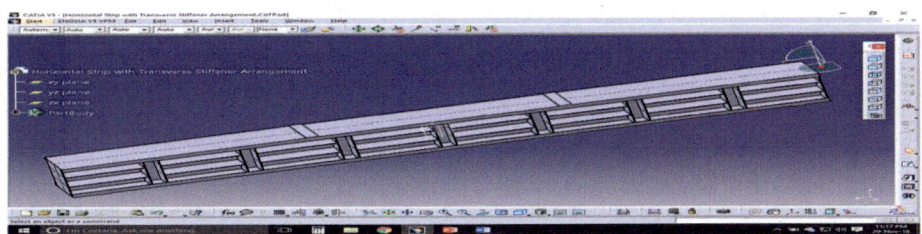

Fig. 8. Pattern 5. Horizontal strips with transverse stiffener arrangement

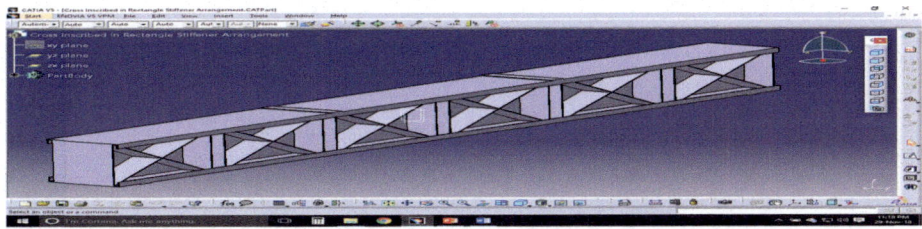

Fig. 9. Pattern 6. Cross inscribed in rectangle stiffener arrangement

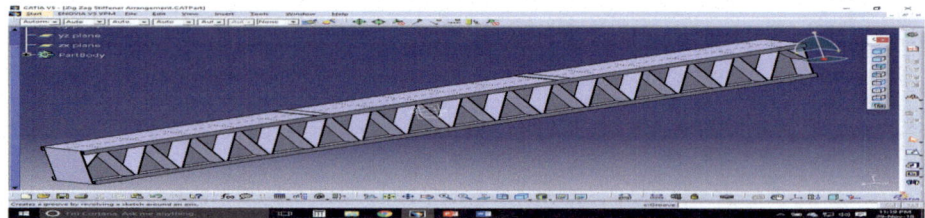

Fig. 10. Pattern 7. Zig zag stiffener arrangement

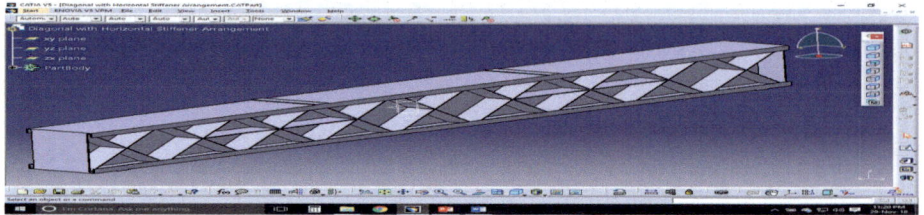

Fig. 11. Pattern 8. Diagonal with horizontal strip stiffener arrangement

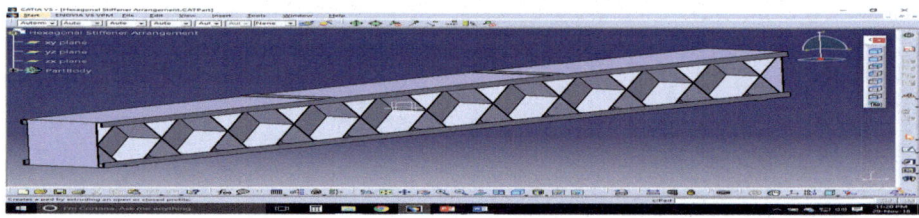

Fig. 12. Pattern 9. Hexagonal stiffener arrangement

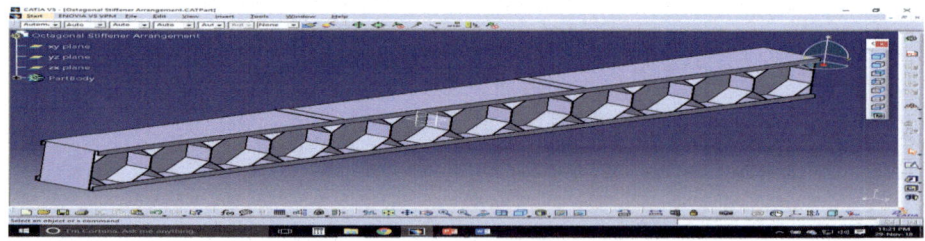

Fig. 13. Pattern 10. Octagonal stiffener arrangement

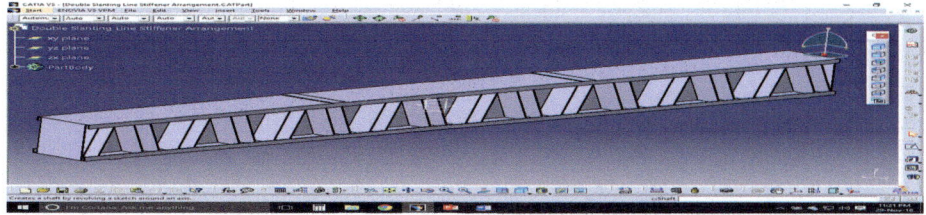

Fig. 14. Pattern 11. Double slanting line stiffener arrangement

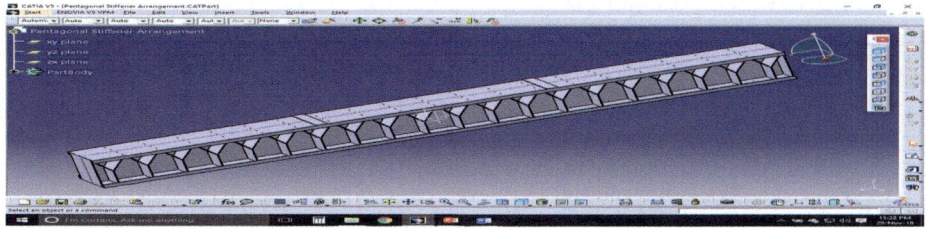

Fig. 15. Pattern 12. Pentagonal stiffener arrangement

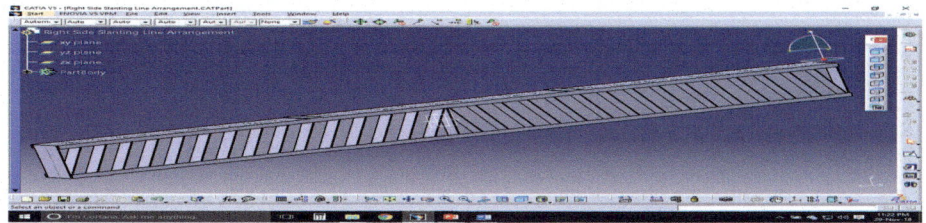

Fig. 16. Pattern 13. Right side slanting line stiffener arrangement

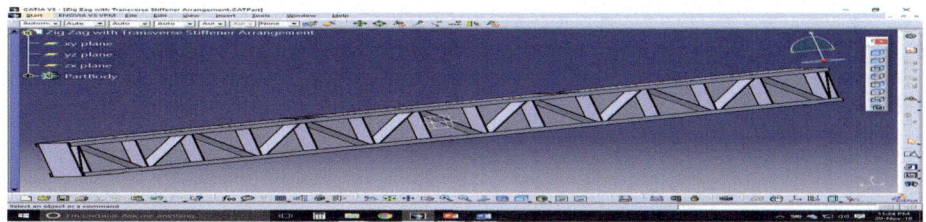

Fig. 17. Pattern 14. Zig-zag with transverse stiffener arrangement

6 Result

Analysis is done with the help of finite element program ANSYS. The result are shown in the table below with constant load and two point loading condition and constant thickness also (Table. 1).

Table 1. Analysis result of plate girder in cold formed material

Sr. no.	Name	Load (KN)	Stress (Mpa)	Deformation (mm)	Strain (mm/mm)
1	IS code design standard stiffener arrangement	40	251.74	2.3988	0.0012986
2	Diagonal stiffener arrangement	40	260.81	2.4952	0.001312
3	"N" type stiffener arrangement	40	252.32	2.5463	0.0013019
4	Diagonal with longitudinal stiffener arrangement	40	254.52	2.5625	0.0013055
5	Diagonal with transverse stiffener arrangement	40	249.62	2.4135	0.0012803
6	Horizontal strips with transverse stiffener arrangement	40	252.17	2.5964	0.0013004
7	Cross inscribed in rectangle stiffener arrangement	40	254.38	2.4724	0.0012958
8	Zig-zag stiffener arrangement	40	253.57	2.5102	0.0012869
9	Diagonal with horizontal strip stiffener arrangement	40	252.34	2.5141	0.0012998
10	Hexagonal stiffener arrangement	40	257.46	2.2554	0.0012915
11	Octagonal stiffener arrangement	40	259.54	2.1713	0.001301
12	Double slanting line arrangement	40	252.00	2.6016	0.0013062
13	Pentagonal stiffener arrangement	40	252.88	2.5606	0.0013009
14	Right side slanting line arrangement	40	255.48	2.5448	0.0013087
15	Zig zag with transverse stiffener arrangement	40	252.47	2.5383	0.0012979

7 Conclusion

1. Result shows that the octagonal stiffener arrangement gives minimum deformation 2.1713 mm compared to all other stiffener arrangements.
2. From result it is concluded that the octagonal stiffener arrangement gives minimum deformation at maximum load so this stiffener arrangement is recommended for plate girder design using cold form material.
3. The maximum stresses for all different stiffener arrangement are in permissible limit.

References

1. Degtyarev, V.V., Degtyareva, N.V.: Numerical simulations on cold-formed steel channels with longitudinally stiffened slotted webs in shear. Thin-Walled Struct. **129**, 429–456 (2018)
2. Xiea, M., Chapmanb, J.C., Hobbsb, R.E.: A rational design model for transverse web stiffeners. J. Constr. Steel Res. **64**, 928–946 (2008)
3. Šakalys, G., Daniūnas, A.: Numerical investigation on web crippling behavior of cold-formed C-section beam with vertical stiffeners. Proc. Eng. **172**, 1102–1109 (2017)
4. Torabian, S., Fratamico, D.C., Schafer, B.W.: Experimental response of cold-formed steel Zee-section beam-columns. Thin-Walled Struct. **98**, 496–517 (2016)
5. Keerthan, P., Mahendran, M.: Improved shear design rules for lipped channel beams with web openings. J. Constr. Steel Res. **97**, 127–142 (2014)
6. Lee, S.C., Lee, D.S., Yoo, C.H.: Design of intermediate transverse stiffeners for shear web panels. Eng. Struct. **75**, 27–38 (2014)
7. Gotluru, B.P., Schafer, B.W., Peköz, T.: Torsion in thin-walled cold-formed steel beams. Thin-Walled Struct. **37**, 127–145 (2000)
8. Wang, L., Young, B.: Design of cold-formed steel channels with stiffened webs subjected to bending. Thin-Walled Struct. **85**, 81–92 (2014)

Displacements in Thick Cantilever Beam Using V Order Shear Deformation Theory

Girish Joshi, Sagar Gaikwad, Ajay Dahake[(⊠)], and Amardip Girase

Civil Engineering Department, G. H. Raisoni College of Engineering
and Management, Pune, India
ajaydahake@gmail.com

Abstract. Paper presents the study of displacements in cantilever thick beam via 5th order function of study of shear deformation when exposed to a cosine loading. The theory is based upon the elementary theory of beam by considering shear deformation effects applying function of 5th order using variables of thickness. This study gratified the zero shear stress condition on top and bottom of the beam. As the deflection is more definite in cantilever sections, the cantilever beam is considered here. For obtaining equilibrium equations a well-known source of virtual work is used. To demonstrate the worth of the theory, the longitudinal and axial displacements are worked out for beam which is thick in nature, when subjected to cosine load as such type of load is very common in aerospace and marine structures. Outcomes are likened with the other theories.

Keywords: Cantilever · Thick beam · V order

1 Introduction

Beams and plates are more common terms in civil, mechanical, aerospace and marine engineering. As the thickness required for resisting moment, shear force and deflection criteria as compared to its length is much more, the beams are known as thick or deep beam. The in-plane and longitudinal displacements are carried out. Bernoulli-Euler [1, 2] recognized the most commonly used classical or elementary theory of beam. Galileo investigated the progressive work on beam concept initially, upto Saint Venant [3]. It suits for the analysis of slender beams as negligence of the transverse shear deformation. Due to underestimation of bending in dense beams when bending due to shear is predominent, there is need of higher order shear deformation theories.

The first order shear deformation theory (FSDT) is developed by Timoshenko [4] considering the refined effects such as rotatory inertia and distortion in shear. This theory assumes the shear strain as straight line throughout the depth, so there is a necessity of factor of shear correction expressing shear deformation.

The numerous methods of development of other studies based on the reduction of the three-dimensional problems of elastic bodies are discussed by Ghugal and Shimpi [5]. Krishna Murty [6] has presented the shear deformation theory considering third order function is thickness coordinate. It is satisfying the shear stress free boundary conditions on top and bottom plane of beam and thus there is no requirement of shear correction factor. The hyperbolic shear deformation theory for the static and dynamic

© Springer Nature Singapore Pte Ltd. 2020
V. K. Gunjan et al. (Eds.): *ICRRM 2019 – System Reliability, Quality Control, Safety, Maintenance and Management*, pp. 238–243, 2020.
https://doi.org/10.1007/978-981-13-8507-0_35

analysis of thick beams presented by Ghugal and Sharma [7, 8]. Ghugal and Dahake [9], Dahake and Ghugal [10] and Jadhav and Dahake [11] investigated the trigonometric shear deformation theory for flexural analysis of thick simple and cantilever beams with cosine and linearly varying loads.

A fifth-order shear deformation theory for static bending and elastic buckling of P-FGM beams have presented by Ghumare and Sayyad [12]. Ghugal and Gajbhiye [13] carried out analysis considering bending effect of thick plates by present theory.

2 Mathematical Modelling

The beam under consideration occupies in $0 - x - y - z$ Cartesian coordinate system the region: $0 \leq x \leq L;$ $-b/2 \leq y \leq b/2;$ $-h/2 \leq z \leq h/2.$

2.1 Development of the Theory

The displacement field used to present refined beam theory is of the following form:

$$u(x, z) = u_0 - z\frac{dw}{dx} + z\left[2 - \frac{4}{3}\left(\frac{z}{h}\right)^2 - \frac{16}{5}\frac{z^4}{h^4}\right]\phi(x) \tag{1}$$

$$w(x, z) = w(x)$$

where,

u = axial displacement
w = slanting displacement.

2.2 Equilibrium Equations

Using the equations for strains and stresses and applying the principle of virtual work, equilibrium equations are obtained as follows:

$$\int_{x=0}^{x=L} \int_{y=-b/2}^{y=b/2} \int_{z=-h/2}^{z=+h/2} \left(\sigma_x \delta\varepsilon_x + \tau_{zx}\delta\gamma_{zx}\right) dx\, dy\, dz - \int_{x=0}^{x=L} q(x)\, \delta w\, dx = 0 \tag{2}$$

The equilibrium equations thus obtained are as follows:

$$EI\frac{d^4w}{dx^4} - A_0 EI\frac{d^3\phi}{dx^3} = q(x), \quad A_0 EI\frac{d^3w}{dx^3} - B_0 EI\frac{d^2\phi}{dx^2} + C_0 GA\phi = 0 \tag{3}$$

where the constants, $A_0 = \frac{12}{7}$, $B_0 = 2.96$, $C_0 = 2.4635$.

3 Example: A Beam with Cosine Load

For validation of the present theory, a cantilever beam is considered subjected to cosine load. The material properties for beam are as modulus of elasticity (E) = 210 GPa, Poisson's ratio (μ)= 0.3 and density (ρ) = 7800 kg/m^3. (Fig. 1)

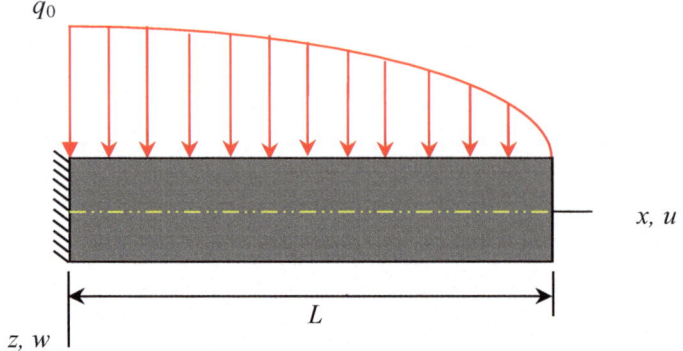

Fig. 1. Cantilever beam with cosine load

Equation for $w(x)$ and $\phi(x)$ are as follows:

$$w(x) = \frac{q_0 L^4}{120 EI} \left\{ \begin{array}{l} \frac{240}{\pi} \left[\frac{8}{\pi^3} \left(\cos \frac{\pi x}{L} - 1 \right) - \frac{1}{6} \frac{x^3}{L^3} + \frac{1}{2} \frac{x^2}{L^2} \right] + \frac{B_0}{C_0} \frac{40}{\pi^2} \frac{E}{G} \frac{h^2}{L^2} \left(\cos \frac{\pi x}{2L} - 1 \right) \\ - \frac{20}{\pi} \frac{A_0^2}{C_0} \frac{E}{G} \frac{h^2}{L^2} \left[-\frac{x}{L} + \frac{1 + \sinh \lambda x - \cosh \lambda x}{\lambda L} \right] \end{array} \right\} \quad (4)$$

$$\phi(x) = -\frac{2}{\pi} \frac{q_0 L}{\beta EI} \left(\cosh \lambda x - \sinh \lambda x - 1 + \sin \frac{\pi x}{2L} \right) \quad (5)$$

The axial displacement and stresses are as follows

$$u = \frac{q_0 h}{Eb} \left\{ \begin{array}{l} -\frac{z}{h} \frac{L^3}{h^3} \frac{1}{10} \left[\frac{240}{\pi} \left[\frac{8}{\pi^3} \left(-\frac{\pi}{2} \sin \frac{\pi x}{2L} \right) - \frac{1}{2} \frac{x^2}{L^2} + \frac{x}{L} \right] + \frac{40}{\pi^2} \frac{B_0}{C_0} \frac{E}{G} \frac{h^2}{L^2} \left(-\frac{\pi}{2} \sin \frac{\pi x}{2L} \right) \right] \\ -\frac{20}{\pi} \frac{A_0^2}{C_0} \frac{E}{G} \frac{h^2}{L^2} \left(-1 + \cosh \lambda x - \sinh \lambda x \right) \\ + \frac{z}{h} \frac{A_0}{C_0} \frac{E}{G} \frac{L}{h} \left(2 - \frac{4}{3} \frac{z^2}{h^2} - \frac{16}{5} \frac{z^4}{h^4} \right) \left(\sinh \lambda x - \cosh \lambda x + 1 - \sin \frac{\pi x}{2L} \right) \end{array} \right\} \quad (6)$$

4 Analytical Results

In this section, the results for in-plane and displacement are given in the following non-dimensional form (Figs. 2, 3 and 4), Table 1.

$$\bar{u} = \frac{Ebu}{q_0 h}, \quad \bar{w} = \frac{10Ebh^3 w}{q_0 L^4}.$$

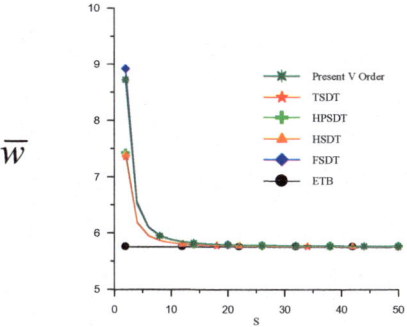

Fig. 2. Graph for maximum transverse displacement ($\bar{w}$) of beam at ($x = L$, $z = 0$) with aspect ratio S.

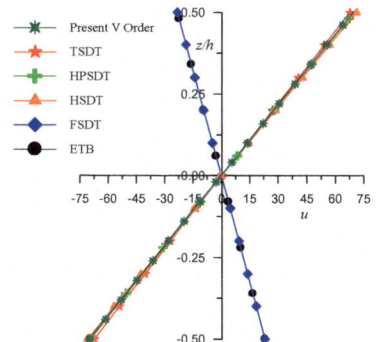

Fig. 3. Graph for axial displacement ($\bar{u}$) through the thickness at ($x = L$, z) for aspect ratio 4.

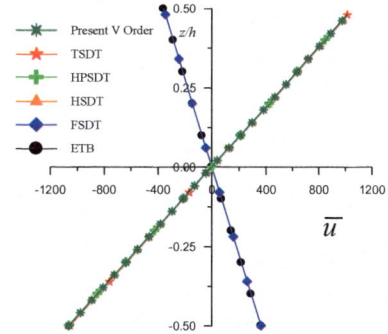

Fig. 4. Graph for axial displacement ($\bar{u}$) of beam at ($x = L$, z) for aspect ratio 10.

Table 1. Axial Displacement ($\bar{u}$) at ($x = L$, $z = h/2$), Transverse Deflection ($\bar{w}$) at ($x = L$, $z = 0.0$) of the Beam for Aspect Ratios.

Source	Aspect Ratio	Model	$\bar{u}$	$\bar{w}$
Present	4	V Order	−69.7185	6.5128
Dahake and Ghugal [10]		TSDT	−67.5978	6.1738
Ghugal and Sharma [8]		HPSDT	−69.9772	6.1846

(continued)

Table 1. (*continued*)

Source	Aspect Ratio	Model	$\bar{u}$	$\bar{w}$
Krishna Murty [6]		HSDT	−71.1524	6.1786
Timoshenko [4]		FSDT	23.1543	6.5444
Bernoulli-Euler		ETB	23.1543	5.7541
Present	10	V Order	−1060.7537	5.8771
Dahake and Ghugal [10]		TSDT	−1055.4521	5.8231
Ghugal and Sharma [8]		HPSDT	−1061.4004	5.8243
Krishna Murty [6]		HSDT	−1064.3386	5.8237
Timoshenko [4]		FSDT	361.7857	5.8806
Bernoulli-Euler		ETB	361.7857	5.7541

5   Conclusions

The transverse displacement by the present theory is overestimate the other refined shear deformation theory and underestimate the FSDT.

1. The inplane displacements and their distributions through the thickness of beam is nearly matching with other refined shear deformation theory for aspect ratio of 4 whereas it is excellent agreement with other higher order refined theories for aspect ratio 10. It is also seen that ETB and FSDT are changes the quadrant.
   This authorizes the worth and reliability of V order shear deformation theory.

References

1. Bernoulli, J.: Curvatura laminae elasticae. Acta Eruditorum Lipsiae **1694**, 262–276 (1744). (Also in Jacobi Bernoulli Basileensis Opera (2 vols.), 1, (LVIII), p. 576,, (1694), (1744)
2. Bernoulli, J.: Explicationes, annotations et additions. Acta Eruditorum Lipsiae 1695, 537–553 (1695). (Also in Jacobi Bernoulli Basileensis Opera (2 vols.), 1(LXVI), p. 639., (1695), (1744)
3. de Saint Venant, B.: Memoire sur la flexion des prismes. Journal de Mathematiques Pures et Appliquees, (Liouville),2(1), pp. 89–189 (1856)
4. Timoshenko, S.P.: On the correction for shear of the differential equation for transverse vibrations of prismatic bars. Phil. Mag. **41**(6), 744–746 (1921)
5. Ghugal, Y.M., Shmipi, R.P.: A review of refined shear deformation theories for isotropic and anisotropic laminated beams. J. Reinf. Plast. Compos. **20**(3), 255–272 (2001)
6. Krishna Murty, A.V.: Towards a consistent beam theory. AIAA J. **22**(6), 811–816 (1984)
7. Ghugal, Y.M., Sharma, R.: A hyperbolic shear deformation theory for flexure and vibration of thick isotropic beams. Int. J. Comput. Methods **6**(4), 585–604 (2009)
8. Ghugal, Y.M., Sharma, R.: A refined shear deformation theory for flexure of thick beams. Latin Am. J. Solids Struct. **8**, 183–193 (2011)
9. Ghugal, Y.M., Dahake, A.G.: Flexure of simply supported thick beams using refined shear deformation theory. Int. J. Civil Environ. Struct. Constr. Archit. Eng. **7**(1), 99–108 (2013)

10. Dahake, A.G., Ghugal, Y.M.: A trigonometric shear deformation theory for flexure of thick beam. Proc. Eng. **51**, 1–7 (2013)
11. Jadhav, V.A., Dahake, A.G.: Bending analysis of deep beam using refined shear deformation theory. Int. J. Eng. Res. **5**(3), 526–531 (2016)
12. Sayyad, A.S., Ghugal, Y.M.: Bending, buckling and free vibration of laminated composite and sandwich beams: A critical review of literature. J. Compos. Struct. **171**, 486–504 (2017)
13. Ghumare, S.M., Sayyad, A.S.: A new fifth-order shear and normal deformation theory for static bending and elastic buckling of P-FGM beams. Latin Am. J. Solid Struct. **14**, 1893–1911 (2017)
14. Ghugal, Y.M., Gajbhiye, P.D.: Bending analysis of thick isotropic plates by using 5th order shear deformation theory. J. Appl. Comput. Mech. **2**(2), 80–95 (2016)

Influence of Soil Structure Interaction of RC Building Considering With and Without Infill Strut Panel - A Review

P. S. Bhurse[✉] and S. S. Sanghai

Department of Civil Engineering, G. H. Raisoni College of Engineering,
Nagpur, India
{bhurse_priya.ghrcemtechstr, sanket.sanghai}@raisoni.net

Abstract. During past years, a lot of research has done by considering the performance of RCC structure with and without infill walls experimental as well as mathematically. Soil structure interface plays an essential character in leading the impact of RC structure through the seismic performance. Persistence of this paper is to examine the various special effects by considering soil-structure interface on a bare frame and infill strut panel with surface must be studied. The soil is model by using Drucker-Prager nonlinear models in ANSYS 15.

Keywords: Infill walls · Soil structure interaction ·
Equivalent diagonal strut method

1 Introduction

The infill wall is mostly used to increase initial stiffness and strength of reinforced concrete building construction. The arrangement of infill strut panel and Reinforced Concrete edge are mostly used in the structure, where the section is predisposed to seismic movement. Infill wall is commonly used as a partition element because of various suitable aspects like lighter in load, straightforwardness in structure, good visual view, protecting material goods, etc., even so in structural analysis, only the effect of mass is measured and its structural physical appearance such as strength, stiffness are usually neglected. Even still it gives sizeable lateral stiffness to the bare framed structures, these existed not measured in the preceding edition of Indian standard code of seismic activity resilient project is IS1893:2002 (Part 1) [1]. The method using which the reaction of the soil impacts the wave of the building besides the wave of the building impacts the reaction of the soil is termed as SSI. In this case, neither the fundamental deformation nor the ground deformation is self-governing of both added. However, in the new publication of Indian standard for a seismic activity resilient project IS 1893:2016(Part 1), [1] various provisions for infill walls are specified. The situation that Unreinforced infill wall can be modeled as an equivalent oblique strut, which if not considered results in unequal structure. The ends of the equivalent strut treated as pin jointed connected to the RC frame and influence of the opening on a width of the equivalent diagonal strut also stated. Existing literature provides a study of the various effect of SSI for tall symmetric and the irregular

© Springer Nature Singapore Pte Ltd. 2020
V. K. Gunjan et al. (Eds.): *ICRRM 2019 – System Reliability, Quality Control, Safety, Maintenance and Management*, pp. 244–251, 2020.
https://doi.org/10.1007/978-981-13-8507-0_36

building, different layer of soil for settlement also, studies Infill walls constructed for different materials, considering with and without opening situations, with and without infill walls, various researchers checked the performance of infill wall by using a shaking table test to the estimated fundamental time period, base shear. But there is very hardly any number of researchers to consider the influence of SSI along with and without infill walls.

2 Infill Wall

Infill wall is unknown, but the mutual masonry wall, brick wall, etc. For the study, we have measured the structural strength, stiffness of infill walls. By using IS 1893:2016 (Part-1), [1], govern the breadth and depth of the infill brace panel. Unreinforced infill wall can be there demonstrated as an equivalent sloping brace method and end of the strut is treated as a pin joint or hinge. According to clause 7.9 pg. no. 25 IS1893:2016 (Part 1), [1],

$$W_{ds} = 0.175 \, \propto_h^{-0.4} L_{ds} \tag{1}$$

Where,

$$\propto_h = h \left\{ \sqrt[4]{\frac{E_m.t.\sin2\theta}{4.E_f.I_c.h}} \right\} \tag{2}$$

E_m = Modulus of resistance of the material of the unreinforced brickwork infill
E_f = Modulus of resistance of the material of the RC moment resisting structure
I_C = Moment of inertia of the adjacent column
t = Width of masonry infill walls
θ = The angle of the diagonal strut with the parallel
h = Height of URM infill walls
W_{ds} = Breadth of equivalent diagonal strut
L_{ds} = Sloping distance of infill strut panel (Fig. 1)

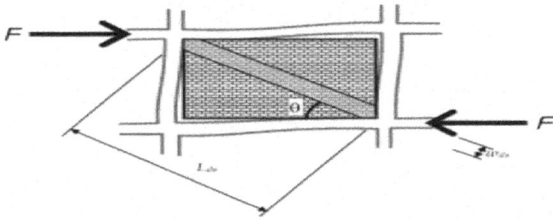

Fig. 1. The equivalent diagonal strut of URM infill wall, [1],

The failure of infilled frame can be categorized in five methods:

1. Sliding shear failure
2. Beam-column junction failure
3. Conner crushing failure
4. Diagonal shear cracking failure
5. Plastic hinge formation

To avoid that mode of failure in infill panel we provide a diagonal strut. Baghi et al. [2], they examine the activities of with and without infill partition by experimentally with Eurocode and observed that infill walls can considerably increase the weight resonant capability of RC edges structure. Perrone et al. [3], evaluated the ductility of masonry infilled RC frame and studied the capability curves, failures methods masonry infilled frame and analyzed the effect of power-driven things of masonry infill. Furtado et al. [4] investigate seismic behavior of 3-storey infilled RC frame structure by using numerical modeling to increase abilities of structure to find their estimated nature. Benavent et al. [5], they investigate experimentally seismic behavior of RC structure which is damaged by the earthquake and retrofitting in masonry walls by using a shaking table test. Choi et al. [6], experimental test on Turkish RC MRF infill walls along with URM infilled walls for multi bays also, investigate the in-plane behavior of infill walls for an experimental model of the diagonal strut. Chrysostomou et al. [7], resistance earthquake load is documented along with the activities of the infill panel and leaping edge. They predict In-plane failure modes, strength, stiffness, and displacement of the infilled frame. Compare methods with experimental method results. Mondal et al. [8], considered the influence of vertical irregularities and thickness of URM infill walls using robustness property of the structural system that can improve the sustainable building to inadequate collapse. Nonlinear static analysis is performed. Shan et al. [9], experimental investigate failures of RC frame with infill walls. Study how the reduce infill failure modes and ductility of RC frame also; as a consequence develop failure confrontation capability of the edge. Furtado et al. [10], improve the numerical modeling which is based on Rodrigues et al. paper, evaluated the behavior of infill wall by using OpenSee system also, Calibrate in-plane analysis is the main purpose of the test. Furtado et al. [11], considering five infill RC edge confirmed by changed biographers through changed arrangement self-control is evaluated by using OpenSee.

3   Soil Structure Interaction

The reaction of loam affects the wave of structure and wave of structure affect the reaction of loam. Loam structure interface categorized three categories below as-

3.1   Kinematic Interaction

Substance into the loam does not move the earthquake ground motion is called kinematic interaction.

3.2 Inertial Interaction

The whole weight of framework transmission the inertial force to the loam is called as inertial interaction.

Loam structure interface is model with the concept of elastic half-space theory.

1. *Direct Method*
   High rise substance system and unrestrained soil mass is model composed of the suitable boundary element.

2. *Substructure Method*
   High rise structure and substance system is model separately with suitable consideration of load transmission from high rise structure to the substance system. Cruz et al. [12], the actual hampering of the structure of essential method reduces through growing structure altitude, also higher mode increases modal frequency. Vasilev et al. [13], improve, confirm and put on in simulated an in effect hybrid method and seismic reaction of loam structure interface. Pulikanti et al. [14], FEM is used for modeling soil-pile-structure interface. They check behavior of the soil model using applying transit loading with the interface element effect between soil and pile system. Hosseinzadeh et al. [15], In this paper they compared results between building code and shake table results.5, 10, 15, and the 20-floor building used for analysis & design of the dynamic test. Celebi et al. [16], they used nonlinear FEM method for seismic response of building with consideration of SSI and for modeling of soil medium using Mohr-Coulomb prototypical further down straight strain state of affairs.

4   Demonstration with a Study on SDOF System

The finite element modeling is done for high rise structure along with foundation system using FEM software ANSYS 15.0. The soil material properties are applied form material library in ANSYS for different linear or nonlinear soil model and structure.

4.1 Effect of SSI on With and Without Infill Walls

Many researchers have studied the effect of SSI on structure considering different soil model. Also, many researchers have studied seismic behavior of with and without infill walls by experimentally. But very hardly any researchers study the SSI effect with making an allowance for infill walls.

In the existing evaluation, the analysis of loam modeled by using ANSYS15.0 (Tables 1 and 2).

The soil dimension is modeled as solid element with dimensions as length & width as five times the parallel dimension of the structure and depth of soil should be at least three times depth of substance. In this paper, high rise structure and foundation system are modeled by using the direct method. Dead load and live load is given as per respectively. The dead load includes self-weight and wall loads.

Table 1. Information of the building

Sr. no.	Information	Value
1	Floor height	4.5 m
2	Structure dimension	4 m × 4 m
3	Cross section for beam	230 × 300 mm
4	Cross section for strut	465 × 345 mm
5	Mat foundation	300 mm
6	Soil volume	20 × 20 × 14 m
7	Stiffness of frame	3.0476×10^6 N/m

Table 2. Material properties

Properties	Structure	Soil
Material	Concrete (M25)	Soft Soil
Young's Modulus E (Pa)	2.5×10^{10}	2.5×10^7
Poisson Ratio	0.25	0.15
Density, P (Kg/m^3)	2500	1900
Cohesion, C(kN/m^2)	–	23
Internal Friction Angle (∅)	–	23°

Following models are studied.

Bare frame	Bare frame with infill strut without SSI
Type I	Type II
Bare frame with SSI without infill strut-	Bare frame with SSI and infill strut
Type III	Type IV

4.2 Total Deformation Model in ANSYS

Figures 2, 3, 4 and 5.

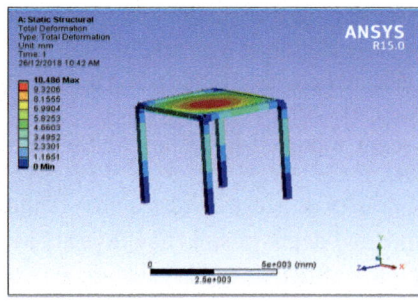

Fig. 2. Bare frame

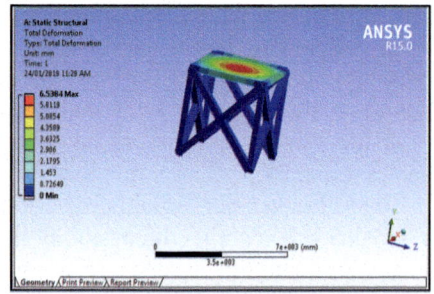

Fig. 3. Bare frame with infill strut without SSI

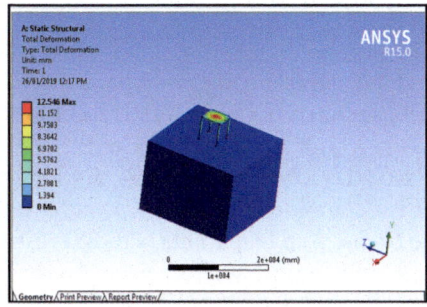
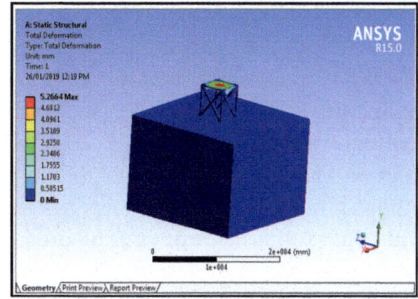

Fig. 4. Bare frame with SSI without infill strut   **Fig. 5.** Bare frame with SSI and infill strut

4.3   Modeling Element Section in ANSYS-15

The ANSYS framed framework is model with 2 node Beam element BEAM188, it has six degrees of freedom at each node. Slab surface used SHELL181 also, have six DOF at each node. And foundation with SOLID 186, interface with the element is CONTA174 & TARGE170; SURF154 is used for various loads and surface effect application in 3D analysis of structure. The soil is modeled with SOLID65 and Drucker-Prager model is used for nonlinear material of soil activities.

5   Results

Soil model is considered has hard layers so it has reserved the movement in any direction because we provide stable boundary condition at bottom of soil model. If soil horizontal side is vertical to X-axis, deformation in X-axis is reserved & deformations in Y & Z are unrestricted to move. Deformation in X, Z direction is calculated for dynamic loading. Base shear in Y, Z direction are shown in below Table 3.

Table 3. SDOF system

Parameters	Type I	Type II	Type III	Type IV
Displacement (mm)	11.680	6.538	12.546	5.266
Time Period (Sec)	0.0166	0.0123	0.9377	0.5727
Frequency (Hz)	60.064	80.89	1.0664	1.7459
Base Shear (kN)	40.63	37.24	17.16	34.63

Time period occurs more in the Type I as related to the Type II. Then in Type III is more as related to Type IV. Total displacement occurred to the column exceeding consideration is more in Type I as compared to the Type IV.

6 Concluding Remarks

This paper presents a review of the studies done going on the various impact of considering soil-structure interface on infill strut panel. RC building with respect to the loading of 4.5 m height with floor dimension 4 m × 4 m is analyzed for effect of soil structure interaction by using Drucker-Prager model nonlinear in Ansys-15 with and without infill walls. From exceeding results, it is determined that the bare frame with infill strut results superior as compared to a bare frame. And bare frame with SSI results superior as compared to a bare frame with SSI & infill strut. Therefore we can say that bare frame with SSI & infill struts results are much superior to another model.

References

1. I.S 1893[PART1]: Criteria for earthquake resistant design of the structure (2016)
2. Baghi, H., et al.: Behavior of reinforced concrete frame with masonry infill wall subjected to a vertical load. Eng. Struct. **171**, 476–487 (2018)
3. Perrone, D., Leone, M., Aiello, M.A.: Non-linear behavior of masonry infilled RC frames: influence of masonry mechanical properties. Eng. Struct. **150**, 875–891 (2017)
4. Furtado, A., et al.: Prediction of the earthquake response of a three-storey infilled RC structure. Eng. Struct. **171**, 214–235 (2018)
5. Benavent-Climent, A., Ramírez-Márquez, A., Pujol, S.: Seismic strengthening of low-rise reinforced concrete frame structures with masonry infill walls: Shaking-table test. Eng. Struct. **165**, 142–151 (2018)
6. Choi, H., Sanada, Y., Nakano, Y.: Diagonal strut mechanism of URM wall infilled RC frame for multi bays. Proc. Eng. **210**, 409–416 (2017)
7. Chrysostomou, C.Z., Asteris, P.G.: On the in-plane properties and capacities of infilled frames. Eng. Struct. **41**, 385–402 (2012)
8. Mondal, G., Tesfamariam, S.: Effects of vertical irregularity and thickness of unreinforced masonry infill on the robustness of RC framed buildings. Earthq. Eng. Struct. Dynam. **43**(2), 205–223 (2014)
9. Shan, S., et al.: Experimental study on the progressive collapse performance of RC frames with infill walls. Eng. Struct. **111**, 80–92 (2016)
10. Furtado, A., Rodrigues, H., Arêde, A.: Modelling of masonry infill walls participation in the seismic behavior of RC buildings using OpenSees. Int. J. Adv. Struct. Eng. (IJASE) **7**(2), 117–127 (2015)
11. Furtado, A., Rodrigues, H., Arêde, A.: Calibration of a simplified macro-model for infilled frames with openings. Adv. Struct. Eng. **21**(2), 157–170 (2018)
12. Cruz, C., Miranda, E.: Evaluation of soil-structure interaction effects on the damping ratios of buildings subjected to earthquakes. Soil Dynam. Earthq. Eng. **100**, 183–195 (2017)
13. Vasilev, G., et al.: Soil-structure interaction using BEM–FEM coupling through ANSYS software package. Soil Dynam. Earthq. Eng. **70**, 104–117 (2015)
14. Pulikanti, S., Ramancharla, P.K.: SSI analysis of framed structure supported on pile foundations-with and without interface elements. Frontiers in Geotechnical Engineering (FGE) 1.3 (2014)

15. Hosseinzadeh, N., Davoodi, M., Roknabadi, E.R.: Comparison of soil-structure interaction effects between building code requirements and shake table study. J. Seismolog. Earthq. Eng. **11**(1), 31–39 (2009)
16. Celebi, E., Göktepe, F., Karahan, N.: Non-linear finite element analysis for prediction of seismic response of buildings considering soil-structure interaction. Nat. Hazards Earth Syst. Sci. **12**(11), 3495–3505 (2012)

A Corrosion Study of Steel in Ferro-cement

Kapil Gupta[✉], Mo Irshad Kazi, Mahendra Mane, and Ajay Dahake

Civil Engineering Department,
G. H. Raisoni College of Engineering and Management, Pune, India
Kapil7.1995@yahoo.com, ajaydahake@gmail.com

Abstract. Ferro-cement can be define as the compound material which consist of cement, sand, water, admixtures, galvanizes steel mesh and skeleton steels. It is a cement mortar which is reinforced with closely space steel mesh wire. Ferro-cement is one form of reinforced concrete which differs from conventional reinforced concrete or prestressed concrete primarily by the manner in which the reinforcing elements are arranged and dispersed. Ferro-cement element are not only very thin but also the cover provided to elements very small which may make them more susceptible to corrosion. A number of literatures have been studied and analyzed to understand the factors that can affect the durability of ferro-cement and RC structures with references to corrosion. The present paper describes experiment conducted on ferro-cement panels for corrosion assessment and its effect over the durability of panel. From study it was found that normal corrosion of steel requires long time so acceleration corrosion testing is must to save time and to get effective result for prediction of service life period.

Keywords: Ferro-cement · Corrosion · Panels · Acceleration corrosion

1 Introduction

Ferro-cement popularity is increasing day by day due to its various advantages like constituent material are readily available in most of the countries, low material and construction cost [5]. Initial the development of ferro-cement was not good as compared to development of reinforced concrete and prestressed concrete due to expensive cost of small size mesh which is basic material for ferro-cement. Though as time pass new innovation and technology made it possible to have effective size of mesh at very economical price. Behavior ferro-cement is much different from our conventional reinforced concrete in term of strength, performances and applications so that it should be classified as the separate material.

There is increasing use of thin walled cement composites like ferro-cement and other laminated composites in various applications where a tough and strong protective shell is required in both existing and new structures [1, 2, and 4]. But in this case a very thin cover to the reinforcement i.e. 3 mm is provided is being questioned frequently. Satisfactory answers to this question does not exists due to lack of both adequate research work and information. Through the studies it has been revealed that with use galvanized steel wire mesh, significant amount of corrosion damage can occur within the time period of natural exposure if none of the extra measures are taken [1, 2]. Also in recent

© Springer Nature Singapore Pte Ltd. 2020
V. K. Gunjan et al. (Eds.): *ICRRM 2019 – System Reliability, Quality Control, Safety, Maintenance and Management*, pp. 252–256, 2020.
https://doi.org/10.1007/978-981-13-8507-0_37

model code of International Ferrocement Society have identified that information available on this area is not good enough to provide general recommendation [5].

There are considerable literature available on the topic as corrosion of steel reinforcement in conventional concrete but it is reluctance in applying the total findings to the ferro-cement because of difference in both material [1, 2, and 3]. Comparing with conventional concrete overall thickness of ferrocement hardly exceeds 25 mm, the reinforcing wire mesh is distributed properly and evenly with very thin cement motar protection, and coarse aggregate are completely removed from matrix [2, 3, 7]. By this differences there can be change in mechanics of corrosion process and its output may differ considerably. So this bring need to address question regarding to durability getting affected by corrosion of steel wire mesh reinforcement in thin walled composites and to search a ways to increase its long term durability and performance.

Natural corrosion processes in steel embedded in cement mortar matrix require decades to cause significant damage, so an acceleration corrosion technique is required to save time period in experimental study purpose [6]. There are various technique which are used for acceleration corrosion process like alternate wetting/drying in Nacl Solution method[2], impressed current method [6], glavanostatic method [1] etc. The factors which affect the durability ferro-cement and make susceptible to corrosion are as:-

(1) The mesh reinforcement cover is very small.
(2) Surface area of steel mesh is very high and the cross sectional area is very small.
(3) Mesh reinforcement used in ferro-cement are galvanized (zinc coating) which can cause hydrogen gas bubbles during hydration [7] (Fig. 1).

Fig. 1. Ferrocement panel

2 Experimental Investigation

2.1 Test Program

The present aim of experimental program is to perform a systematic corrosion study of ferro-cement panel with different cement with curing conditions and its assessment through half-cell potential meter. As time period is limited for experimental study acceleration corrosion technique is implemented.

A total 18 ferro-cement roof panels were casted which were divided into two groups. First group specimen are casted with OPC (Birla) and second group specimen are casted with PSC (JSW HD) which are green cement. All the specimen were reinforced with 2 layers of Square welded steel mesh having diameter 1 mm.

2.2 Test Specimen Specification

The specimen are casted with following Specification.
SIZE: 330 mm × 100 mm.
THICKNESS: 20 mm.
COVER: 5 mm.
WATER CEMENT: 0.5.
CEMENT MORTAR: 1:2.5.
FINE AGGREGATE: Passing through Sieve No. 7 (2.63 mm).
CEMENT: OPC 53 grade (Birla) and PSC 53 grade (JSW Hd).
REINFORCEMENT: Galvanized welded steel wire mesh having diameter 1 mm (18 gauge) and square grid 17 mm. Steel mesh to be cut in size of 310 mm × 90 mm.
SKELETAL STEEL: No.

2.3 Formwork and Casting

The ferro-cement panels have to be casted horizontally in rectangular ply mould having inner size 330 mm × 100 mm with a thickness 20 mm. for casting purposes the wire mesh is to be cut in size of 320 mm × 90 mm from wide roll of mesh in order to have cover and to prevent early corrosion of steel mesh. After cutting steel mesh, it was to be flatten with help of hammer. For assessment of corrosion, single color plastic coated copper wire is to be joint to each steel mesh which can be seen in Fig. 2 below.

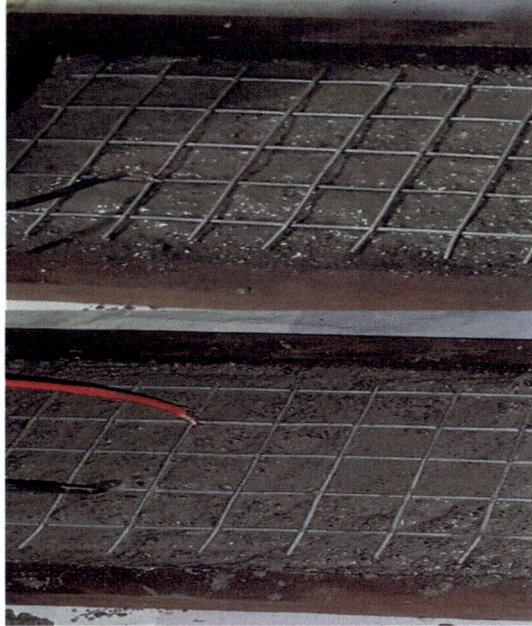

Fig. 2. Copper wire jointed to mesh reinforcement

2.4 Curing Conditions

In order to achieve acceleration corrosion the specimen will be cured in different concertation solution of sodium chloride as 0%, 3% & 5%. Condition for curing will be alternate wetting and drying will be implemented. The period of wetting cycle will be 4 days (96 h) and that for drying cycle will be 3 days (72 h).

2.5 Testing of Panels

For measurement of corrosion half-cell potential meter is used. Half- cell which is to be used should be confining to ASTM-876C. Half-cell potential meter consist of micro-voltmeter and copper/copper sulphate as reference electrode. Corrosion is to measure at 1 and 7 day.

3 Experimental Results

After demolding of specimen it was allowed to cure in portable water for 1 day and after that specimen were cured in respective Nacl solution and result obtain with half-cell potential meter are (Tables 1 and 2).

Table 1. Spontaneous reading at day 1.

Test specimen	Half-cell potential meter (mvolts)		
	0%	3%	5%
Specimen 1 (opc)	212	225	228
Specimen 2 (psc)	198	192	218

Table 2. Spontaneous reading at day 7.

Test specimen	Half-cell potential meter (mvolts)		
	0%	3%	5%
Specimen 1 (opc)	217	220	240
Specimen 2 (psc)	210	207	235

4 Conclusions

(1) Corrosion assessment by half-cell potential meter which is generally used for RC structure can also be used for corrosion prediction in thin laminate like ferro-cement.
(2) The result obtain will be helpful for further studies regarding its effect on strenght of ferro-cement panel.

References

1. Mansur, M.A., Maalej, M., Ismail, M.: Study on corrosion durability of ferro-cement. ACI (American Concrete Institude) Mater. J. **105**(1), 28–34 (2008)
2. Nedwell, P.J., Damola, O.O., Stevens, N.: Corrosion in ferro-cement. In: 11th International Symposium on Ferrocement and Textile Reinforced Concrete 3rd ICTRC, pp. 267–273. University of Manchester (2012)
3. Jin, M., Gao, S., Jiang, L., Jiang, Y., Wu, D., Song, R., Wu, Y., He, J.: Continuous monitoring of steel corrosion condition in concrete under drying/wetting exposure to chloride solution by embedded MnO_2 sensor. Int. J. Electrochem. Sci. **13**, 719–738 (2018)
4. Quraishi, M.A., Nayak, D.K., Kumar, R., Kumar, V.: Corrosion of reinforced steel in concrete and its control: an overview. J. Steel Struct. Constr. **3**(1), 1–6 (2017). ISSN 2472-0437
5. Balaguru, P.N., Soroushian, P., Arockiasamy, M., Daniel, J., Mansur, M., Paramasivam, P., Wecharatana, M., Ahmad, S.H., Gale, D.M., Mobasher, B., Reddy Robert, D.V., Williamson, B., Zellers Gordon, C., Batson, B., Hackman, L., Naaman, A.E., Shah, S.P., Zollo, R.F., Castro, J.O., Iorns, M.E., Nanni, A., Swamy, N., Zubieta, R.C.: State of the Art Report on Ferrocement (Reported by ACI Committee 549), ACI 549R-97
6. Bhalgamiya, S., Tivadi, G., Jethva, M.: Techniques for accelerated corrosion test of steel concrete for determine durability. IRJET J. (2018). ISSN 2395-0072
7. Akhtar, S., Daniyal, Md., Quraishi, M.A.: A review of corrosion control methods in ferrocement. J. Steel Struct. Const. (2015). https://doi.org/10.4172/2472-0437.1000103

Effect of Prop in Cantilever Thick Beam Using Trigonometric Shear Deformation Theory

Ajay Dahake[(✉)], Sandeep Mahajan, Akshay Mane, and Sharif Shaikh

Civil Engineering Department, G. H. Raisoni College of Engineering
and Management, Pune, India
ajaydahake@gmail.com

Abstract. Many researchers have carried out the analysis for regular boundary conditions like simply supported, fixed and cantilever. An attempt has been made to compare the effect of deflection in cantilever and propped cantilever beam as it is more important in case of cantilever beams. As many parts of spacecraft, airplane are made up of aluminum which is also thick in nature, we have considered the thick beam and material as aluminum. As the effect of warping is more pronounced in case of thick beam and which is neglected in elementary theory of beam, the refined shear deformation theories are used for validation. Theory includes the sinusoidal function in terms of thickness coordinate for consideration of shear deformation effects. The theory fulfilled the condition of zero shear stresses on the top and bottom of the beam. To prove the results of the theory, the transverse deflection is carried out for thick aluminum beams, loading considered as varying load for both the cases.

Keywords: Cantilever · Propped cantilever · Thick beam ·
Trigonometric shear de-formation

1 Introduction

Propped cantilever beams are used in many structural elements wherever anyone feels the dangerous situation. Simply supported beams are often used for design purpose due to not achieving the fixity of the support at field. Ghugal and Dahake [1, 2] have studied flexure of simply supported thick beams with cosine load using refined shear deformation theory and flexure of cantilever thick beams using trigonometric shear deformation theory subjected to the varying load. Comparison of various shear deformation theories for the free vibration of thick isotropic beams is presented by Sayyad [3]. Thermo-mechanical vibration of orthotropic cantilever and propped cantilever nano-plate using generalized differential quadrature method presented by Ghadiri et al. [4]. Nonlinear vibration of the cantilever FGM plate based on the third order shear deformation plate theory is investigated by Hao and Zang [5]. Reshmi and Indu [6] have detected crack of propped cantilever beam using dynamic analysis. Flexural vibrations in propped cantilever are also studied by Peek [7]. Krisna Murty [8] have provided third order shear deformation theory.

© Springer Nature Singapore Pte Ltd. 2020
V. K. Gunjan et al. (Eds.): *ICRRM 2019 – System Reliability, Quality Control,
Safety, Maintenance and Management*, pp. 257–262, 2020.
https://doi.org/10.1007/978-981-13-8507-0_38

2 Development of Theory

The beams are considered as cantilever and propped cantilever to verify the effect of transverse deflection at same location. The beam material is as follows (Table 1 and Fig. 1):

Table 1. Properties of Aluminum 6061-T6, 6061-T651 [9]

Physical properties	Quantity
Density	2700 kg/m³
Ultimate tensile strength	310 MPa
Modulus of elasticity	**68.9 GPa**
Notched tensile strength	324 MPa
Ultimate bearing strength	607 MPa
Bearing yield strength	386 MPa
Poisson's ratio	**0.33**
Fatigue strength	96.5 MPa
Shear modulus	**26 GPa**
Shear strength	207 MPa

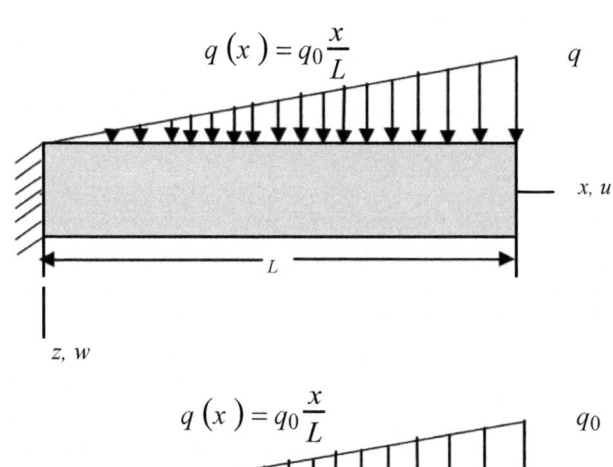

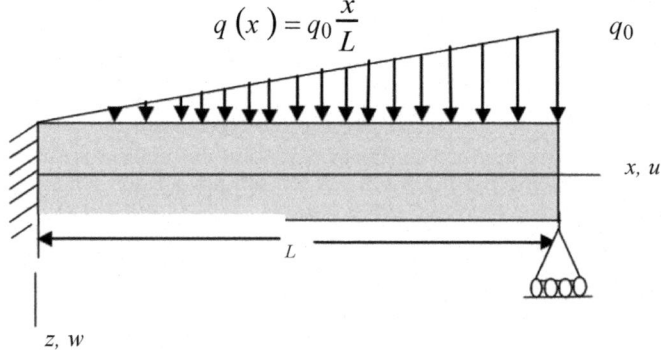

Fig. 1. Cantilever and propped cantilever beam

2.1   The Displacement Field

The displacement field of the present trigonometric shear beam theory is of the following form:

$$u(x, z) = -z \frac{dw}{dx} + \frac{h}{\pi} \sin \frac{\pi z}{h} \phi(x) \tag{1}$$

$$w(x, z) = w(x)$$

in which u is the axial displacement in x direction and w is the transverse dis-placement in z direction. The sinusoidal function is associated with the shear stress distribution through the thickness of the beam. The function ϕ represents rotation of the beam at neutral axis, which is an unknown function to be determined.

We obtain the coupled equilibrium equations and associated boundary conditions of the beam by employing the fundamental lemma of calculus of variations. The equilibrium equations thus obtained are as follows:

$$\frac{d^3 w}{dx^3} = \frac{24}{\pi^3} \frac{d^2 \phi}{dx^2} + \frac{Q(x)}{EI}, \frac{d^3 w}{dx^3} = \frac{\pi}{4} \frac{d^2 \phi}{dx^2} - \beta \phi \tag{2}$$

$$w(x) = \frac{q_0 L^4}{120 EI} \left\{ \frac{x^5}{L^5} - 10 \frac{x^3}{L^3} + 20 \frac{x^2}{L^2} + 6 \frac{E}{G} \frac{h^2}{L^2} \left[\left(\frac{x}{L} + \frac{(\zeta(x) - 1)}{\lambda L} \right) - \frac{x^2}{L^2} \right] \right\} \tag{3}$$

$$\text{where } \zeta(x) = (\sinh \lambda x - \cosh \lambda x) \tag{4}$$

3   Transverse Displacement

The results for transverse displacement presented in the following non-dimensional form (Table 2 and Figs. 2, 3).

Table 2. Transverse displacement at $(x = 0.75L)$ of the beam for aspect ratio (AR) of 2 to 20.

AR	ETB	FSDT	HSDT	HPSDT	Present TSDT
Cantilever beam					
2	7.2686	7.8275	11.8363	11.8484	11.8218
4	7.2686	7.4083	8.4243	8.4249	8.4222
6	7.2686	7.3307	7.7843	7.7842	7.7835
8	7.2686	7.3035	7.5592	7.5591	7.5589
10	7.2686	7.2909	7.4548	7.4547	7.4546
12	7.2686	7.2841	7.3980	7.3979	7.3979
14	7.2686	7.2800	7.3637	7.3636	7.3636
16	7.2686	7.2773	7.3414	7.3414	7.3414
18	7.2686	7.2755	7.3262	7.3261	7.3261
20	7.2686	7.2741	7.3152	7.3152	7.3152

(continued)

Table 2. (*continued*)

AR	ETB	FSDT	HSDT	HPSDT	Present TSDT
Propped cantilever beam					
2	0.3076	4.6635	3.2955	3.2981	3.2910
4	0.3076	4.2442	1.0608	1.0603	1.0604
6	0.3076	4.1666	0.6433	0.6429	0.6432
8	0.3076	4.1394	0.4967	0.4964	0.4967
10	0.3076	4.1269	0.4287	0.4285	0.4287
12	0.3076	4.1200	0.3918	0.3916	0.3918
14	0.3076	4.1159	0.3695	0.3694	0.3695
16	0.3076	4.1132	0.3550	0.3549	0.3550
18	0.3076	4.1114	0.3450	0.3450	0.3451
20	0.3076	4.1101	0.3379	0.3379	0.3379

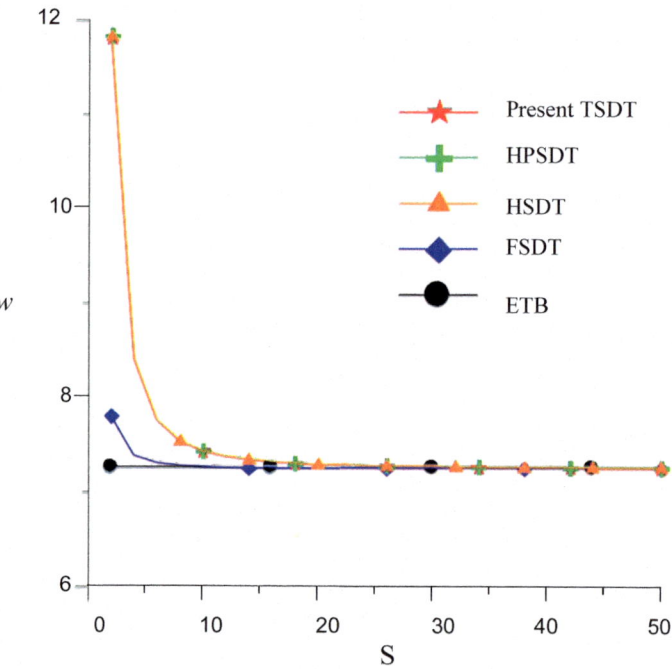

Fig. 2. Transverse displacement of cantilever beam

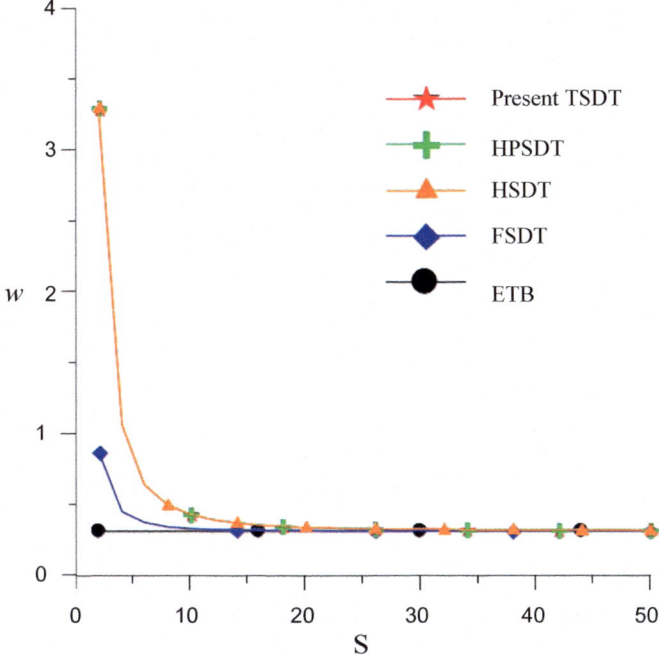

Fig. 3. Transverse displacement of propped cantilever beam

$$\bar{w} = \frac{q_0 L^4}{10 E b h^3 w}$$

4   Conclusions

It is observed that the transverse deflection is considerably reduced from 11.82 to 3.29 by present TSDT, hence it is proposed to have propped cantilever beam in-stead of cantilever beam for design purpose of thick beam as the section require-ments reduces considerable.

References

1. Ghugal, Y.M., Dahake, A.G.: Flexure of simply supported thick beams using refined shear deformation theory. Int. J. Civil Environ. Struct. Constr. Archit. Eng. **7**(1), 99–108 (2013)
2. Ghugal, Y.M., Dahake, A.G.: A trigonometric shear deformation theory for flexure of thick beam. Proc. Eng. **51**, 1–7 (2013)
3. Sayyad, A.S.: Bending, buckling and free vibration of laminated composite and sandwich beams: a critical review of literature. J. Compos. Struct. **171**, 486–504 (2017)

4. Ghadiri, M., Shafiei, N., Alavi, H.: Thermo-mechanical vibration of ortho-tropic cantilever and propped cantilever nanoplate using generalized differential quadrature method. Mech. Adv. Mater. Struct. **24**(8), 639–646 (2016)
5. Hao, Y.X., Zhang, W.: Nonlinear vibration of the cantilever FGM plate based on the third-order shear deformation plate theory, pp. 522–527 (2010). https://doi.org/10.1063/1.3452226
6. Reshmi, R., Indu, V.S.: Crack detection of propped cantilever beam using dynamic analysis. Int. J. Tech. Res. Appl. **3**(4) 274–278 (2015). www.ijtra.com, e-ISSN 2320-8163
7. Peek Jr, R.L.: Flexural vibrations of a propped cantilever, pp. 609–636 (1963)
8. Krishna Murty, A.V.: Towards a consistent beam theory. AIAA J. **22**(6), 811–816 (1984)
9. Properties of Aluminum 6061-T6, 6061-T651. http://www.aerospacemetals.com

Optimal Structural Design of Diagrid Structure for Tall Structure

Chittaranjan Nayak$^{(\boxtimes)}$, Snehal Walke, and Suraj Kokare

Department of Civil Engineering, VPKBIET, Baramati, India
cbnnayak@gmail.com, snehalwalke87@gmail.com,
kokaresuraj91@gmail.com

Abstract. Diagrid systems have strong structural efficiency than other system like braced system. Diagrid system have unique geometric configuration used for various heights of tall structure. Geometric configurations and grid geometrics of diagrid structure depends on the heights and angle of the diagrids. Joints of diagrid structure are more complicated than conventional structure therefore Construction of diagrid structure is complicated than conventional structure. In early days tall buildings have importance due to its architectural view and diagrid structure have better architectural view as well as good structural stability. In recent forms in tall structure have complex shapes like tapered, twisted and tilted. This paper includes required data, model, Earthquake and Wind analysis of Braced Tube Structure and diagrid structure with Circular, Square and Rectangular plan. Then by keeping same plan area and structural data for circular, square and rectangular plan, Earthquake and Wind analysis result of both Braced Tube and Diagrid Structures is carried out and by comparing the braced tube structures results and diagrid structures results conclusions drawn from the present investigation.

Keywords: Diagrid steel-frame · Structural design · Structural robustness · Sustainability · Tall buildings

1 Introduction

In the today's trend population has rapid growth and the space is limited for the building. Therefore there is need to develop tall buildings for development of city. Diagrid system is economical and has good lateral resisting system and carrying gravitational loads. Important point are taken into consideration while designing tall structure is lateral loads due to the wind and earthquake, diagrid system have better latter load resisting system due to inclined column located at the periphery of the structure [2]. Diagrid System and braced tube structure are as follows generally for tall structure braced tubes system is used for 100 storeys John Hancock Center of 1970 in Chicago. But due to various structural problems in braced tube structure, diagrid performs better than conventional system [3]. In conventional system columns are spaced widely by diagonal braces and diagonal braces create wall like characteristics. By increasing the spacing in column, increases the opening for structure for door and windows [6]. Normally purpose of beam and column are to control the bending action

© Springer Nature Singapore Pte Ltd. 2020
V. K. Gunjan et al. (Eds.): *ICRRM 2019 – System Reliability, Quality Control, Safety, Maintenance and Management*, pp. 263–271, 2020.
https://doi.org/10.1007/978-981-13-8507-0_39

resulting due to gravitational and lateral loads and due cantilever in structure crates large shear effect therefore due to these problems braced systems overcomes problem stiffening the perimeter frames in their own planes. ETABS is a sophisticated, yet easy to use, special purpose analysis and design program developed specifically for building systems. ETABS features an intuitive and powerful graphical interface coupled with unmatched modeling, analytical, and design procedures, all integrated using a common database [12]. Although quick and easy for simple structures, ETABS can also handle the largest and most complex building models, including a wide range of geometrical nonlinear behaviors, making it the tool of choice for structural engineers in the building industry. All structural members are considered as per IS 800:2007 considering all load combinations. Dynamic along wind and across wind are considered for analysis of the structure. Comparison of analysis results in terms of time period, storey displacement, storey shear and storey drift with conventional building [20]. Diagrid structural system is adopted in tall buildings due to its structural efficiency and flexibility in architectural planning. Due to inclined columns lateral loads are resisted by axial action of the diagonal compared to bending of vertical columns in framed tube structure [20].

2 Methodology

It includes gathering required data, preparing model of both Braced Tube structure and Diagrid Structures with Circular, Square and Rectangular plan, applying load cases to both Braced Tube and Diagrid Structures using ETABS. Storey displacement, story drift and base shear were then obtained after analysis of both Braced Tube and Diagrid Structures with Circular, Square and Rectangular plan. Then comparison between both Braced Tube and Diagrid Structures with Circular, Square and Rectangular plan was made and conclusions were drawn based on the result of analysis. IS 800:2007, IS 1893:2002 (Part I), IS 875:1987 (Part I), IS 875:1987 (Part II) and IS 875:1987 (Part III) were used. For structures number of storeys, plan area, storey heights loads on structure seismic data, wind data, materials used and section properties is kept same varying only on shapes. Methodology Presents Earthquake and Wind analysis result of both Braced Tube and Diagrid Structures and their comparisons. Collection of relevant research data from national and international journals is presented. Conclusions will be drawn based on the analysis. The following design characteristics are considered for diagrid structure and braced tube structure.

Table 1 Required structural data for braced tube structure and diagrid structure. Table 1 shows the all structural data required for analysis of diagrid structure and braced tube structure. In this table by keeping plan area and all structural data same (1296 m^2) for rectangular plan, square plan and circular plan wind analysis and earthquake analysis is carried out (Fig. 1).

Table 1. Parameters in detail for diagrid structure and braced tube structure

Sr No	Description	Data/Value	Sr No	Description	Data/Value
1	Number of storey	60	6	Wind data (a) wind speed	39 m/s
2	Plan area	1296 m^2		(b) Terrain category	2
3	Storey height	3.0 m		(c) Structure class	B
4	Loads on structure (a) Wall load (Cladding)	6 KN/m		(d) Risk coefficient k_1	1
	(B) Parapet wall load	3 KN/m		(e) Topography factor k_3	1
	(c) Roof live load	1.5 KN/m^2	7	Material used (a) Concrete	M20 for deck slab, M40 for infill concrete
	(d) Floor live load	3 KN/m^2		Steel	fy310 for fill and infill tube section, fy250 for deck section
	(e) Floor finish	1 KN/m^2		Rectangular plan size	54 m × 24 m
5	Seismic data (a) Seismic zone	III, Z = 0.16		Square plan size	36 m × 36 m
	(b) Response reduction factor	5		Circular plan size	40.62 m Diameter
	(c) Importance factor	1	8	Sections used (A) Beam	ISB 475X475X50
	(d) Soil type	Type II Medium Soil		(B) Inner column	(a) storey 1–21 Infill column 2200X2200X50
	(e) Structure type	Steel frame structure			(b) storey 22–33 Infill column 2000X2000X50 (c) storey 34–45 ISB 2000X2000X75 (d) storey 46–60 ISB 1800X1800X50
				(C) Peripheral column	(a) storey 1–40 Infill column 900X900X75 (b) storey 41–60 ISB 900X900X75
				(D) Bracing	ISB 850X850X50
				(E) Deck slab	120 mm thick

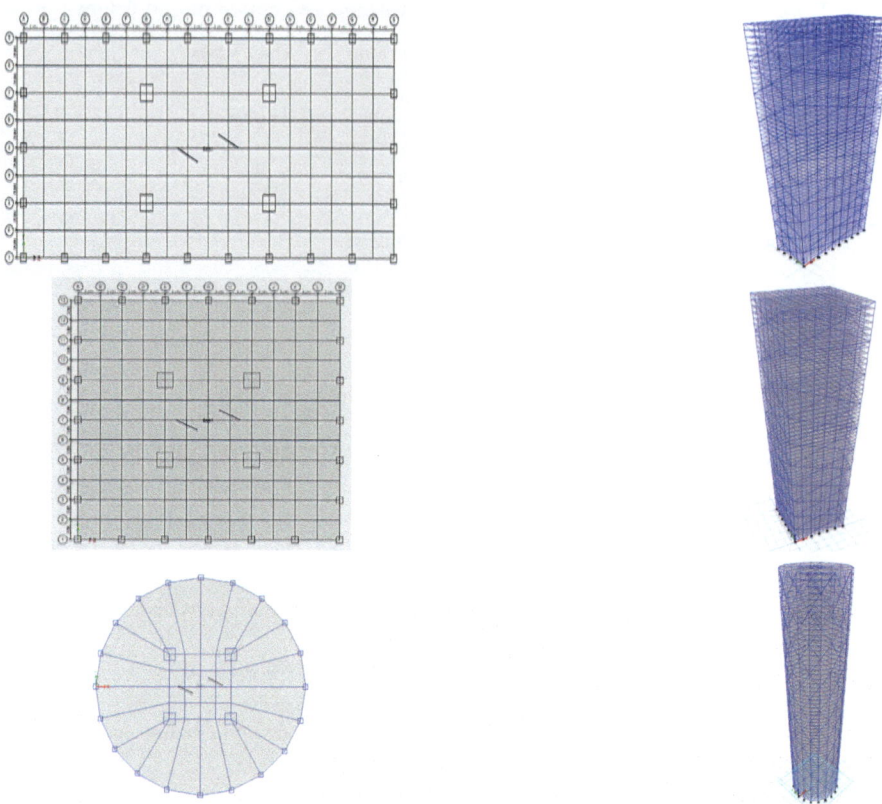

Fig. 1. Model and configuration of rectangular, square and circular braced tube structure

3   Results and Discussion

3.1   Earthquake Analysis Result

From Table 2 it is concluded that in earthquake analysis diagrid structure has less storey displacement, storey drift than braced tube structure and base shear for diagrid structure is greater than braced tube structure (Graphs 1, 2, 3, 4, 5 and 6).

Table 2. Comparison of braced tube structure and diagrid structure (earthquake analysis results in Y and X direction)

Shape of structure	Braced tube structure						Diagrid structure					
	Circular plan		Square plan		Rectangular plan		Circular plan		Square plan		Rectangular plan	
Directions	Y	X	Y	X	Y	X	Y	X	Y	X	Y	X
Storey displacement (mm)	57.6	56.2	73.1	72.8	108.8	57.8	45.8	46.6	43.9	43.9	58.9	32.1
Storey drift (kN)	0.000563	0.000554	0.000595	0.000595	0.000878	0.000571	0.0006 56	0.000671	0.000472	0.000471	0.000548	0.000485
Base shear(kN)	7029.6259	7032.8214	6261.0379	6264.4996	5941.6315	7467.7997	10654.9725	10724.3015	13242.7011	13242.7011	10170.2155	18425.5402

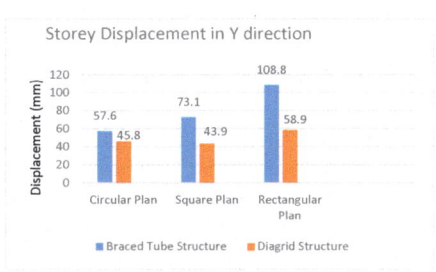

Graph 1. Storey displacement in Y direction (earthquake analysis)

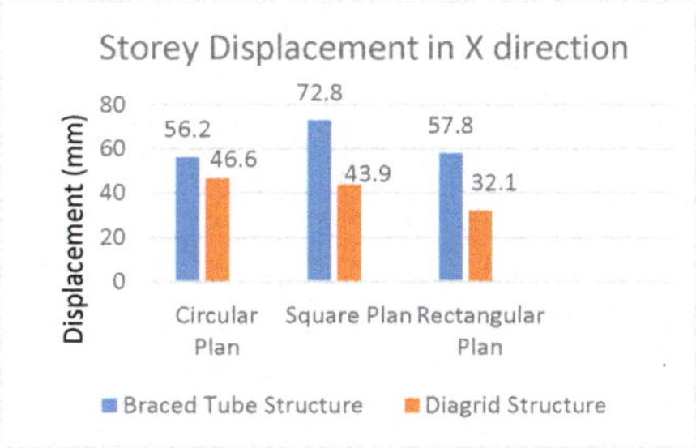

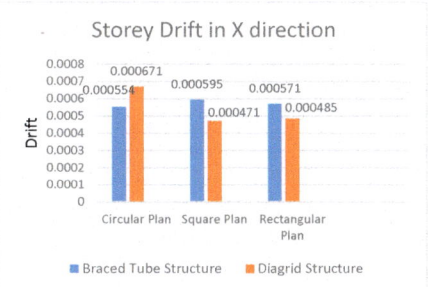

Graph 2. Storey displacement in X direction (earthquake analysis)

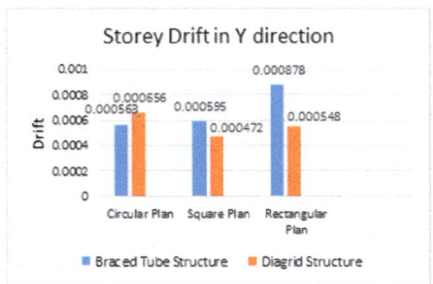
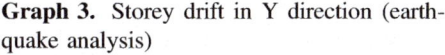

Graph 3. Storey drift in Y direction (earthquake analysis)

Graph 4. Storey drift in X direction (earthquake analysis)

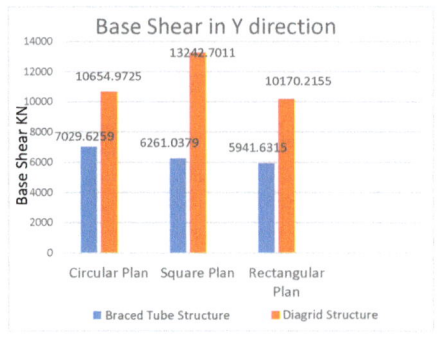

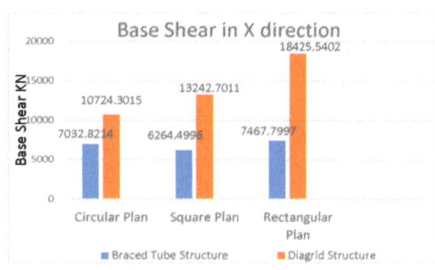

Graph 5. Base Shear in Y direction (earth-quake analysis)

Graph 6. Base Shear in X direction (earth-quake analysis)

3.2   Wind Analysis Result

From Table 3 it is concluded that in wind analysis diagrid structure has less storey displacement, storey drift than braced tube structure (Graphs 7, 8, 9 and 10).

Table 3. Comparison of braced tube structure and diagrid structure (wind analysis results in Y and X direction)

Shape of structure	Braced tube structure						Diagrid structure					
	Y	X	Y	X	Y	X	Y	X	Y	X	Y	X
Storey displacement (mm)	85.2	56.9	103.8	68.2	226.8	47.3	44.9	30.3	28.9	19	70.9	10.2
Storey drift	0.000749	0.000609	0.000826	0.000591	0.001661	0.000493	0.000673	0.000606	0.000447	0.000408	0.000627	0.00046

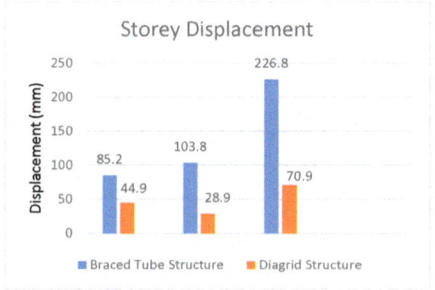
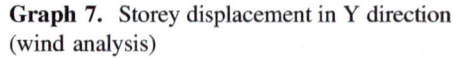

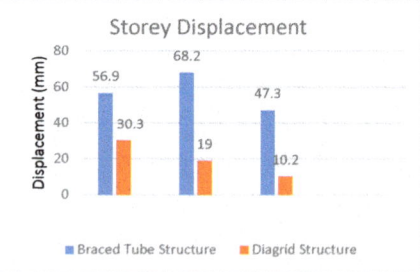

Graph 7. Storey displacement in Y direction (wind analysis)

Graph 8. Storey displacement in X direction (wind analysis)

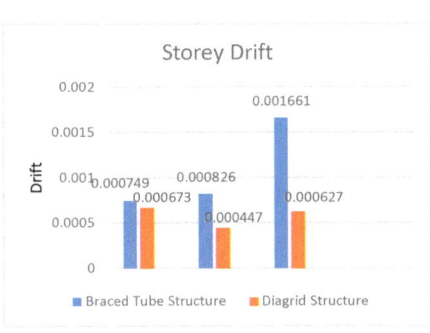

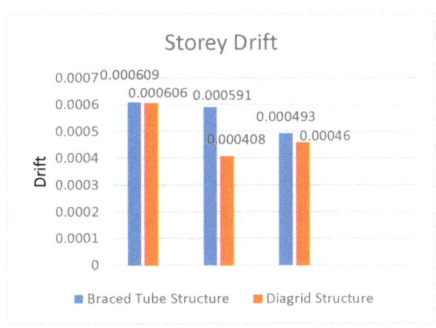

Graph 10. Storey drifts in X direction (wind analysis)

Graph 9. Storey drifts in Y direction (wind analysis)

4 Conclusion

After carrying out analysis by using ETABS 2015 software for Braced tube structure and Diagrid structure, various parameters like storey displacement, storey drift, and base shear are compared. Following conclusions are made.

[1] From all storey displacement and storey drift result we can say that all the storey drift (<H/500), where H is total height of building and storey displacement (<0.004 * h), where h is storey height values are within limits.

[2] Braced tube structure in Earthquake analysis circular plan has least storey displacement (y direction-57.6 mm, x direction-56.2) and storey drift (y direction-0.00563, x direction-0.00554) values compared to square plan has storey displacement (y direction-73.1 mm, x direction-72.8 mm) and storey drift (y direction-0.00595, x direction-0.000595) and rectangular plan has storey displacement (y direction-108.8 mm, x direction-57.8) and storey drift (y direction-0.00878, x direction-0.000571).

[3] Braced tube structure in Earthquake analysis circular plan has least storey displacement (y direction-85.2 mm, x direction-56.9 mm) and storey drift (y direction-0.00749, x direction-0.00609) values compared to square plan has storey displacement (y direction-103.8 mm, x direction-68.2 mm) and storey drift (y direction-0.00426, x direction-0.000591) and rectangular plan has storey displacement (y direction-226.8 mm, x direction-47.3) and storey drift (y direction-0.0001661, x direction-0.000493)

[4] Rectangular plan has maximum base shear value (y direction-5941.62 KN, x direction-7467.79 KN) compared to square (y direction-6261.03 KN, x direction-6264.41 KN) and circular plan (y direction-7029.62 KN, x direction-7032.82 KN) for Braced tube structure.

[5] Diagrid structure in Earthquake analysis square plan has least storey displacement (y direction-43.9 mm, x direction-43.9 mm) and storey drift (y direction-0.000472, x direction-0.00471) values compared to circular plan has

storey displacement (y direction-45.8 mm, x direction-46.6 mm) and storey drift (y direction-0.00656, x direction-0.000671) and rectangular plan has storey displacement (y direction-58.9 mm, x direction-32.1 mm) and storey drift (y direction-0.00548, x direction-0.000485).

[6] Diagrid structure in wind analysis square plan has least storey displacement (y direction-28.9 mm, x direction-19 mm) and storey drift (y direction-0.000447, x direction-0.00408) values compared to circular plan has storey displacement (y direction-44.9 mm, x direction-30.3 mm) and storey drift (y direction-0.00673, x direction-0.000603) and rectangular plan has storey displacement (y direction-70.9 mm, x direction-10.2 mm) and storey drift (y direction-0.00627, x direction-0.00046).

[7] Rectangular plan has maximum base shear value (y direction-10170.21 KN, x direction-18425.54 KN) compared to square (y direction-13242.7 KN, x direction-13242.70 KN) and circular plan (y direction-10654.97 KN, x direction-10724.30 KN) for diagrid structure.

[8] From above results from earthquake and wind analysis of storey drift, storey displacement and base shear we can conclude that circular plan performs better than square and rectangular plan in braced tube structure.

[9] From above results from earthquake and wind analysis of storey drift, storey displacement and base shear we can conclude that square plan performs better than circular and rectangular plan in diagrid structure.

[10] Diagrid structure has least storey displacement and storey drift values and maximum base shear values compared to Braced tube structure so we can say that Diagrid structure performs better than Braced tube structure.

References

1. Jani, K., Patel, P.V.: Analysis and design of diagrid structural system for high rise steel buildings. In: Chemical, Civil and Mechanical Engineering Tracks of 3rd Nirma University International Conference on Engineering (NUiCONE 2012), pp. 92–100 (2013)
2. Panchal, N.B., Patel, V.R., Pandya, I.I.: Optimum angle of diagrid structural system. Int. J. Eng. Tech. Res. (IJETR) 2(6), 150–157 (2014). ISSN 2321-0869
3. Prashant, T.G., Badami, S.S., Gornale, A.: Comparison of symmetric and asymmetric steel diagrid structures by non-linear static analysis. Int. J. Res. Eng. Technol. (IJRET) 04(05), 486–492 (2015). eISSN 2319-1163, pISSN 2321-7308
4. Thomas, F.M., Issac, B.M., George, J.: Performance evaluation of tall buildings with steel diagrid system. In: 2nd International Conference on Science, Technology and Management (ICSTM), September 2015, pp. 2242–2256 (2015)
5. Nimisha, P., Namitha, K.: Structural comparison of diagrid building with tubular building. Int. J. Eng. Res. Technol. (IJERT) 5(04), 57–60 (2016). ISSN 2278-0181
6. Shah, M.I., Mevada, S.V., Patel, V.B.: Comparative study of diagrid structures with conventional frame structures. Int. J. Eng. Res. Appl. (IJERA) 6(5), 22–29 (2016). ISSN 2248-9622, (Part - 2)
7. IS 800:2007: Indian Standard Code of Practice for General Construction in Steel, Bureau of Indian Standards, New Delhi

8. IS 1893:2002 (Part I): Indian Standard Criteria for Earthquake Resistant Design of Structures, Bureau of Indian Standards, New Delhi
9. IS 875:1987 (Part II): Indian Standard Code of Practice for Design loads (Other than Earthquake) for buildings and structures, Bureau of Indian Standards, New Delhi
10. IS 875:1987 (Part III): Indian Standard Code of Practice for Design loads (Other than Earthquake) for buildings and structures, Bureau of Indian Standards, New Delhi
11. Boake, T.M.: Diagrids, the new stability system: combining architecture with engineering. School of Architecture, University of Waterloo, Cambridge, ON, Canada
12. Moehle, J.P.: Performance-based seismic design of tall buildings in the U.S. In: The 14th World Conference on Earthquake Engineering, Beijing, China, 12–17 October (2008)
13. Kim, J., Jun, Y., Lee, Y.H.: Seismic performance evaluation of diagrid system buildings. In: 2nd Specialty Conference on Disaster Mitigation, Manitoba, June 2010
14. Jani, K., Patel, P.V.: Analysis and design of diagrid structural system for high rise steel. In: Chemical, Civil and Mechanical Engineering Tracks of 3rd Nirma University International Conference on Engineering, Procedia Engineering, vol. 51, pp. 92–100 (2013)
15. Alaghmandan, M., Pehlivan, N.A., Elnimeiri, M.: Architectural and structural development of tall buildings. In: ATINER Conference Paper Series No. ARC 2013: 0758, ISSN 2241-2891, 16 December 2013
16. Mazinani, I., Jumaat, M.Z., Ismail, Z., Chao, O.Z.: Comparison of shear lag in structural steel building with framed tube and braced tube. Struct. Eng. Mech. **49**(3), 297–309 (2014)
17. Revankar, R.K., Talasadar, R.G.: Pushover analysis of diagrid structure. Int. J. Eng. Innov. Technol. (IJEIT) **4**(3), 168–174 (2014). ISSN 2277-3754
18. Patil, D., Naveena, M.P.: Dynamic analysis of steel tube structure with bracing system. IJRET: Int. J. Res. Eng. Technol. **04**(08) (2015)
19. Chandwani, V., Agrawal, V., Gupta, N.K.: Role of conceptual design in high rise buildings. Int. J. Eng. Res. Appl. (IJERA) **2**(4), 556–560 (2012). ISSN 2248-9622
20. Varsani, H., Pokar, N., Gandhi, D.: Comparative analysis of exo-skeleton, diagrid and conventional structural system for tall building. Int. J. Adv. Res. Eng. Sci. Technol. (IJAREST) **2**(3) (2015). ISSN(O) 2393-9877, ISSN(P) 2394-2444
21. Harsha, S., Raghu, K., Narayana, G.: Analysis of tall buildings for desired angle of diagrids. IJRET: Int. J. Res. Eng. Technol. eISSN 2319-1163, pISSN 23s21-730
22. Asadi, E., Adeli, H.: Seismic performance factors for low-to mid-rise steel diagrid structural system, **27**(15) (2018)
23. Asadi, E., Li, Y., Heo, Y.A.: Seismic performance assessment and loss estimation of steel diagrid structures. J. Struct. Eng. **144**(10) (2018)

Determination of Optimum Height of Small Prototype Model of Chimney Operated Solar Power Plant

P. J. Bansod$^{(\boxtimes)}$

G. H. Raisoni College of Engineering and Management, Wagholi, Pune, India
premendra.bansod@raisoni.net

Abstract. This paper presents a research on a small model of solar chimney that was fabricated for taking necessary experimental readings in western region of India. The operating parameters considered are temperature, pressure, velocity, mass flow rate and geometrical parameters such as chimney height and collector diameter are considered. Influence of chimney height is studied on above mentioned parameters. The draught produced at the base of the chimney largely depends on the height of the chimney and subsequently is directly proportional to the power produced by the turbine; therefore the optimum height of chimney is first determined for present area of the collector. Even though the power produced by the turbine is directly proportional to the available draught which in turn is proportional to the height of the chimney, but it is found that after optimum height has reached the increase in power produced is not that significant.

Keywords: Solar chimney · Optimum height · Solar energy ·
Power generation · Operating parameters

1 Introduction

Solar chimney power plant is an old technology but it is not explored much for power generation. The First Pilot power plant was set in Manzanares Spain in 1981 with the help of funds provided German ministry of research & technology of 50 kW capacities. The chimney of this plant was 194.6 m long and 5.08 m of radius [1]. *Mullet* studied overall efficiency, design and performance of SCPP by considering Manzanares plant as reference [2]. Pasumarthi et al. performed experimental and theoretical performance of SCPP by doing mathematical modeling of various parts [3]. *Ganon et al.* carried an ideal air standard cycle analysis of SCPP. The authors studied limiting performance, ideal efficiencies and relationship between main variables [4]. *Von Backstrom et al.* investigated through a representative tall solar chimney bracing wheels [5]. Bilgen et al. studied solar chimney power plant for high latitude [6]. *Pretorius et al.* did sensitivity analysis of the operating and technical specifications of a solar chimney power plant [7]. *Zohu et al.* performed experimental study of temperature field in a solar chimney power plant [8]. *Maia et al.* performed theoretical evaluation of the influence of geometric parameters and materials on the behavior of the airflow in a

© Springer Nature Singapore Pte Ltd. 2020
V. K. Gunjan et al. (Eds.): *ICRRM 2019 – System Reliability, Quality Control, Safety, Maintenance and Management*, pp. 272–278, 2020.
https://doi.org/10.1007/978-981-13-8507-0_40

solar chimney [9]. *Zhou et al.* carried analysis of chimney height for solar chimney power plant. The maximum power output of 102.2 kW was obtained for the optimal chimney height of 615 m, which is lower than the maximum chimney height with a power output of 92.3 kW [10]. *dos Santos Bernardes et al.* analyzed some available heat transfer coefficients applicable to solar chimney power plant collectors [11]. *Petela* had performed the thermodynamic analysis of a simplified model of the solar chimney power plant [12]. *Nizetic et al.* provided a simplified analytical approach for evaluating the optimal ratio of pressure drop across the turbine in solar chimney power plants [13]. *Lorente et al.* showed how to use constructal design to distribute solar chimney power production on available land area most [14]. *Al-Dabbas* performed the first pilot demonstration of solar chimney power plant in Jordan [15]. *Cao et al.* performed simulation of a sloped solar chimney power plant in Lanzhou [16]

2   Experimental Setup

The main basis for the selection of chimney height and size was to construct small setup of SCPP so that it can be convenient to take reading at any suitable locations. The suitable size available was 38.1 mm (1.5 in.). Accordingly cylindrical chimney shape is selected of small heights of 0.5 m, 1.0 m, and 1.5 m 2.0 m. Pipe was cut in two pieces of 0. 5 m each, one piece of 1 m. diameter of 0.038 m was same for all the three pieces. The collector was fabricated with the help of support wooden stand by fixing crystal clear glass of 4 mm thick. The triangular and rectangular cut glass was attached on the wooden frame of the same shape with the help of Teflon tape. The micro wind turbine generator with small fan (turbine) has output voltage of 1 V to 18 V and DC output current of 1A to 3A with a rated speed of 200–6000 rev/min. was used in this setup. The diameter of motor is 24.5 mm and 32.2 mm height. Output shaft of 2 mm diameter with a shaft length of 13.5 mm was used. The blades of turbine has diameter of 32 mm along with aperture blades of 1.95 mm (Table 1 and Fig. 1).

Table 1. Size of SCPP considered for experimental setup

Height of chimney	Collector diameter	Chimney diameter at base	Chimney diameter at exit
0.50 m to 2.0 m	1.8 m	0.038 m	0.038 m

3   Mathematical Modeling

3.1   Energy Input/Output for Collector

According to the principle of radiation a solar collector converts available solar radiation 'G' falling on the Collector surface A_{coll}, into heat output Q. The steady state collector efficiency η_{coll} can be expressed as ratio of the heat output and G times Collector area

Fig. 1. Photograph of experiment setup [17]

$$\eta_{coll} = Q/GA_{coll} \tag{1}$$

Where,

η_{coll} – Efficiency of collector, Q – Heat output, G – Solar radiation available, A_{coll} – collector surface.

3.2   Chimney Efficiency

The chimney efficiency is given by

$$\eta_{ch} = gHc/C_{pa}Ta \tag{2}$$

Where,

η_{ch}-Chimney efficiency, g-Gravitational force, Hc-Chimney height, C_{pa}-Specific heat at constant pressure, Ta-Ambient Temperature.

3.3   Total Pressure

Total pressure produced in the collector area and chimney is given by

$$\Delta P_{total} = \rho gHc\Delta T/Ta \tag{3}$$

Where,

ρ-Air density produced inside chimney, ΔT-Temperature difference from inlet to outlet from collector to chimney.

4   Results and Discussions

For determination of optimum height of SCPP the readings are recorded for different duration of the day. Average values of various process parameters and its variation with different heights of cylindrical (PVC) chimney are considered for determination of optimum height of chimney. The details process parameters considered for different height of SCPP along with their average values are shown in Table 2.

Table 2. Average values of process parameters for different heights of chimney

S. no	Process parameters (with avg. of readings from 10 am to 5 pm)	0.5 m height	1.0 m height	1.5 m height	2.0 m height
1	Mass flow rate (kg/s)	11.03	13.742	18.876	18.906
2	Total power (P tot) W	4.660	12.620	32.895	37.600
3	Maximum velocity (V_{max})	0.953	1.190	1.645	1.641
4	Avg. Temp. Diff. (ΔT) °C	3.73	4.20	4.80	3.90
5	Max. theoretical power (Pwt, max) W	1.36	3.69	9.24	10.13
6	Collector efficiency (ηcoll) %	59.08	58.73	59.02	59.12

The values shown in Table 2 are plotted for different heights of chimney which considered for experimentation. The graphs are drawn for these parameters as shown in the Figs. 2, 3, 4 and 5.

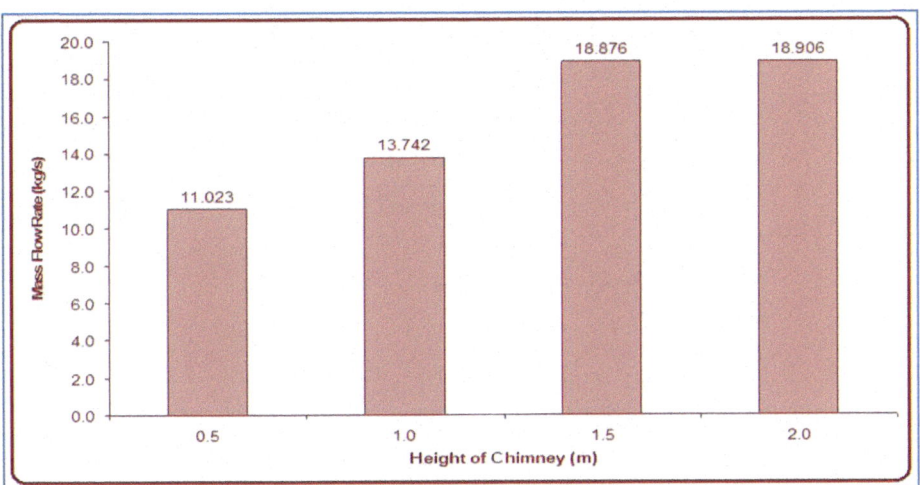

Fig. 2. Height vs Mass flow rate

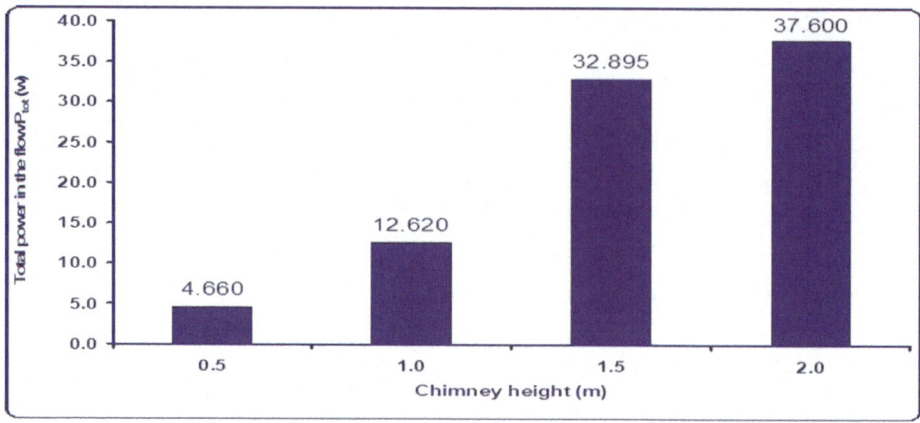

Fig. 3. Height vs Total power

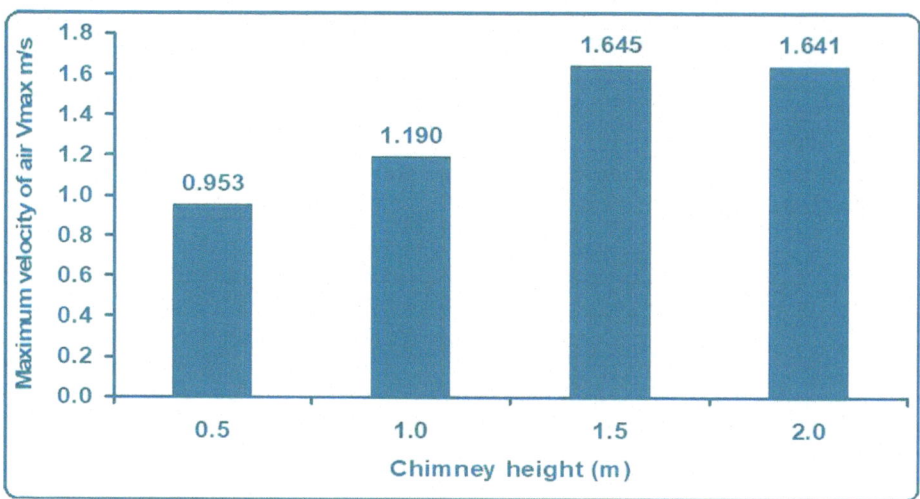

Fig. 4. Chimney height vs Maximum velocity

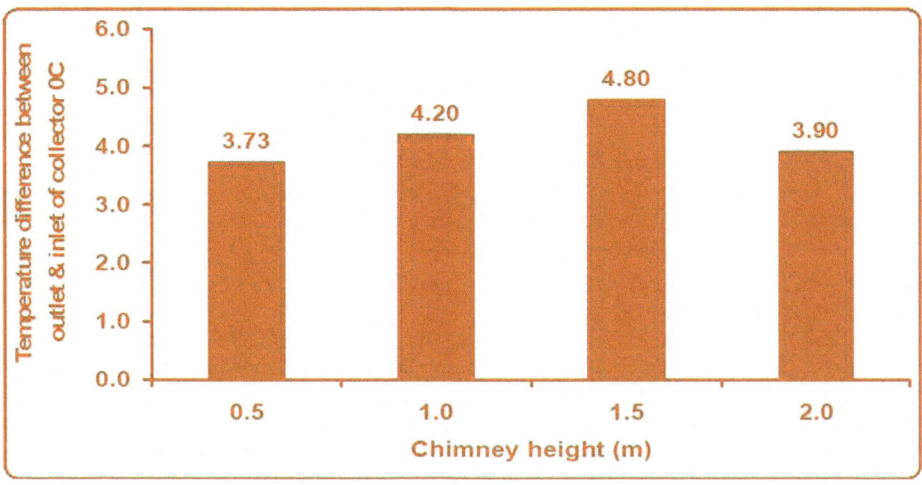

Fig. 5. Chimney height vs Temperature difference

5 Conclusions

- From the results and discussion it is concluded that the prototype model developed for experimentation shows the optimum height of cylindrical chimney as 1.5 m.
- If 2.0 m long chimney is used, there is marginal increase in power output but the cost of construction for marginal power output of chimney is not economical.
- If there is small increase in power output then it is suggested that unnecessary increase in height of chimney will increase the construction cost and finally increase in unit cost of power generated, which is not desirable.
- From the present work it can be justified that for every SCPP depending upon collector area, the height of chimney should be calculated and the power plant should be developed accordingly.

References

1. Haaf, W., Friedrich, K., Mayr, G., Schlaich, J.: Solar chimneys part i: principle and construction of the pilot plant in Manzanares. Int. J. Sol. Energy **2**, 3–20 (1983)
2. Mullet, L.B.: The solar chimney-overall efficiency, design and performance. Int. J. Ambient Energy **8**, 35–40 (1987). https://doi.org/10.1080/01430750.1987.9675512
3. Pasumarthi, N., Sheriff, S.A.: Experimental and theoretical performance of a demonstration solar chimney model part II: experimental and theoretical results and economic analysis. Int. J. Energy Res. **22**, 443–461 (1998)
4. Gannon, A.J., von Backstrom, T.W.: Solar chimney cycle analysis with system loss and solar collector performance. Int. J. Sol. Energy Eng. **132**, 132–137 (2000)
5. von Backstrom, T.W., Bernhardt, A., Gannon, A.J.: Pressure drop in solar power plant chimneys. Int. J. Sol. Energy Eng. **123**, 165–169 (2003)

278   P. J. Bansod

6. Bilgen, E., Rheault, J.: Solar chimney power plants for high latitudes. Int. J. Sol. Energy **79** (5), 449–458 (2005)
7. Pretorius, J.P., Kroger, D.G.: Sensitivity analysis of the operating and technical specifications of a solar chimney power plant. J. Sol. Energy Eng. **129**, 171–178 (2007)
8. Zhou, X., Yang, J., Xiao, B., Hou, G.: Experimental study of temperature field in a solar chimney power setup. Appl. Therm. Eng. **27**, 2044–2050 (2007)
9. Maia, C.B., Ferreira, A.G., Valle, R.M., Cortez, M.F.B.: Theoretical evaluation of the influence of geometric parameters and materials on the behavior of the airflow in a solar chimney. Comput. Fluids **38**, 625–636 (2009)
10. Zhou, X., Yang, J., Xiao, B., Hou, G., Xing, F.: Analysis of chimney height for solar chimney power plant. Appl. Therm. Eng. **29**, 178–185 (2009)
11. dos Santos Bernardes, M.A., Von Backstrom, T.W., Kroger, D.G.: Analysis of some available heat transfer coefficients applicable to solar chimney power plant collectors. Sol. Energy **83**, 264–275 (2009)
12. Petela, R.: Thermodynamic study of a simplified model of the solar chimney power plant. Sol. Energy **83**, 94–107 (2009)
13. Nizetic, S., Klarin, B.: A simplified analytical approach for evaluation of the optimal ratio of pressure drop across the turbine in solar chimney power plants. Appl. Energy **87**, 587–591 (2010)
14. Lorente, S., Koonsrisuk, A., Bejan, A.: Constructal distribution of solar chimney power plants: few large and many small. Int. J. Green Energy **7**, 577–592 (2010)
15. Al-Dabbas, M.A.: The first pilot demonstration: solar updraft tower power plant in Jordan. Int. J. Sustain. Energy **31**, 399–410 (2012)
16. Cao, F., Zhao, L., Guo, L.: Simulation of a sloped solar chimney power plant in Lanzhou. Energy Convers. Manag. **52**, 2360–2366 (2011)
17. Bansod, P.J., Thakre, S.B., Wankhade, N.A.: Expermentational data analysis of chimney operated solar power plant. Int. J. Mech. Eng. Tech. **7**, 225–231 (2016)

Analysis of Single Curvature Arch Dams Using Finite Element Method

Deepak Patil[1]([⊠]) and Shrikant Charhate[2]

[1] Department of Civil Engineering, Pillai's HOC college of Engineering
and Technology, Rasayni, University of Mumbai, Mumbai, MS, India
deepakpatil813@gmail.com
[2] Department of Civil Engineering, Amity School of Engineering
and Technology, Amity University, Mumbai, MS, India
scharhate@mum.amity.edu

Abstract. The arch dam is a massive water retaining structure made up of concrete. The buckling of arch dam belongs to theory of buckling of shells. This paper deals with finite element analysis of constant angle concrete arch dam with full reservoir conditions. Keeping the central angle 120° and valley profile constant at an angle of 30° with the vertical; dam is analyzed for maximum deflections and maximum principal moments for varying thickness across the dam profile. Von Mises stress theory is applied to find maximum stress factors in membrane and flexure. The parametric investigation of constant angle arch dam reveals that the maximum stresses and displacements are minimized with increasing base thickness and vice versa with aspect ratio.

Keywords: Finite element method · Constant angle · Deflections ·
Stress factors · Von Mises stress theory

1 Introduction

In the analysis of concrete arch dam the circular or parabolic shape of arch dam does not affect the strength, but the B/H ratio has a greater effect (Zingoni et al. 2013). The problem of dam buckling belongs to fundamental subjects of shell. A lot of theoretical and experimental research has been undertaken in past many decades (Papadakis 2008). The foundation and abutments are considered as the finite elements, while arch is considered as an arch cantilever system (Lin and Su 2002). Trial load and finite element methods are the main existing stress analysis methods for arch dams. Difficulties involved in trial load method are overcome by finite element method (Zhing et al. 2010). It is worldwide accepted that shape of dam has a great influence on economy and safety of an arch dam. Arch dams are designed by trial and error method. If the condition satisfies the specification, then the dam is adopted else revised.

© Springer Nature Singapore Pte Ltd. 2020
V. K. Gunjan et al. (Eds.): *ICRRM 2019 – System Reliability, Quality Control,
Safety, Maintenance and Management*, pp. 279–284, 2020.
https://doi.org/10.1007/978-981-13-8507-0_41

The difficulty and details of the design, project and application parts as well as the supervision and control stages are troublesome. Particularly the modeling of the finite elements of dams with arch form is a time consuming and exhausting process. Numerous analytical (Kartal et al. 2015), numerical (Ohmachi and Jajali 1999, Oliveira and Faria 2006, Sevim et al. 2014) and experimental studies (Masserzarea et al. 2000, Wang 2007, Sevim et al. 2011, Sevim et al. 2012) have been carried out to obtain the structural behavior of dams. Since last decades due to development in technology there is an increasing trend towards use of software operations to ensure ease of process (Altunisik et al. 2018). Mittrup and Hartmann studied on software development for structural controls of dams. In this study we first focused on arch dams of single curvature with constant central angle. Dam is analyzed for maximum displacements and maximum principle stresses in membrane and flexure.

2   Scope of Present Study

The present paper deals with the finite element analysis of single curvature constant angle arch dams set in a trapezoidal valley. Horizontal circular curved arches are selected for analysis. The dam height is referred as H and span of the dam between left and right abutments as B. Thickness of dam is referred as t. The B/H referred as aspect ratio. The thickness of dam is varying with maximum at base and zero at the crest. The dependence of parameters such as B/H ratio and thickness of dam on the behavior of dam is studied thoroughly. The findings of the parametric analysis of such dams can be useful in studying the behavior of double curvature arch dams. Thus the study of maximum displacement and maximum stress factors for various parameters considered is the only concern of present work.

3   Finite Element Modeling

Finite element modeling of arch dam is done by using general purpose fortran77 program. The poisons ratio assumed is 0.15 ($\vartheta = 0.15$), Young's modulus of elasticity $E = 0.25 \times 10^8$ KN/m^2, Density of material $\rho = 24$ KN/m^3. Thickness of dam adopted at the base 0.1H, 0.15H and 0.2H. Heights of dam 100 m considered with various spans with B/H ratio = 2, 3, 4. All the nodes along the abutments and base nodes of the shell were completely restrainedd (all the boundary nodes are fixed). The angle made by vertical abutments with the horizontal is 30°. The dam is discritised into finer mesh of size (8 × 8), (16 × 16) & (32 × 32). The central angle for all the dam cases is kept 120° meaning there by half the central angle is 60°.

The results of the analysis were reviewed in respect of the maximum deflection at top and development of the maximum stress factors (Von Mises criterion). The investigation relates to static behavior of dam with consideration to full reservoir condition. The finite element analysis of various cases as formulated above is conducted by thin plate Krichoffs formulation with the plates oriented in a global space.

4   Numerical Results and Discussion

Parametric investigation in terms of maximum displacements and maximum stress factors are presented below for a dam height of 100 m.

4.1   Maximum Displacements

Table 1 below shows the maximum deflection for 100 m height dam for various parameters considered (i.e. thickness and B/H ratio), the results of displacements are graphically presented in Fig. 1(a–c).

Table 1. Results of maximum displacements in meters for various parameters (H = 100 m)

B/H ratio	Mesh size								
	8 × 8			16 × 16			32 × 32		
	Base thickness of dam(t)								
	0.1H	0.15H	0.2H	0.1H	0.15H	0.2H	0.1H	0.15H	0.2H
2	0.0418	0.0235	0.0141	0.0450	0.0249	0.0148	0.0463	0.0255	0.0151
3	0.0817	0.0501	0.0330	0.0889	0.0527	0.0350	0.0917	0.0542	0.0360
4	0.1430	0.0834	0.0535	0.1570	0.0922	0.0588	0.1600	0.0958	0.0615

Table 2 below shows the maximum and minimum membrane and flexural stresses for arch dam with different mesh sizes. The results of stress factors are graphically presented in Figs. 2(a and b).

With reference to Fig. 1(a–c) it is observed that mesh size influences the deflections, manifested such that finer the mesh, deflection goes on increasing. Deflections reduce with increased base thickness and displacement variations with respect to B/H ratio are almost linear for 0.2H. (Patil and Charhate 2018).

4.2   Maximum Membrane and Flexural Stresses

Table 2 above and Figs. 2(a and b) represents the maximum membrane and flexural stress factors. From the figures it is studied that stress factors reduces with increased base thickness and there is linear variation with respect to B/H ratio. Finer the mesh size, stress factors are increased.

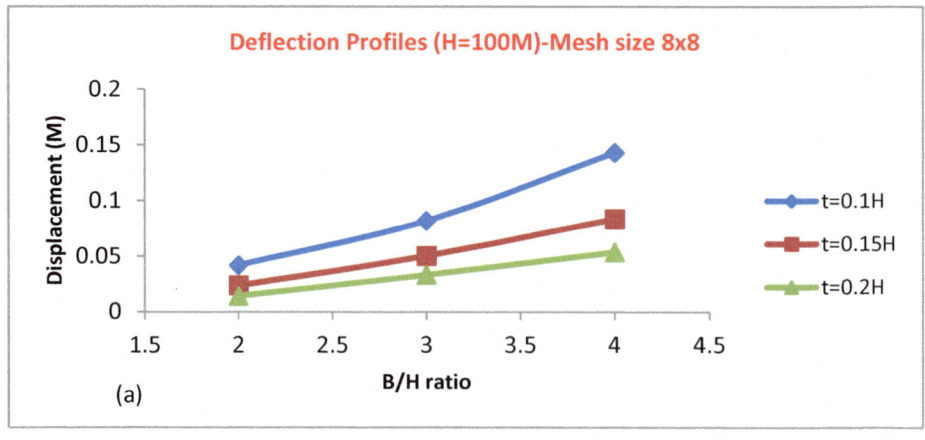

(a)

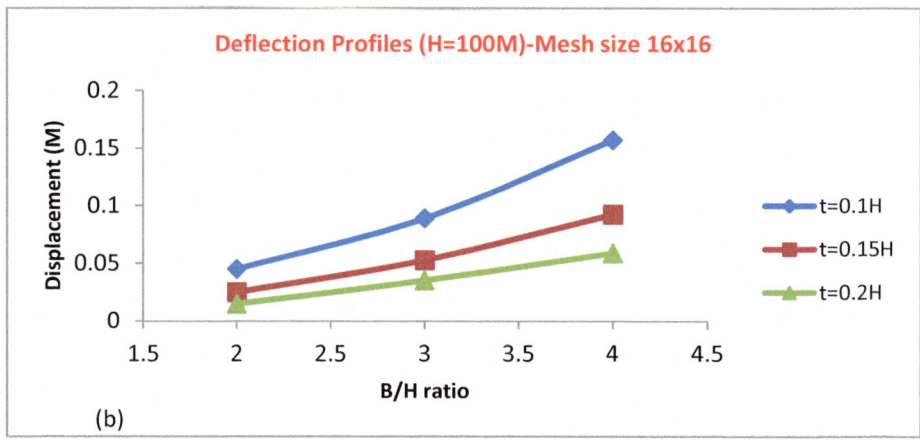

(b)

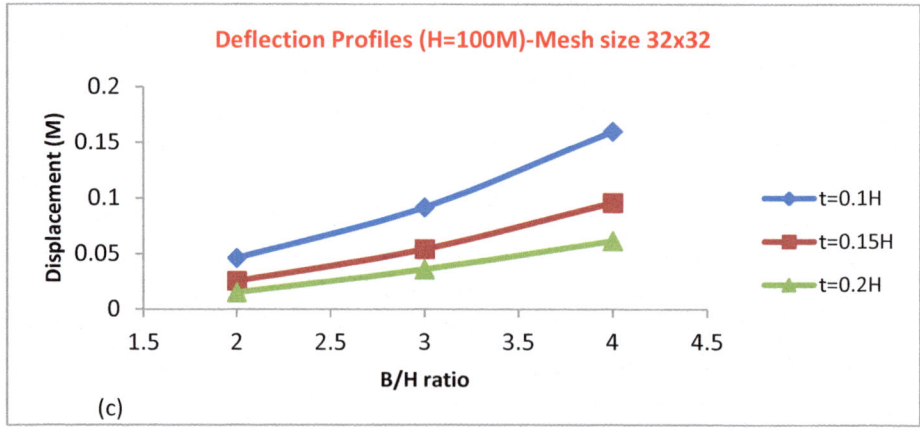

(c)

Fig. 1. (a–c) Displacement plots for constant angle arch dam for varying thickness

Table 2. Results of maximum membrane and flexural stresses (KN/m^2)

B/H Ratio & base thickness	Mesh size					
	8 × 8		16 × 16		32 × 32	
	Membrane	Flexure	Membrane	Flexure	Membrane	Flexure
2(t = 0.1H)	6.65E+03	6.03E+04	7.81E+03	9.19E+04	7.45E+03	1.13E+05
3(t = 0.1H)	1.06E+04	1.20E+05	1.10E+04	1.64E+05	1.14E+04	1.90E+05
4(t = 0.1H)	1.45E+04	1.81E+05	1.54E+04	2.41E+05	1.57E+04	2.74E+05
2(t = 0.15H)	3.90E+03	8.83E+04	4.22E+03	1.26E+05	4.41E+03	1.49E+05
3(t = 0.15H)	6.31E+03	1.72E+05	6.68E+03	2.21E+05	6.90E+03	2.48E+05
4(t = 0.15H)	8.52E+03	2.50E+05	9.40E+03	3.13E+05	9.77E+03	3.45E+05
2(t = 0.2H)	2.52E+03	1.00E+05	2.76E+03	1.38E+05	2.90E+03	1.61E+05
3(t = 0.2H)	4.02E+03	2.02E+05	4.04E+03	2.50E+05	4.25E+03	2.57E+05
4(t = 0.2H)	5.37E+03	2.84E+05	6.03E+03	3.45E+05	6.40E+03	3.74E+05

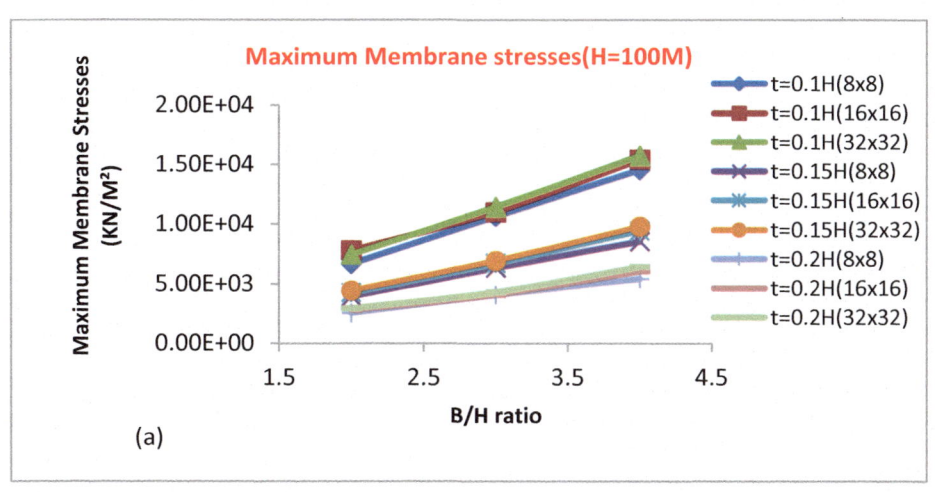

(a)

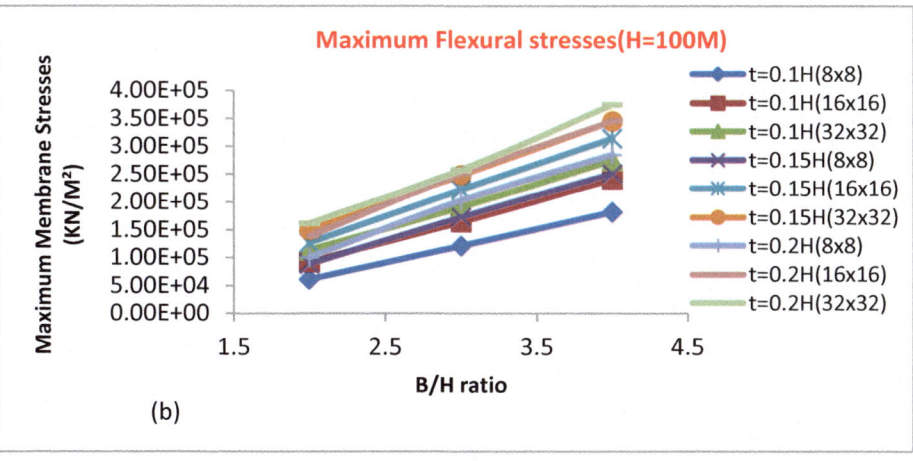

(b)

Fig. 2. (a & b) Plot for maximum stress factor for constant angle arch dams

5 Conclusions

In this study parametric investigation of constant angle concrete arch dam is undertaken with various parameters as mentioned earlier. An attempt was to study the static behavior of dam in terms of maximum displacements and principle stresses and moments. These principle stresses were converted into maximum stress factors using Von Mises theory. A number of significant observations have been made. The geometric properties of arch dam such as B/H ratio, thickness of dam has greater effect on behavior of dam. As fineness of mesh increases deflection as well as stress factors increases. As expected in theoretical study of finite element analysis that structural response increases with fineness of mesh. From the parametric investigation it is revealed that with increase in aspect ratio B/H, maximum deflections and flexural stresses increases and membrane stress factors decreases.

References

Zingoni, A., Mudenda, K., French, V., Mokhothu, B.: Buckling strength of thin-shell concrete arch dams. Thin Walled Struct. **64**, 94–102 (2013)

Papadakis, G.: Buckling of thick cylindrical shells under external pressure: a new analytical expression for the critical load and comparison with elasticity solutions. Int. J. Solids Struct. **45**, 5308–5321 (2008)

Lin, S., Su, D.: Stress analysis for arch dams on complicated rock foundation using trial load methods. J. Yangtze River scientific Res. Inst. **19**(5), 27–30 (2002)

Kehong, Z., Tong-Chun, L.: Coupled analysis of arch dam using trial load technique and displacement compatibility based on finite element method. J. Earth Sci. ASCE (2010)

Altunisik, A.C., Kalkan, E., Basaga, H.B.: Development of engineering software to predict the structural behavior of arch dams. Adv. Comput. Des. **3**(1), 87–112 (2018)

Kartal, M.E., Bayratkar, A., Caruslu, M., Karabulut, M., Basaga, H.B.: Investigation of RCC dams considering viscous boundary conditions. In: 2nd International Sustainable Building Symposium, Ankara, Turkey (2015)

Ohmachi, T., Jalali, A.: Fundamental study on near field effects on earthquake response of arch dams. Earthq. Eng. Eng. Seismolog. **1**(1), 1–11 (1999)

Patil, D., Charhate, S.: Finite Element Analysis of Constant radius and Constant angle Arch Dams. Int. J. Civil Eng. Technol. **9**(12), 666–688 (2018)

Oliveira, S., Faria, R.: Numerical simulation of collapse scenarios in reduced scale tests of arch dams. Eng. Struct. **28**, 1430–1439 (2006)

Sevim, B., Altunisik, A.C., Bayratkar, A.: Construction stages analysis using time dependent material properties of concrete arch dams. Comput. Concr. **14**(5), 599–612 (2014)

Masserzarea, J., Leib, Y., Eskandari-S Hiria, S.: Computation of natural frequencies and mode shapes of arch dams as an inverse problem. Adv. Eng. Softw. **31**, 827–836 (2000)

Wang, H., Li, D.: Experimental study of dynamics damage of an arch dam. Earthq. Eng. Struct. **36**, 347–366 (2007)

Sevim, B., Altunisik, A.C., Baytakar, A., Akkose, M., Calayir, Y.: Water length and height effects on the earthquake behavior of arch dam reservoir foundation system. J. Civil Eng. **15**, 295–303 (2011)

Sevim, B., Altunisik, A.C., Baytakar, A.: Experimental evaluation of crack effects on the dynamic characteristics of a prototype arch dam using ambient vibration tests. Compt. Concr. **10**(3), 277–294 (2012)

Assessment of Heavy Metal Contamination in Groundwater Around Dumping Site

G. M. Hattekar[✉] and V. S. Pradhan

Department of Civil Engineering, Jawaharlal Nehru Engineering College,
Aurangabad 431001, Maharashtra, India
gayatri89hatteakar@gmail.com, vizpradhan@gmail.com

Abstract. In Aurangabad, lack of capital and appropriate technology for environment-friendly waste management practices has been made the dumpsite at Naregaon for the disposal of Municipal Solid Waste (MSW). Unplanned dumping leads groundwater contamination of nearby area, as rainwater percolates through MSW forms leachate that consists of decomposing organic matter combined with other heavy metals. pH, Electrical Conductivity (EC) and GIS-based spatial distribution of heavy metals such as Copper (Cu), Lead (Pb), Cadmium (Cd), Arsenic (As), Manganese (Mn) and Chromium (Cr) of Thirty (30) groundwater samples around dumping area have been determined. Based on heavy metal concentration, we concluded that our study area as a whole is critically polluted.

Keywords: Heavy metals · MSW · Leachate · Groundwater

1 Introduction

As water is essential for life and groundwater is the major source for drinking, domestic and irrigation purposes in both rural and urban areas. Threats to the groundwater from unlined and uncontrolled landfills exist in many parts of the world [1]. Aurangabad city is one of the major industrial centres in central Maharashtra. City is facing serious environmental degradation and public risk due to uncontrolled disposal of MSW [3]. The waste collected from Aurangabad city is transported to Naregaon dumping ground [2]. Solid waste disposal represents a significant source heavy metal which leads the contamination in groundwater is documented as severe environmental pollution and therefore the study of this problem is important [4]. GIS is an effective tool for collection, storage, management and retrieval of a multitude of spatial and non-spatial data. The objectives of this study is to map using GIS and describe the distribution of some selected heavy metals like Cu, Pb, Cd, As, Mn and Cr.

2 Study Area

Aurangabad city is located at coordinate 19°52'34' N and 75°20'35' E. The waste collected from Aurangabad city is transported to Naregaon dumping ground which is 6 km away from city limits. The total area of Naregaon waste dumping site is about

© Springer Nature Singapore Pte Ltd. 2020
V. K. Gunjan et al. (Eds.): *ICRRM 2019 – System Reliability, Quality Control,*
Safety, Maintenance and Management, pp. 285–289, 2020.
https://doi.org/10.1007/978-981-13-8507-0_42

46 acres. It is situated at latitude 19° 54'15' North and longitude 75° 23'45' East. Figure 1 shows the location map study area with sampling site.

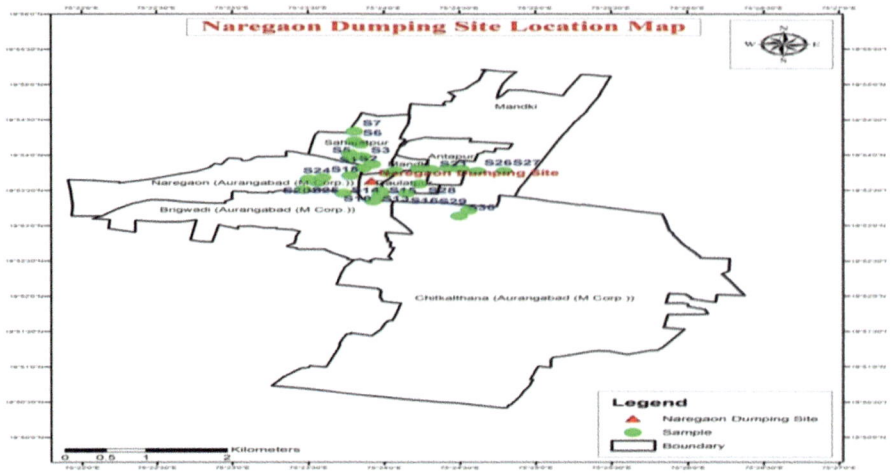

Fig. 1. Location map of Naregaon dumping site showing sampling locations

3   Materials and Methods

Sample Collection and Analysis
The location of sampling points (S1 to S30) were obtained with hand held Global Positioning System (GPS, Germin ETREX 30X model) (Table 1) in 2 km of radius by stratified random sampling. Samples for heavy metal analyzed and collected by standard method. To maintain integrity of water samples pH and EC which are sensitive to environment were measured in situ using portable digital meter. For spatial analysis of heavy metals Q-GIS software has been used.

4   Results

The summary of the results of laboratory analyses conducted on the samples are in Table 1. pH value indicates that water around dumping site is slightly acidic to alkaline in nature. The overall EC values varied between 213 to 3550 µS/cm. Figure 2 shows the spatial distribution of investigated heavy metals (Cu, Pb, Cd, As, Mn and Cr) were found to be high in concentration exceeding their permissible limit which makes it unsuitable for drinking. Figure shows the concentration of heavy metals were maximum for sample S10, S13, S14 and S15.

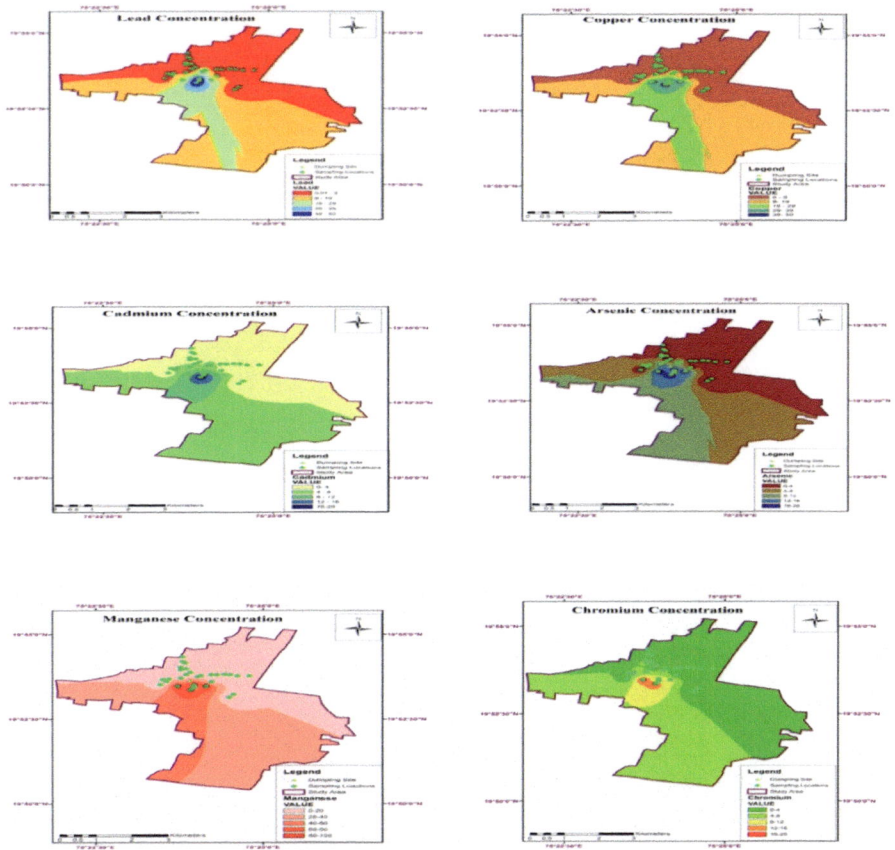

Fig. 2. Spatial distribution of heavy metals

Table 1. Samples showing concentration of heavy metals

Sample number	Latitude	Longitude	pH	EC (µS/cm)	Cu (mg/l)	Pb (mg/l)	Cd (mg/l)	As (mg/l)	Mn (mg/l))	Cr (mg/l)
S1	19° 53' 48.81' N	75°23' 51.60' E	8	3250	1	1	5	1	5	1
S2	19° 53' 57.33' N	75°23' 46.74' E	7.5	3550	5	10	5	5	20	10
S3	19° 53' 58.89' N	75°23' 51.64' E	7.5	720	0.04	0.9	0.03	1	0.3	0.03
S4	19° 53' 1.31' N	75°23' 46.56' E	7.5	900	5	1	1	1	5	0
S5	19° 54' 9.29' N	75°23' 50.63' E	7.5	820	1	1	1	1	5	0.8
S6			8	960	1	1	0	1	1	1

(*continued*)

Table 1. (*continued*)

Sample number	Latitude	Longitude	pH	EC (μS/cm)	Cu (mg/l)	Pb (mg/l)	Cd (mg/l)	As (mg/l)	Mn (mg/l))	Cr (mg/l)
	19° 54' 11.83' N	75°23' 48.44' E								
S7	19° 54' 20.01' N	75°23' 48.34' E	7.5	2020	1	1	0.8	1	5	1
S8	19° 53' 44.95' N	75°24' 1.98' E	7.5	1091	20	50	5	10	20	5
S9	19° 53' 46.83' N	75°24' 4.44' E	8	726	10	5	10	10	5	5
S10	19° 53' 25.87 N	75°23' 58.63' E	8	1593	20	50	20	10	20	5
S11	19° 53' 52.62' N	75°23' 55.45' E	8.3	521	0.1	1	0.05	1	1	0.1
S12	19° 53' 48.35' N	75°24' 13.74' E	6.3	293	5	1	1	0.58	1	0.8
S13	19° 53' 21.43' N	75°23' 55.69' E	7.8	1973	50	50	20	20	100	20
S14	19° 53' 27.69' N	75°23' 43.90' E	7.3	1921	50	20	10	20	100	20
S15	19° 53' 29.82' N	75°23' 58.60' E	8.1	733	50	50	10	20	100	10
S16	19° 53' 27.84' N	75°24' 9.09' E	7.2	312	50	20	10	20	100	10
S17	19° 53' 42.51' N	75°23' 46.75' E	8.2	512	10	5	5	20	10	5
S18	19° 53' 10.12' N	75°23' 58.55' E	7.3	623	10	5	10	10	5	1
S19	19° 53' 40.75' N	75°23' 35.79' E	7.5	1021	5	10	5	5	20	5
S20	19° 53' 36.636 N	75°23' 33.62' E	7.3	1129	1	10	5	1	10	5
S21	19° 53' 48.179' N	75°24' 18.74' E	8	792	5	1	0.69	0.23	10	0.89
S22	19° 53' 49.092' N	75°24' 25.44' E	8.3	1022	0.89	1	0.68	1	5	0.21
S23	19° 53' 48.527 N	75°24' 30.57 E	8	552	1	0	1	0.36	0.69	0
S24		75°23' 30.16' E	8.2	213	1	5	0	5	10	1

(*continued*)

Table 1. (*continued*)

Sample number	Latitude	Longitude	pH	EC (μS/cm)	Cu (mg/l)	Pb (mg/l)	Cd (mg/l)	As (mg/l)	Mn (mg/l))	Cr (mg/l)
	19° 53' 39.724' N									
S25	19° 53' 30.62' N	75°23' 27.92' E	7.5	993	0.82	0.94	0.52	0.65	1	0.85
S26	19° 53' 46.57' N	75°24' 37.11' E	8	1121	0	1	0	0	1	0
S27	19° 53' 46.54' N	75°24' 47.68' E	8.2	757	0	0	0	0	0	0
S28	19° 53' 35.79' N	75°24' 14.24' E	7.5	1621	1	5	1	1	5	1
S29	19° 53' 13.59' N	75°24' 33.74' E	7.8	933	1	1	0	0	5	0
S30	19° 53' 8.57' N	75°24' 29.75' E	7.5	312	0	1	0	0	1	0
BIS	–	–	6.5–8.5	750–2250	0.05	0.01	0.01	0.01	0.1	0.05
WHO	–	–	6.5–8.5	1000–2000	1.0	0.05	0.01	0.05	0.1	0.05

5 Conclusion

The study has shown that the groundwater sources within 2 kms radius of Naregaon dumping site are contaminated due to heavy metals.

This very large extent is due to the dispersion of chemical constituents from leachates produced at landfill.

It is therefore recommended that the dumpsite condition be improved to minimize the effects on the environment or it be relocated to another area away from residences.

References

1. Shenbagarani, S.: Analysis of groundwater quality near the solid waste dumping site. J. Environ. Sci. Toxicol. Food Technol. **4**(2), 01–05 (2013)
2. Teta, C., Hikwa, T.: Heavy metal contamination of ground water from an unlined landfill in Bulawayo, Zimbabwe. J. Health Pollut. **7**(15), 18–27 (2017)
3. Rajkumar, N., Subramani, T., Elango, L.: Groundwater contamination due to municipal solid waste disposal – a GIS based study in Erode city. Int. J. Environ. Sci. **1**(1), 39–55 (2010)
4. Majhi, A., Biswal, S.K.: Application of HPI (heavy metal pollution index) and correlation coefficient for the assessment of ground water quality near ash ponds of thermal power plants. Int. J. Sci. Eng. Adv. Technol. **4**(8), 395–405 (2016)

Author Index

Printed by Printforce, the Netherlands